PROJECT PARTNERING FOR THE DESIGN AND CONSTRUCTION INDUSTRY

PROJECT PARTNERING FOR THE DESIGN AND CONSTRUCTION INDUSTRY

RALPH J. STEPHENSON, P.E., P.C.
Consulting Engineer
Mt. Pleasant, Michigan

A Wiley-Interscience Publication

JOHN WILEY & SONS, INC.

New York • Chichester • Brisbane • Toronto • Singapore

Library of Congress Cataloging in Publication Data:

Stephenson, Ralph J., 1922–
    Partnering and alternative dispute resolution : in the planning,
design and construction business / Ralph J. Stephenson.
        p.   cm.
    Includes index.
    ISBN 0-471-10716-6 (cloth : alk. paper)
    1. Construction industry—Management.   2. Strategic alliances
(Business)   3. Conflict management.   I. Title.
HD9715.A2S73   1996
624'.068'4—dc20   95-39866

CONTENTS

PREFACE

This book on project partnering is dedicated to my family and to the wonderful opportunities their help and support have offered. Especially, I thank my wife and partner, Betty, for her sparks of enthusiasm and constant optimism. Partnering really works.

For all the help in finding, writing, reading, typing, proofing and the million other details that are the ingredients of a forerunner effort, a special thank you goes to the great friends who contributed in numerous ways to make my work possible: George Wilkinson, who got me started on partnering; Joe Neussendorfer, who had faith in what could be said about it; David Schock, writer, critic, and hard-working editor; Phil Bennett, trainer and educator; Ron Hausmann, a spark for change; John Spittler, professional improver; Kathy Neville, faithful helper and advisor; Kim Fricke, who encourages partnering; Marla Janness, who loves education, writing, and construction; Jessica Bell, a trusty, dependable practitioner; Marlo Wiltse, a new teacher with good sense; and Peter and Susan White, a pair of exceptional learners.

Of course, none of my work in partnering would have been possible without a source of practice in the current techniques of partnering. For this I thank my partnering clients. They are helping create an environment in which partnering in the truest sense is being used to sustain and create a constantly improving planning, design, and construction business and profession. They are an important part of a business world that believes in trust, honesty, hard work, and competence.

RALPH J. STEPHENSON
September 30, 1995

PROJECT PARTNERING FOR THE DESIGN AND CONSTRUCTION INDUSTRY

INTRODUCTION: THINGS THAT GO RIGHT

This book begins with a story: a once-upon-a-time story about a construction professional who has devoted much of his career working in the midst of conflict to get jobs done. This man is wary and a bit weary: he used to like his work, but these days something's missing—it's no longer fun.

Jepson Construction Company project manager, Robert Whitaker, is in charge of the Vinegar Hill power plant job. He, other members of his firm, and other stakeholders on the project spent what he considered an undue amount of time before field work began in partnering sessions. Bob is all too familiar with construction disputes: The last three jobs he has had with Jepson have wound up in legal battles, one in court, the other two in binding arbitration. In the first, the mechanical contractor went bankrupt with the job unfinished. The owner sued Jepson and forced it to finish the project at a much higher cost. In the second, a concrete supplier went to arbitration with Jepson for payment that had been withheld correctly in accordance with its contract with Jepson. In the third, Jepson went to binding arbitration with the owner to recover delay and penalty damages resulting from serious problems with incomplete contract documents.

Although he knows something needs to change fundamentally in the design and construction industry, Bob is not too sure if partnering is the answer; it seems too soft and mushy, too touchy-feely. He has seen psychological and philosophical concepts come and go, each one touted as the next revolution in management and construction. Still, the ideals behind partnering appeal to his strong sense of fair play.

Well, he cannot waste time about that. Today, he and the concrete supplier, Truckcrete, have to put nearly 700 yards of concrete in place for a new retention basin. It is the largest pour of the job. Forms and resteel are set, the crews are ready, and he sees the first ready-mix trucks pull onto the job site. He also sees the concrete subcontractor owner and project manager by her car talking on the cellular phone. She looks distressed. Bob has an instinct for things that are going awry and it is sounding an alarm. The sub, Diane Lowens, finishes her call and starts toward him.

"Here it comes," says Bob to himself. And indeed it does. Diane quickly explains that the primary transformer at her plant has just blown up and put the plant totally out of commission. A new transformer is on the way but will not be available and in place for at least a week. Bob sees certain disaster since the first pour is starting from the trucks that did get to the job site. His first thought is to stop the pour, but it is already too late to avoid cold joints. Next, he has fleeting visions of alerting his legal staff, but they certainly cannot get the immediate problem solved. He hesitates, but Diane continues to explain that her partner has called in every favor he has ever done for all the other concrete contractors in the four-county area. The result—within an hour trucks from four other suppliers will be rolling into the site, each with concrete made to the exact design mix required.

Diane and her partner are as good as their word. Even the drivers from the other companies get into the doing-a-favor act, each trying to better the next in delivery service and truck placement for finishing crew convenience. By 4 P.M. the pour is complete and the finishing crews are hard at work winding down the job. Diane brings out her Thermos and offers Bob a cup of coffee. He accepts and together they discuss the successful conclusion of a day that started in the other direction.

"You know, I thought this whole retention basin was in critical trouble when I saw you on the phone this morning," observes Bob over the rim of his coffee cup.

"Jerry, my partner, knew what was at stake and he was determined not to let anything stop today's work," Diane declared emphatically. "He understands how important it is. And besides, we signed an agreement to maintain both the letter and the spirit of the contract. Jepson and Truckcrete are partners on this job."

Bob returns to his office and sits thinking. He turns to his desk, opens the file drawer, and takes out the job charter prepared in the partnering meeting. There is his signature. Above it and to the right are Jerry's and Diane's. He cannot even remember what Jerry looks like. As he reviews the moral agreement they all signed, several phrases jump out at him: "on time and within budget," "keep site safe and clean," "resolve conflicts quickly and at the level where they originate," "be a good neighbor," "enjoy your work." Maybe, he thinks, there *is* something to this partnering idea after all.

Bob has cause to reexamine that charter 15 or 20 times more during the course of the project. In one case the pipefitters identified a change that they said could be made both to better the facility performance and to save money. Better yet, it was a change welcomed by the owner and the design engineer.

Initially, Bob thought they just wanted to recover an additional chunk of their costs, but after a calm discussion with the sub, he realized that the goal had more far-reaching consequences. Appropriately, he brought in the owner and the engineers and the modification was both okayed and praised.

Again and again he sees instances where the job could have, and normally would have, become contentious. Subcontractors settled disputes among themselves without escalating issues. When there was a legitimate disagreement, Bob's boss and Jepson's vice president told him to consider calling in a standing neutral, an independent construction expert who listened and made recommendations. Bob did, and everyone involved in the dispute agreed to be bound by the neutral's findings.

Season by season the work progressed. By the time the project closed out after 30 months of steady work, Bob was older, wiser, and richer. In particular, Bob had developed a group of friends on the job. He knew that he could trust these people no matter what—all because they had agreed to be partners. Additionally, he himself had become an advocate of partnering and a well-respected authority on partnering among his peers. What he had seen in the field at Vinegar Hill convinced him that partnering was not a fad and that it was not even a new idea. It was a return to the sunnier days of construction when a person's word was his or her bond, when no one disparaged the idea of a reasonable profit for a job well done, and where people made an effort to get along. For Bob, Vinegar Hill was a sweet experience, one that changed his life. Construction was once again both fun and a professional calling.

So, is Bob a figment of my imagination? Well, the names and places have been changed to protect the innocent, but the experience is one I have seen often in the last few years. I, too, was skeptical about partnering. I had come across an early wave of partnering enthusiasm and thought, "This is just too pat and naive." I remember addressing an annual session of a local construction association in 1991. I recommended to the members attending that they wait awhile, perhaps five years, before considering the use of partnering on their projects. There were several in the audience who are not going to let me forget what I said. Well, I have changed my mind. I have since told them that I have revised my early estimates to this extent: If they aren't partnering, they are missing one of the most unique and gratifying experiences they could have in the planning, design, and construction business. I strongly believe that partnering can so enhance our jobs that it gives back to us the spirit we saw when we first considered construction as a career.

It is in that frame of reference that I address this book to you. You will find here a brief history of the construction industry as we know it, a schema of the work and organizations we must deal with, the kinds of profit we have a right to expect from our work, an assessment of the need for partnering, and a how-to manual showing how to practice partnering. My goal in this book is to give you the understanding that will make a difference in the industry. Along the way, you might learn a great deal about yourself; I know I have.

1

THE EVOLUTION OF
MODERN CONSTRUCTION

Contemporary construction practices have their roots in the mid-nineteenth century when the technological tempo of the Industrial Revolution accelerated design and construction change. Increasing market demands for new and improved products and a manufacturing shift from home industries to large mechanized corporate entities intensified the need for new construction. This construction climate forced dramatic changes in building practices within the world's market-driven economies.

Additional impetus was given planning, design, and construction change by several major technical advances. These included increased use of advanced mathematical techniques, emergence of structural products capable of spanning long, open areas, improvements in exterior cladding, and increasing urban land values, making high-rise structures economical and desirable. Other factors contributing greatly to ongoing construction industry change included innovations in mechanical and electrical systems and improved education and training of planners, designers, and builders. These improvements led to refinements in translation techniques that allowed construction drawings to be prepared from which builders could bid building jobs competitively but equitably.

Construction quickly became a business, an industry, and a profession. Construction practitioners began to organize into design and building disciplines as the Chicago World's Fair, the Schlesinger and Meyer Store (later Carson, Pirie Scott), the Empire State Building, and other forerunner projects were designed and built. Many of these early pioneer efforts still stand and serve current needs. Construction professionals seemed to concentrate their efforts on significant architectural, engineering, and construction advance-

4

ments. There were disputes, jealousies, and conflicts within the building professional's closed circle. The issues were resolved among the practitioners with very little advice or interference from outside forces. The high standing of the construction professional encouraged a feeling of trust by owners, contractors, designers, and planners. Contemporary building teams got along reasonably well, solving their own problems.

Out of this increasingly complex but well-defined practice evolved the discipline-oriented operational organizations of the late 1920s. Separation of design and construction offices into related but separate disciplines or functions became commonplace. This, in turn, called for producing construction documents distinct and separate from the actual construction activity. Full-service design and build companies[1] often took on single-source responsibility for a total facility. They, too, were divided organizationally into separate operations, providing services such as real estate, financing, land planning, building design, construction, and facility operation and maintenance. These operations were usually managed by department or division managers reporting to a single executive group.

The onset of a sizable economic depression in the late 1920s literally brought all design and construction to a halt. A very slow recovery period followed and until the late 1930s, when the sounds of war began to rumble, increased construction volume was not to be seen in the U.S. economy. However, the Depression had highlighted the need to better integrate aesthetics and mechanistic concepts into the economy of the society that had to buy and use these facilities. The 1930s were, oddly enough, marked by significant struggles between engineers and architects. The conflict between the two led to a reluctant and uneasy merging of architecture with science and technology. The need for the merger of philosophies was made vividly real by the situation that in 1933 found as many as 85% of all architects and engineers unemployed. By the end of World War II, in the late 1940s, most architects and engineers in the construction disciplines were either in the armed forces or were working on military construction, and we saw the start of a modest improvement in design and construction business volume.

The 1930s and 1940s saw design and construction of facilities move into a commonsense pattern that produced functionally effective design better suited to the needs of the user than had been the case previously. This approach was required by the growing technical business sectors, in which new facilities were being constructed that were forced through competition to better match needs with functions. Economy of construction was a major objective of the post–World War II era. The corresponding design simplification effort was aimed at taking unneeded and often very expensive features out of facilities

---

[1] Those construction businesses that accept the full responsibility for designing, building, and often owning, leasing, and operating facilities. Such organizations usually provide single-point responsibility to the client. Design and build firms originated as a business form late in the nineteenth century.

designs. Some of these perceived excesses were removed from the structure and the technical systems. Most, however, came out of the ornamental or decorative features of the building.

Meanwhile, the various parties to construction were getting along with each other, despite their differences of opinion, by holding onto a strong belief in the value of accepting the construction project team as a group of highly trained people working as informal partners. The individuals had unwritten rules that seldom were violated and most often followed. The shift of design emphasis during this period had many benefits and some shortcomings. It forced architects and engineers to address continually the problem of producing truly well-functioning, economical structures. Buildings took on a new look that to a great extent resulted from this economic and technical disciplining. The new look in commercial and industrial buildings was generally applauded, even though building appearances were often excessively severe and cold.

Blending of disciplines continued full force in the late 1940s and early 1950s. During this period, a new awareness of the need for further merging of the efforts of construction professionals began to appear. It was reflected by extensive interdisciplinary efforts in urban land planning and in large regional shopping facilities design. Those who actually constructed facilities that were designed by architects and engineers began to make their presence felt. Much of this influence came from the increased educational level experienced after World War II. Many construction tradespeople and technicians who had served with the armed forces were encouraged by government subsidies to return to college and continue their education in the theoretical aspects of design and building. There they were exposed to the complexities of construction economics as they affected planning and design. This combination of practical training in the trades, often in the military, and the technical and business education at college provided the basis for a management system in which these future construction contractors and managers were to play a major role.

Design and construction practitioners became more and more technically dependent on the efforts of manufacturers, fabricators, and constructors. In addition, the rules governing design and construction became more complex and difficult to interpret. These rules were expressed in many ways—by building codes, ordinances, labor guidelines, governmental regulations, and many other decrees of all kinds at all levels. As the construction business grew, prospered, and changed between the 1950s and the 1990s, related business also grew, prospered, and changed. The result is that today, construction is a business and profession that has collected an increasing retinue of followers providing support functions. Support functions are activities which people perform that bring resources to the point of use. Support actions are vital to ensuring proper execution of material and equipment placement in its permanent position. The other main category of activities are performed by people in *ex'e'cutive*[2] positions. Ex'e'cutive activities are those that put the available resources in place.

In construction the primary product is a completed facility. It might be an office building, a school, a stadium, a park, or any of literally hundreds of other types of facilities created to improve our environment. Organizations that design and construct most buildings are those that perform the basic functions of planning, programming, design, construction, and public protection. Within these basic actions exist a variety of supportive and ex'e'cutive positions that must be filled by tradespeople, technicians, planners, engineers, architects, and other specially trained and educated professionals. These people are drawn to the construction business because of its excitement and ever-present challenge of visible, complex work that contributes mightily to a civilized way of life. However, the design and construction professional is now facing a serious and powerful deteriorating influence on his and her work. Beginning in the mid-1960s, our society began to embrace the concept of outside, third-party decision making to solve problems resulting from differences in opinion. This attitudinal shift accelerated the trend of technical design toward external liable involvement. It also encouraged an increased use of mechanical and electronic systems to draw, design, construct, and control the work of an industry that previously had been almost totally people operated.

Rapid growth of construction volume and complexity has continued to stimulate an increase in the potential for destructive conflict and risk passing to those not positioned to take such risks. The high growth rate has concurrently generated a powerful litigious movement working inside and outside the construction business. Unfortunately, this litigious movement has had a damaging negative influence which has added little, if any, value to the end product of planner, designer, and constructor. The reason is simple: Expending excessive resources to determine who, if anyone, is right or wrong, and to what degree, usually adds nothing of quality or significance to the end-of-construction product. Issue resolution is commendable; expending excessive resources is not. Overreactive and superfluous conflict resolution efforts add significantly to the cost of design and construction by forcing key personnel to engage in lose–lose[3] fighting. Fortunately, most planners, architects, engineers, contractors, owners, and users want to improve society by the synergistic application of their combined talents and their training and education. This is what this book is all about.

Even when the design and construction practitioner does focus on professional performance, he or she may be thrown into conflict by forces that appear to be outside the control of those being damaged. Most times the problems that are encountered in a day-to-day construction environment can be solved by intelligent people working among themselves toward a mutual solution (see Chapter 5 and Appendix D for a review of design and construc-

---

[2] *Ex'e'cutive* is used here as intended in "the executive carries out, or executes, a duty."
[3] A conflict in which there are no winners.

tion problems). Resolution of disputes also consumes much more of the construction professional's time than is usually justified. The mere filing of a legal claim by a contractor escalates the claim amount since a settlement in binding resolution is often for only 20 to 50% of the amount claimed. From the filing point, the time required to follow the legal settlement process consumes valuable time that can usually be spent more profitably in other areas of the organization's work.

Another debilitating effect of excessive conflict is the erosion of income potential normally available in the professional's fee and allocated to achieving program, design, and construction excellence. When a construction company project manager and an estimator are required to spend unreimbursed time documenting a job claim and reestimating the impact of the claim event, they are not applying resources to improve performance on present or future work. This will diminish profit potential on present and future work. Destructive conflict has seriously harmed the abilities of many small and midsized design firms to survive. Some of these organizations are apt to be driven into bankruptcy or be forced to sell to other organizations just to survive the drain of their operating capital into the sump of dispute resolution. The effect of excessive conflict is seen most dramatically in the reluctance of the architects, engineers, contractors, and owners to retain other firms that seem to attract litigation and binding third-party settlements. For instance, construction firms engaged in current disputes which they hope will be resolved by binding arbitration or litigation may have difficulty getting sufficient bonding to maintain the type and size of workload they need to survive. Or they might find usually regular customers abandoning them.

In design, an architectural and engineering firm engaged in a dispute about a design problem might see a sizable increase in its professional liability insurance. This, in turn, can seriously diminish the opportunity to obtain new work and finance its operations. Firms in external conflict are put in a difficult competitive position irrespective of the cause or merit of the dispute in which they are engaged. Another dramatic change in the construction industry has been a transition since the early 1950s to increased demands for higher technical abilities and increased proof of competence of the participants in the process of generic construction.[4] This is especially so in the builder segment of construction.

Immediately after World War II, it was unusual to find college graduates or registered professionals in contractors' offices, and there were practically none in their field staffs. This began to change around 1960. The shift was stimulated by increased use of project delivery systems encompassing many of the early stages of program writing, financing, real-estate development, design, construction, and ownership and operation of the completed facility.

---

[4] The field of business practice that encompasses all phases of the construction industry, including programming, planning, designing, building, operating, and maintaining facilities, described best as the full set of activities shown in the line of action (see Chapter 2).

Those who embraced such generic construction techniques found themselves taking a dominant position in the business. Others who began practicing the expanded concepts in part found that their operations were much more in demand. Many of the pure, but innovative, general contractors of those early expansion days were somewhat dismayed to find that although their construction skills were still in demand for certain kinds of project delivery systems (e.g., hard-money or lump-sum projects), they were losing jobs to firms that were able to offer extended services. Owners were increasingly interested in single-source responsibility and management among their professional constructors.

In addition, the professional architectural and engineering societies increasingly accepted the concept of construction as a true technical discipline. These groups saw the need for more attention to construction actions as an extension of the concept of the protection of public health, welfare, and safety. The definitions of responsibility were stretched, expanded, and the field was further regulated. Today, many system components[5] in a facility, including chillers, elevators, escalators, boilers, roof systems, windows, generators, and similar elements, are designed, manufactured, and often installed by parties separate from the architectural or engineering designer of record. Often, these components are so complex that they demand an exceptionally high level of technical excellence in their use. Many times, they are smaller pieces of full systems specified by performance only.

This sizable increase in the demand for well-educated, highly qualified, and publicly licensed persons working as contractors is being accommodated by the professional and technical societies. Today, large numbers of functional groups address the special interests of construction-related employees. Examples can be seen in the construction functional groups of the national and state societies of Professional Engineers and in the various construction divisions of the American Society of Civil Engineers. Today, we see generic construction as a visible but vulnerable sector of our business society. We also see how this visibility and vulnerability have stimulated heavy inroads into the management and execution of planning, design, and construction by outside parties, some of whom attempt to capitalize on claims against the design and construction professional in legal actions.

It is the unwanted and unneeded incursion into design and construction by outside parties that we strive to minimize through the use of partnering. Legitimate, properly placed, competent business advice has always been an absolute necessity in the successful practice of construction. For instance, the legal profession plays a very important role when advising on contract terms and contract language during the predesign and preconstruction negotiating period. Good legal direction here is critical to the preparation of a sound and reliable contract document. Additionally, legal advice is of help when a dispute

---

[5] In construction, an assemblage or combination of things or parts forming a unitary and functional whole.

boils over into full-fledged binding arbitration or litigation. Here the position of participants can be endangered by a wrong action outside the prescribed limits of legal procedures.

It is not in these proper areas of participation that partnering concepts seek to eliminate the unneeded and undesired outsider. Instead, partnering focuses on the intelligent use of well-prepared legal documents to minimize or eliminate a need and the potential for binding arbitration or litigation. Partnering is a method of settling disputes before they erupt into destructive and costly conflict—conflict that can be resolved only by outsiders. It is a tried and effective technique for placing the responsibility for issue resolution in the appropriate time slot and on those best equipped to effect the resolution. Successful partnering requires that those in leadership positions understand how the construction industry is structured and organized. The history of design and construction is of value because it allows insights into trends of the future. To fully understand how design and construction is to be shaped so that it uses partnering and other similar concepts effectively, we must understand how the past has produced today's business form.

The keys to understanding the systems of generic construction practices are available to anyone who looks closely at the structure, organization, and components of the system. This understanding is vital to the career of all design and construction professionals seriously interested in improving their abilities.

2

KEYS TO THE SYSTEM: UNDERSTANDING CONSTRUCTION

SECTION A: THE STRUCTURE AND ORGANIZATION OF GENERIC CONSTRUCTION

Generic construction: The field of business practice that encompasses all phases of the construction industry, including programming, planning, designing, building, operating, and maintaining facilities. Described best as the full set of activities shown in the line of action.

A careful look at our construction business today reveals several well-defined components that have always been a part of the practice of generic construction. Understanding these components helps construction professionals better understand their technical work, organizational structure, and management work, authority, and responsibilities. The three main components of generic construction are:

X   *A line of action:* the general sequence followed in bringing a project on line from conception to full operation (a detailed explanation of line-of-action elements appears in Chapter 8, Section B)

Y   *A set of functions:* the specialized actions necessary to accomplish the total process as shown in the line of action

Z   *A set of participants:* the people, organizations, and working groups that combine and manage the functions by which the objectives of the project design and construction process are achieved

The three main components, X, Y, and Z, can be further divided into subelements that allow us to identify clearly the responsibilities, authority, and duties of each party involved in the project. Subelements usually contained in each of the three components are listed below.

Actions (What Must Be Done)

- Conceive and communicate
- Program and articulate
- Approve
- Design
- Construct
- Turn over
- Operate and maintain

Functions (Skills That Must Be Applied)

- Programming
- Planning
- Marketing and sales
- Engineering
- Architecture
- Estimating
- Administration
- Real estate
- Financing
- Legal
- Construction
- Leasing
- Property management

Participants (Those Who Take The Action)

- Conceiver
- Translator
- Constructor
- Operator
- Regulator
- User

The graphic representation of the three dimensions—action, function, and participants—is shown in Figure 2.1. The macro matrix shows a line of action along the horizontal or X axis. Functions are shown on the vertical or Y axis, and participants are shown on the depth or Z axis. Combining the components

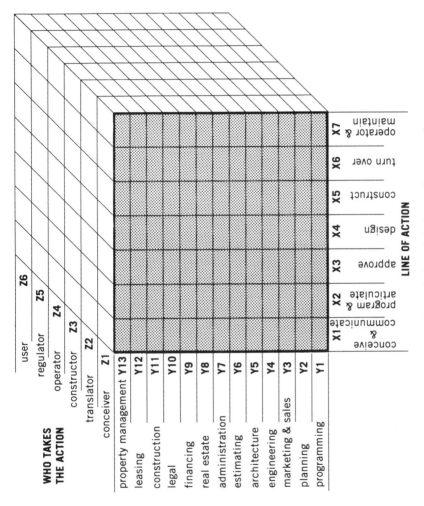

Figure 2.1 Macro matrix boundaries of design and construction.

13

es the seven phases on the line of action, the 13 functional
, participants that are key to carry a planning, design, and
:ct from its beginning to its end.
compartments in the macro matrix. The competent manage-
,al can combine these 546 compartments into a cohesive project
value-added potential far exceeds the total of the value-added
ᵣ , each compartment produces separately. *The ability to blend the
produc, ,ch compartment into a successful whole is one of the main ingredients of successful project management.* The macro matrix is considered to be
the fundamental diagram of the construction industry since it contains most
major action, function, and participant components needed in today's building profession.

SECTION B: ANALYZING THE PROJECT ORGANIZATION

Planning, design, and construction organizations are expected to produce
a specific end product such as a set of construction documents or a completed facility. The production process has specific beginning and ending
points, and those involved have identifiable individual and organizational
missions, goals, and objectives. The end product of this organization's work
is called a project.

There are two basic types of management used within the planning, design,
and construction organization: functional or ongoing management and project management.

- *Functional management* is responsible for directing business actions designed to perform continuous specialized activities. These activities are
  managed in a manner to ensure effective and successful ongoing operations that accomplish a specialized activity or duty. Examples of functional
  departments are accounting, architecture, legal, structural engineering,
  estimating, purchasing, maintenance, and similar ongoing functions. Functional management provides technical and administrative support to project managers and superintendents.
- *Project management* is responsible for directing business and technical
  actions to achieve defined objectives for a specific project. Project operations consist of related, discrete project work having a definable beginning
  and ending. Project managers are usually provided with ongoing services
  such as accounting, engineering design, estimating, and others by the
  functional operation groups. Managing the production process is called
  project management.

A list of the two kinds of actions that are managed by the generic construction professional is shown below.

Functional	Project
Engineering design	Cogsdill Elementary School
Architectural design	Franklin Center Office Building
Estimating	Bengst Research and Development
Specification writing	Stoddard Warehouse
Legal	Olanta Data Systems, Ltd.
Accounting	
Personnel	
Shop	
Yard	

The functional manager is concerned with specific disciplines and the application of special knowledge to those disciplines. The project manager is concerned with combining functions to produce a product. In construction it is a completed facility.

Functional and project management groups both require internal and external resources applied to their respective continuous or project-related work. Within the two types of management we usually find two subtypes of action: ex'e'cutive and supportive. Successful implementation of the organization usually requires excellence in both. The ex'e'cutive group is usually responsible for actually *doing* the direct actions needed to achieve the total organization's mission, goals, and objectives. Members of this group exert day-to-day control of project expense and income flows and are those who provide and manage line operations. Supportive divisions are usually responsible for actions that bring resources to the point of use by the ex'e'cutive project group. They manage and direct the main administrative and staff functions so critical to smooth running and effective executive line operations. Below are shown some of the descriptions that often are applied to supportive and ex'e'cutive elements.

Supportive	Ex'e'cutive
Administration	Operations
Nonproduction	Production
Overhead costs	Direct costs
Staff	Line
Ongoing	Project
Backup	On line
Front end	Closing
Burden	Costs

There are an immense number of participant interrelations possible within the two basic kinds of management—functional and project—and the two fundamental kinds of possible action—supportive and ex'e'cutive. A major factor contributing to partnering success is an understanding of the interrelations between the people and organizations that are stakeholders or other types of participants in the project.

Within management and management action, there are five key business and technical relations, some of which are present on all projects and in all organizations:

1. Formal functional
2. Reporting
3. Informal
4. Staff
5. Temporary

The definitions and use of these relations should be clearly understood by the various responsible project staff as early as possible, even before a partnering effort is made (for a detailed discussion of the five relations, see Chapter 7, Section C).

A successful partnering effort is, in the main, dependent on a smoothly functioning ex'e'cutive operation. The majority of the people who will make partnering commitments at a charter meeting, at an issue resolution conference, or in a project evaluation session will be ex'e'cutive managers. However, a strong support or administrative staff must be available to perform the logistical duties needed to feed the line personnel. As in a military operation, the line operations of planning, design, or construction organizations are logistically dependent on a well-managed flow of supplies, material, and personnel to maintain their prime efforts.

It is essential that key support personnel be aware of what partnering is, how they fit into the management and action picture, how they might contribute to writing and implementing the charter mission and objectives, and how they assist in achieving success from the use of the issue resolution policy and the evaluation system. To emphasize a key element in partnering—as managerial participants—we must know the nature of the organizations involved and understand how to encourage high-value contributions from supportive and ex'e'cutive personnel as well as from functional and project staff.

SECTION C: LEGAL COMPONENTS
OF THE CONSTRUCTION PROJECT

The legal agreement among a construction project team is a pivotal document around which positive or destructive conflict flows. The nature of the flow is dependent primarily on how clearly and soundly the agreement conditions are written. The legal structure dictating how a design and construction project should be implemented is derived from the conditions spelled out in the contract agreement. This discussion will center on the nature of the agreement rather than the legal language of the contract. The two are, however, closely interrelated and interdependent.

A well-written legal agreement between parties to a planning, design, and construction project should contain four major components, each of which is crucial to ensuring the health of the entire program of work:

1. Agreement premises
2. Authority limits
3. Payment method
4. Scope of services

Two important generic construction agreements which contain the four components are the consulting services contract and the construction services contract. A consulting services contract, often called a professional services contract, defines the relationship between a client and an advisor who is retained for a fee. A construction services contract defines the relationship between a client and a contractor who has agreed to do a specific amount of construction work for a specific amount of money.

The various combinations possible using the four basic components for the two kinds of contracts provide an opportunity to tailor consulting and construction services to the specific needs of the client or principal. They also allow the client, consultant, and the constructor to receive or offer services in a manner best suited to their talents and to tailor a business form to their needs. A combination of the four components used in each contract defines the type of project delivery system to be used. Combinations possible in consulting services contracts are shown in Table 2.1. In a construction services contract, components A and B are the same as for the consulting contract. However, component C, payment methods, and component D, scope of services, are considerably different. These components indicate the change in position of a construction manager, or consultant who is advising a client, as compared to the construction professional who is constructing a facility for a client. Construction services contract components are shown in Table 2.2.

SECTION D: PROJECT DELIVERY SYSTEMS[1]
AND THEIR CHARACTERISTICS

The project delivery systems described by the contract components above define the business forms by which the conceiver, owner, user, designer, constructor, and regulator, working together, are to produce a facility and convey it to the ultimate owner and user. The construction professional can formulate

---

[1] Methods of assembling, grouping, organizing, and managing project resources so as best to achieve project goals to objectives.

TABLE 2.1   Consulting Services Contract Components

A. Agreement premises
　　1. Totally negotiated: broad range of criteria used in selection
　　2. Partially qualified: moderate range of criteria used in selection
　　3. Totally qualified: very narrow range of criteria used in selection
B. Authority limits
　　1. As agent, can commit fully for a principal
　　2. As limited agent, can commit partially, and as defined, for a principal
　　3. As contractor, commits to consult only for a defined scope and at a specified
　　　　price
C. Payment methods
　　1. Fixed total (includes direct costs, overhead, profit, and all other costs)
　　2. Multiplier × payroll costs (includes direct costs and overhead) + fixed
　　　　fee + expenses
　　　　a. With limit or cap on:
　　　　　　(1) Payroll hours
　　　　　　(2) Expenses
　　　　b. With no limit on:
　　　　　　(1) Payroll hours
　　　　　　(2) Expenses
　　3. Multiplier × payroll costs (includes direct costs, overhead, and
　　　　profit) + expenses
　　　　a. With limit or cap on:
　　　　　　(1) Payroll hours
　　　　　　(2) Expenses
　　　　b. With no limit on:
　　　　　　(1) Payroll hours
　　　　　　(2) Expenses
　　4. Percent of total construction cost (includes payroll costs, direct costs,
　　　　overhead, and profit)
　　　　a. Expenses included
　　　　b. Expenses reimbursed separately
D. Scope of consulting services
　　1. Single responsibility
　　　　a. All in-house
　　　　b. In-house and outside consultants
　　2. Split responsibility
　　　　a. In-house, client, and other prime consultants
　　　　b. In-house and client
　　　　c. In-house and other prime consultants

many organizational and operational project models from the material contained in the macro matrix (Figure 2.1), the consulting services contract components, and the construction services contracts components. These together
define how a generic construction project is to be assembled, organized, programmed, designed, built, and occupied. They describe the project delivery
system to be used on the project. The project delivery system is conceived

TABLE 2.2   Construction Services Contract Components

A. Agreement premises
   1. Totally negotiated: broad range of criteria used in selection
   2. Partially qualified: moderate range of criteria used in selection
   3. Totally qualified: very narrow range of criteria used in selection
B. Authority limits
   1. As agent, can commit fully for a principal
   2. As limited agent, can commit partially and as defined for a principal
   3. As contractor, commits only to a defined scope of work and at a specified price
C. Payment methods
   1. Fixed total (covers all costs including direct costs, overhead, and profit)
   2. Time and material + fixed fee
      a. Limit on time-and-material costs [guaranteed maximum price (GMP)] with shared savings
      b. Limit on time-and-material costs (GMP) with no shared savings
      c. No limit on time-and-material costs
   3. Time and material + % fee
      a. Limit on time-and-material costs (GMP) with shared savings
      b. Limit on time-and-material costs (GMP) with no shared savings
      c. No limit on time-and-material costs
   4. Conditional bonuses or penalties
      a. Incentives/disincentives
      b. Liquidated damages
      c. Bonuses
D. Scope of construction services
   1. Single prime contract responsibility—in-house trades and subcontractors
      a. All project trades
         (1) Provide management
         (2) Provide design
         (3) Provide construction labor
         (4) Provide construction material
      b. Limited project trades
         (1) Provide management
         (2) Provide design
         (3) Provide construction labor
         (4) Provide construction material
   2. Split total contract responsibility with owner and/or other prime contractors
      a. All project trades in prime construction contract
         (1) Provide management
         (2) Provide design
         (3) Provide construction labor
         (4) Provide construction material
      b. Limited project trades in prime construction contract
         (1) Provide management
         (2) Provide design
         (3) Provide construction labor
         (4) Provide construction material

and modeled in words, just like a building design is conceived and modeled in wood, plastics, and wire. The word-based model gives a narrative description of the system proposed. The three-dimensional physically based model provides a miniature replica of a building. Both allow nondestructive simulation and testing of various assemblies before a final selection is made.

One of the most elemental word models is the traditional project delivery system. Referring to the consulting services contract components and the construction services contract components, this traditional model often contains the components outlined in Table 2.3, the traditional project delivery system model.

The project delivery system defined in Table 2.3 usually will have an overall unique set of characteristics. Usually:

1. Checks and balances are built in from the start.
2. Construction decisions are based on capital costs.
3. Participant selection is made by cost-competitive bidding.
4. Job control is highly centralized in most stages.
5. Project is being built for owner and/or users.
6. Contract documents are completed before bidding.
7. Bidders are selected from a short list derived from a long list (occasionally, the owner will use the long list only).
8. Bonding is required.
9. Site preparation and other early work is done by owner before building construction starts.
10. Majority of attention is given to the need and want list. Wish list is usually considered a luxury.

Another frequently used project delivery system is the design and build technique. This system, and others like it, are often called nontraditional systems. Design and build places a single organization in charge of both designing and building the facility. If the design and build delivery system is analyzed in the same manner as the traditional system, we find that the components are considerably different. Referring to Tables 2.1 and 2.2, the design and build model can be described by the set of components listed in Table 2.4.

The project delivery system described in Table 2.4 is often called a nontraditional system. Nontraditional systems have certain basic characteristics that suit them to negotiated techniques and concentrated responsibility patterns. Such a system might have these characteristics:

1. Checks and balances evolve as the project proceeds and as the need arises.

TABLE 2.3　Traditional Project Delivery System Model

Consulting Services Contract Type—A2/B3/C1/D1

- A2　*The agreement is based on moderate multivalued competition.* The consulting team is selected from a screened list of those who respond to a request for qualifications (RFQ). The screening is usually by a selection task force and is based on ratings given the respondents from a rating checklist. Final selection is usually by the selection committee through direct interviews of the finalists. Services and fees are usually negotiated with the consultant firm.
- B3　*Authority limit is restricted to action as a contractor only.* The consultant is considered as serving the client by agreeing to provide a well-defined end product, the building construction documents, for a stipulated sum. No agency is intended.
- C1　*Payment for services is a fixed cost, including all direct costs, overhead, and profit.* The total cost is the full amount the contractor is to be paid by the client except for costs of agreed-upon changes to the project.
- D1　*The scope of consulting services is very broad.* Services are provided under a single consulting responsibility using all in-house professional services or a combination of in-house services and outside consultants reporting to the lead or prime consultant.

Construction Services Contract—A3/B3/C1/D1,a1, 3, 4

- A3　*The agreement is based on narrow, multivalued competition.* This selection process usually means that a complete set of working documents is available and has been bid competitively by a set of prime contractors who have been prequalified by such elementary requirements as bondability. Sometimes freedom from lawsuits, current or pending, is considered one of the basic prequalifications.
- B3　*Authority limit is restricted to action as a contractor only.* The contractor is obligated to provide a well-defined end product (the building) for a stipulated sum of money.
- C1　*Payment for services is a fixed cost, including all direct costs, overhead, and profit.* The total contract cost is the full amount the contractor is to be paid by the client, except for costs of agreed-upon changes to the project. This type of project is called a hard-money, lump-sum, or fixed-cost job.
- D1, a1, 3, and 4　*All construction work specified is to be provided by the contractor and the subcontractors.* The management and the provision of all construction labor, materials, and equipment are the responsibility of the prime contractor. The prime contractor is liable for their cost irrespective of who provides them. The payment by the client for these costs is to the prime contractor, who, in turn, reimburses the subcontractors, suppliers, and vendors.

TABLE 2.4   Design–Build or Nontraditional Project Delivery System Model

Consulting Services Contract—A1/B2/C1/D1a or b

The consulting services delivery system of a design–build project is usually determined by the design–build contractor and the consulting team. It may be similar to that described in Table 2.3, a traditional project delivery system model, described previously. Sometimes, however, it is best due to the overlapping design/construction needs of design–build to issue the contract documents in packages. In such a situation the consulting services contract would take the following form:

- A1   *The agreement is negotiated based on a broad range of selection criteria.* Usually, the design–build firm has either in-house design capabilities or works closely with consulting firms that are knowledgeable about the construction practices of the designer–builder. Thus the consultants are selected quickly and usually work under a negotiated agreement.
- B2   *Authority limit is sometimes on a limited-amount agency basis.* The design team may have to commit to limited cost expenditures by its client, the design–build construction organization directing the project team. This is usually the case where cost decisions made are affected by a need to move rapidly in the design process.
- C1   *Payment for services is a fixed cost, including all direct costs, overhead, and profit.* This total cost is the full amount the consultant is to be paid by the client, except for costs of agreed-upon changes to the project. Other payment methods may be designated, particularly if the consulting division is a separate cost center within the total design–build firm.
- D1a or b   *Scope of consulting services.* Services are provided under a single consulting contract responsibility using all in-house professional services, or a combination of in-house services and outside consultants reporting to the lead consultant.

Construction Services Contract—A1/B3/C2a/D1-a1, 2, 4

- A1   *Agreement based on negotiations using multiple selection factors.* Usually, a full set of construction documents is not available, as design and construct proposals are solicited and received. Proposals are usually derived from a set of performance specifications prepared by the prospective client. The basis of selection may include price but will also consider such factors as:
- History on similar projects
- Reputation in the community
- Reputation with subcontractors
- Staff availability for assignment to the project
- Degree to which the work is placed under a single control management
- Financial strength of the design and build organization
- Services available within staff
- Control of desirable property
- Availability of project financing
- Design ability

TABLE 2.4 (*Continued*)

- Construction ability
- Ability to meet desired or mandated minority requirements
- B3 *Authority limit is as a contractor but may be extended into modest agency responsibilities.* Often the design–build contractor is selected and entrusted with limited cost authority because of unique features found in the firm. For instance, long experience with the design–build firm may have developed a trust by the client in its abilities to complete successfully the construction of grain elevators to the exact needs of the client. This would tend to direct selection toward a certain few firms and would be heavily influenced by the confidence the owner has in any one of these firms.
- C2a *Payment for services is on a time-and-material contract with a fixed fee. The total cost is to be within a not-to-exceed cap or limit [guaranteed maximum price (GMP)]. The client and the contractor are to share in the savings below the GMP on a basis to be defined.* The total cost of the facility is that actually incurred in the time-and-material construction process plus the agreed-upon fixed fee. The fixed fee tends to eliminate a temptation to spend more if the fee is tied to the direct costs. The share in the savings is an incentive to both the contractor and the client to build economically.
- D1 a1, 2, 3, and 4 *All construction specified is to be provided by the contractor and subcontractors.* The management, and the provision of all design, construction labor, materials, and equipment, are the responsibility of the prime contractor. Payment for all work is to the design–build contractor, who, in turn, pays the subcontractors (which would include the designers on the project).

2. Construction decisions are based on capital costs, maintenance costs, operating costs, desired project quality, and desired investment return.
3. Lead participant selection is made on professional and technical abilities and on reputation and past performance together with estimated project cost.
4. Job control is somewhat decentralized during early program and design stages, with progressive centralization as the working document and construction phases are approached.
5. Project could be for a variety of conceivers and prime movers, including owners, users, investors, developers, funders, syndicates, governmental agencies (privatization), and groups assembling capital to gain desired returns on investment.
6. Construction is often closely dovetailed with design of the project. Design usually proceeds with construction guidance and advice from a construction discipline.
7. Capital cost is often negotiated from the pro forma base and reduced in stages to a guaranteed maximum price (GMP).

8. Need for bonding is usually minimized or eliminated by careful selection procedures to maximize the probability of success.

9. Site preparation and expense work is often done by various members of the project or program team.

10. Design and construction is heavily influenced by consideration of the needs, wants, and wishes of the participants.

Occasionally, experience with extended project delivery systems encourages an entrepreneur to work in even more compartments of the macro matrix than is usual. This activity and functional expansion may ultimately lead to provision of all services needed from conception to operation, including acquisition of real estate and ownership and maintenance of the occupied facility. In a land development program where the physical facility is to be built by a land developer for a specific tenant—often called *build to suit*—the line of action that is the X axis in the macro matrix may take on the form shown in Figure 2.2. In this model we often find that the grouping of the components fits a scenario where the driving force is the developer or a potential tenant for a building (conceivers). In this delivery system the developer retains total control of the project from conception through occupancy and operation of the facility. Within the model the line-of-action components are shown in three groups:

- Group A, *Plan, design, construct line of action,* shows the required activities from the recognition of need through to the discharge of responsibility. This line of action is nearly identical to the X axis in the macro matrix (Figure 2.1).
- Group B, *Development Phases,* shows the major phases of the project. The terminology is uniquely tailored to the nature of the development business. Often, the grouping content and arrangement will be changed dependent on the development staff, its organization, and its operations.
- Group C, *Who Participates?,* identifies the various parties of a development project by responsibility and authority. Notice that no outside participants are named in the model. Contacts with these parties is maintained by the interior team participants.

The generic construction professional should be well acquainted with the many methods available to him or her by which a facility can be designed and constructed. The delivery systems described above are encountered frequently and have become common practice in the construction business. Each has its place in our economic system. The potential for destructive conflict exists in both the traditional and nontraditional methods irrespective of their contract characteristics. However, the more trust and faith that exists in the relations between planners, architects, engineers, contractors, and owners, the less chance there is for a project to become claim prone. It appears that the more

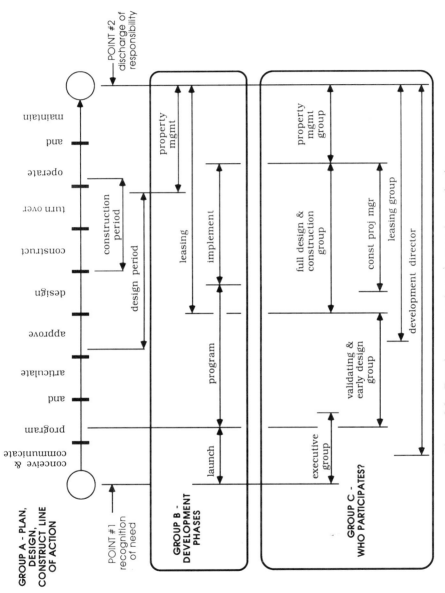

Figure 2.2 Development cycle actions and organization.

25

a set of project contract characteristics favor negotiated agreements, limited agency, payments with caps or limits, and centralization of project control, the less chance there is for debilitating arguments and third-party binding resolutions of these arguments. Partnering can help achieve this within both the traditional and nontraditional methods of design and construction.

3

THE NATURE OF CONFLICT

Conflict is a state of disagreement and disharmony.

Conflict may clarify or muddy the method by which we distinguish good from bad. It follows that conflict can be either, or both, destructive or constructive. Destructive conflict is highlighted by animosity or disagreement, which, in turn, lowers the potential for an individual or an organization to succeed. Positive conflict is hostility that is managed so that its resolution raises the potential for individuals or organizations to succeed. For instance, an argument which, in the heat of angry words, ends with "I'll see you in court" is usually destructive unless followed by a retraction of the threat. A disagreement about late payment by an owner can be positive if resolution of the dispute improves future owner payment practices—or future contractor billing practices—or both.

Conflict in construction is often caused by:

- Lack of understanding that conflicts lead directly to results
- Frustration over a lack of control of events affecting performance
- Differences in goals and objectives of parties in the project
- Lack of understanding about the needs of others also involved in the planning, design, and construction process
- Resentment or dislike resulting from a perceived lack of value added to projects by those responsible for adding value.
- Excessive technical and legal delays to resolution of conflict
- Excessive demands on resources normally depended on to assist in the resolution of conflict

- Greed
- Incorrect assumptions made from biased perceptions
- Demands for higher quality than specified
- Failure to meet commitments
- Insufficient time to make required decisions
- Lack of ability to do the job
- Poor or inadequate training
- Inadequate credentials to do the job
- Indifferent leadership
- Actual or perceived overwork
- Bad blood among participants
- Desire to take advantage of those in weaker positions
- Misplaced attempts to demonstrate who is in charge

The list could be extended almost without end. For instance, in a recent series of 23 meetings designed to explore team techniques and how to get along better on construction projects, the participants identified more than 2800 ways that people cause problems for designers, owners, contractors, users, and regulatory agencies in the planning, design, and construction business. Boiling these down to major dispute categories shows, for this particular set of responses, that there are at least 45 major classes of potential conflict (see Appendix D).

To begin turning conflict into positive channels, it is important to adopt some basic assumptions about people. A good starting point is to trust the concept that *most people are honest, concerned, desirous of challenge, need attention, and welcome help in times of trouble.* This might seem somewhat naive in the face of today's apparent cynicism and dishonesty. However, test after test of this concept has proven its truth—and, most important, this belief is widely held by competent professional construction practitioners. Assuming the best about people has some very practical benefits:

- It allows identification of those who do not subscribe to the concept and who might be potential proponents of destructive conflict.
- It gives confidence to those who accept the concept in guiding their behavior toward others.
- It provides guidelines in informal, moral agreements as well as in structured legal agreements.

Active use of such assumptions plays a very important role in making partnering a successful and flexible system designed to improve the probability of success in the generic construction profession. A common belief about the construction industry and the people involved is that most architects, engineers, and owners are born adversaries of the contractor and all others in the building business, and that contractors are concerned only about profit, even at the expense of fulfilling contract requirements.

This is simply not true. I recently had a phone call from a concerned city project manager whose construction traffic was constantly closing streets in a downtown location. The street closings were driving a small retail candy-maker out of business. She had been doing a good business there for 18 years but the retailer said if her business was disrupted for another three months, as planned in the construction of a new municipal parking deck, it would mean the end of her shop. The parking deck traffic problem was being caused by the large number of truck maneuvers needed to unload and erect precast structural elements with a single crane. This equipment configuration necessitated several forward and reverse moves to allow the use of a single crane and to keep unloading time to a minimum.

The city's project manager asked me what he as the responsible agent of the city, the contractors, and the architect/engineer could do to keep the shopkeeper in business. One solution already thought through by this concerned project manager was to have another crane brought on the job. This would allow the contractor to erect the precast elements in a hoisting pattern that allowed the trucks to move in and out of the site without necessitating a reverse move. Adding another crane to the hoisting assignment would cost the city some of its contingency fund but would additionally help smooth out and maintain auto traffic.

The discussions, not yet complete at this writing, will undoubtedly result in a compromise traffic plan. This plan may cost the city additional dollars from the contingency fund. It might even cost the contractor a very small part of his profit. However, as the project manager put it, "keeping the residents of downtown in business is just as important as bringing in new downtown business through added parking." The contractor felt the same way, as he showed by his actions and cooperative spirit.

This project manager and the project team thought enough of the candy-store owner to want to help keep her in business. The contractors thought enough of the storekeeper, the city's project manager, and their responsibilities to the city's business health to want to help, if financially possible. Of greatest interest to us here, however, is that the collective and cooperative approach by the team management was undertaken in a way that kept potentially dangerous conflicts under control. People attempted to help, not hurt, others.

I have found there are seven actions managers can take to smooth out and resolve potentially destructive conflict before a point of no return is reached. The good manager should remember that before resorting to a hard, unrelenting stance, he or she may be able to make progress by taking some or all of the actions. The actions can be used individually or in various combinations. All are essential to effective dispute resolution.

1. Understand the cause of the conflict.
2. Put yourself in the other person's shoes.
3. Understand the relative importance of resolution versus nonresolution.
4. Become competent in proper application of the technical and professional management tools of our profession.
5. Don't lie—always tell the full truth.
6. Understand thoroughly the obligations you have to society and to your clients, your employer, and your peers.
7. Understand everything you can—not just your own field—and work to be effective in managing intersections of diverse interests.

What is meant by each of these statements? A short parable for each action might lead to a better understanding of their application to construction conflicts and disputes.

Action 1: Understand the cause of the conflict

Case Study 3.1: The Accidental Double Payment

The Corporate Construction Company normally tries to pay, promptly and fully, all properly submitted invoices, relying heavily on the honesty and integrity of those submitting the invoices. Two months ago, John Post, Triangle Computer's president, a consultant and vendor to Corporate, informally mentioned to Bruce Allen, president of Corporate, that a double payment had been made to Triangle on one of their invoices and that he would like to know how Corporate would like Triangle to process the return of the overpayment. Meanwhile, Fred Snow, Triangle's accountant, had routinely inquired of Corporate's accounting manager, Susan Larkin, how she would like to process the return of the overpayment. A method was worked out and the return and crediting made promptly.

The Corporate president, Bruce Allen, meantime had forgotten his informal conversation with John Post, Triangle's president, until just yesterday, when he mentioned it to Corporate's accounting manager, Susan Larkin. This routine follow-up remark, coming late and occurring after the matter had already been taken care of at the originating level, started a whole set of side effects, each of which had to be explained or communicated to others in the communications chain. Some of the side effects were:

1. Bruce Allen of Corporate Construction, not knowing the chronology of the resolution, could not figure why John Post of Triangle was bothering him with an accounting matter that was between two staff departments in their firms.
2. John Post of Triangle was embarrassed that his relations with Corporate, a good customer, had been endangered by his casual remark

about the overpayment. He also expressed poorly concealed anger at Fred Snow, his accountant, for not informing him of what corrective actions were taken.

3. Susan Larkin of Corporate was embarrassed and angry that Triangle's people had brought her apparent error to the attention of her boss, Bruce Allen, when she thought the matter had been settled at the proper level of their respective organizations.

4. Fred Snow was very upset because he perceived that he had taken what he considered an honest and routine action to resolve a minor accounting glitch. His action had suddenly been translated by others, unaware of the true circumstances, into a finger-pointing critique of him, resulting in his chastisement by his boss for "not letting me know what's going on out there."

There are many fallouts from this situation which ended up costing almost as much as the extra payment—all resulting from an innocent but unneeded statement made during a lull in a conversation and forgotten about until some event triggered its being relayed to those most affected. This illustrates a simple case of not understanding the cause of a conflict.

Often, the parties most affected by a series of events leading to conflict do not even know they are involved. A large percentage of disagreements that occur in any planning, design, and construction-related situation have their roots in misunderstandings about the cause of the conflict. And, if those affected are aware of their involvement, they may not understand how or why. An example of obscure communications about conflict can be seen when we yield to the everybody-must-know (EMK) system of keeping a department staff informed. In the EMK communication system the staff is required to attend meetings, absorb information, read documents, communicate with others, and sometimes act on the information, whether or not it is pertinent to their immediate work. The reason often offered by the EMK-style manager is that this is being done so that all people can be aware of all things. The use of such a broadband communications method may lead to incorrect management conclusions about what people actually know about the activities of other departments in the organization.

Case Study 3.2: Carl, Tom, and Anita

Carl, president of the Ajax Company, may assume that Tom knows something about Anita's new project just because Tom and Anita were in a meeting where her new job was discussed. However, at the time of the meeting, Tom was totally preoccupied with closing out his own project. So he used his need-to-know (NTK) filter to block out worries, concerns, and information about Anita's project. When Anita tells Tom that the president, Carl, expects Tom to help her in the project startup, as discussed in the project meeting, Tom is totally at a loss, simply because the EMK technique did not register some important information with him.

Some simple principles of analysis can often put information into the correct perspective and into the right hands to see that it resolves conflict rather than causing it.

1. Try to pinpoint communications targets on a need-to-know (NTK) basis.
2. When the need-to-know is recognized, act on it.
3. Get to the root of conflicts quickly and determine the cause of the dispute, the people involved, and the circumstances surrounding the conflict.
4. Seek firsthand information when collecting information about a problem that has surfaced. Don't depend on gossip, the grapevine, or hearsay to provide your basis of discovering the cause.
5. Seek out the parties most affected and most knowledgeable to use for determining the facts about a conflict.
6. Determine how serious a conflict is before taking action. The best solution may be one of allowing the parties involved to resolve the matter among themselves.

Action 2: Put yourself in the other person's shoes

Case Study 3.3: The Nonmissing Dimension

Tom Beard, a young architect project manager and co-owner of a small architectural engineering firm, Beard and Strate, has just completed preparing bid package #1, Site Preparation and Foundations, for his first fast-track project. B & S had been retained by Charles Lever, the client and owner, to work with Bristol Construction, the prime contractor, and Bristol's key subcontractors to prepare a set of progressively issued bid packages. The prime construction contract with Bristol had been negotiated on a time and material, fixed-fee, guaranteed maximum basis. Bristol's contracts with the subcontractors had been awarded competitively to a selected short list of firms.

About one week after bid package #1 was issued, Bristol's field superintendent, Harold Whirley, called the owner, Mr. Lever. Whirley told him that he had found a major dimension missing in the building layout and that he would have to shut the job down until the architect/engineer provided some answers to resolving the error. Mr. Lever got on the phone immediately to Tom Beard and explained the problem to him with the request that he get out to the job site immediately to find out what was wrong and make certain that the problem was corrected.

Tom drove to the job site, found Harold, the superintendent, who said he really didn't have time to solve Tom's problems but had found the needed dimension after a further search of the site plans. He also stressed to Tom that Tom's staff had to be more meticulous in the preparation of subsequent bid packages or the project was in for a lot of trouble. So the problem had been resolved but many hard feelings were left from the residual of the words spoken and the actions taken.

The true story behind this unfortunate conflict is that years before Harold Whirley had worked for a local architectural firm as a field inspector. He had once been reprimanded for being careless about inadequate dimensional clarifications to contractors who were building jobs he was on for the firm. He had carried this perceived injustice and uncertainty with him from job to job. His continued lack of confidence in his ability to interpret dimensions often caused him to make snap decisions which resulted in scenarios such as this one with Tom Beard. Further, it had often caused him and others great difficulty with those affected by his incorrect actions.

Some of the unfortunate results of the false cry for help included:

1. Mr. Lever's confidence in the architect he had hired took a temporary dip. Restoring the owner's confidence may now require an embarrassing revelation that the cry of help from the contractor was not totally justified.
2. Harold Whirley's ability as a competent superintendent has been placed in jeopardy by his hasty call to the owner. The alternative would have been a further study of the drawings and an initial clarification call to Tom Beard. This would have settled the matter at the originating level and been to the credit of all involved.
3. Tom Beard's confidence in the prime contractor has been considerably diminished and his working relations with Bristol Construction may have been damaged.

This entire sequence might have been avoided if the superintendent had thought back to his earlier problems and put himself in Tom Beard's shoes instead of seeing it as a way of getting even with architects. Also, Mr. Lever, the owner, might have placed himself in the shoes of the young architect and suggested to Harold that he call Tom directly for a technical resolution between the parties immediately affected. Tom Beard now has the opportunity to restore good working relations by:

1. Finding out more about Harold Whirley's past and how he has behaved on jobs previously. Then he can prepare future contract documents that will provide a dimensional clarity and accuracy to Harold that cannot be challenged.
2. Having a short, nonconfrontational discussion with Harold Whirley to see how he can best work with Harold for the remainder of the

project. He should put himself in Harold's position and see what would work for both of them.

Below are some suggestions for gaining a better understanding about how people think and act in any given situation:

1. Determine early if there is actually anything you can do to resolve the situation by employing an empathic approach.
2. Get rid of as many biases as possible you have about the people involved in the situation.
3. Get rid of as many biases about the organization involved as possible.
4. Try hard to find out why the person got into the problem that he or she is in. Then ask yourself if and how you might have gotten into the same position.
5. Determine the ethical and moral implications of your feelings and actions about this particular situation.
6. Be empathic based on what you *see* people are *doing,* and on what you *know* people are *like,* rather than on what you *think* people are *thinking,* and what you *think* they are *like.*

Partnering concepts are designed to keep us empathic while providing guidelines for healthy reaction to situations that may currently be unstable or one sided. However, the basic principles of managing destructive conflict serve to help us achieve the mental discipline to better understand other people's feelings. The application of empathy in any organizational structure is an important part of becoming, and continuing to be a good manager. Often, in a fresh burst of enthusiasm, we resolve to better understand how others feel in the conduct of our day-to-day business affairs. Then, as the days, weeks and months grind on, this resolve is often dissipated in anger, jealousy, fear, fatigue, dislike, or other eroding emotions.

Case Study 3.4: The Unhappy Architectural Graduate

Such a situation might be seen in the case of Fred Liner, a field inspector just out of college, who recently joined the design firm of Witner and Associates. Fred wanted architectural design experience but had to start in the field department if he wanted to work for Witner. His disappointment has caused him to nag Abner Witner, the president, constantly about a transfer. Abner does not realize that his design department head, Laura Kostel, thinks Fred has the potential to be a great architectural designer and had offered Fred a design position. However, when Fred graduated, Laura did not have a place for him in design. To be hired, Fred took the field job and has actually done very well. In fact, his direct superior, Lars

Logan, likes Fred's work so well that he refuses to listen to his requests for a transfer.

Abner, who knows very little about Fred's background, is beginning to think that he is just a chronic complainer. Thus the cycle of disappointment, complaints, delays, nagging, and refusals continues. The result? Fred will ultimately leave or be fired. Laura has lost a potentially good designer and Lars has lost a good field administrator. Abner has two department heads and a young architect angry with him, and Fred has lost considerable time, support, and respect by his misunderstood behavior at the Witner firm. Each of these small failures produced a collective result that damaged each person involved, and all because of the inability or unwillingness of those involved to better understand someone else's point of view.

As with the staff at Witner, we often become tired of trying to understand people. We just plain wear out trying to solve everybody's problems at (we perceive) the expense of our own well-being. This is the point at which conflicts are frequently born. In the flash of a second we may let loose frustrations that create destructive conflicts. These quickly dissolve much of the patient work we did to build a sound, healthy relationship. It may now have to be rebuilt if the damage is serious enough.

Action 3: Understand the relative weights of resolution versus nonresolution   Delays in resolving conflicting positions will usually lengthen the time participants in the conflict can dwell on their hurts. Mature people quickly focus on how they got into such a conflict position in the first place, and from this mental platform try to find a path of resolution. Delays in conflict resolution often allow the ripples of dissent and dislike to spread and to be reinforced by supporting or aggravating points of view. This may then be accompanied by searches for further hurts. An example of how prompt resolution of a conflict can help the people involved can be seen in the experience of Mary Conners and Charmaine Lester.

Case Study 3.5: The Case of the Unavailable Ceramic Tile

Mary Conners is an interior designer in the construction division of a large hotel chain. She has been appointed lead designer for a new middle-priced motel prototype. Currently she is selecting materials and colors for guest-room finishes and discussing these with the production architect of record, Horland and Associates, for the first project to be brought on line in the new model chain. One of the materials that Mary has picked, ceramic tile, is from her personal experience in home interior design and is a particular favorite of hers. The principal in charge of the project for Horland, Charmaine Lester, discovered that Mary's selection had formerly been available in limited quantities for one-of-a-kind high-priced residential interiors. It is, however, no longer on the production market for large-scale use. The tile can be obtained, but only at premium prices and for large production runs.

Charmaine did not have an opportunity to mention her findings and discuss them with Mary before a major project launch celebration to which they were both invited. Charmaine informally mentioned her investigation to Mary Conner's boss, Louise Alder, at the cocktail party. The word quickly got back to Mary the next day that her boss was unhappy with her ceramic tile selections and that the cause of the problem was Miss Lester and her loose tongue. Before the word got to Mary, however, it had passed through several of Mary's peers, being magnified and elaborated with each telling.

Mary was upset and phoned Charmaine, sharply demanding to know what she meant by criticizing her decisions. She claimed the remark about her selection had hurt her work and diminished her stature with her boss and her co-workers. Charmaine quickly asked Mary if they could meet over coffee and talk it over. Mary agreed. The coffee discussion soon brought out the technical reasons for Charmaine's comments. It also resulted in an apology from Charmaine and an offer to explain personally to Mrs. Alder what was meant by the cocktail party remark. The offer was immediately accepted by Mary. A bright spot was that Charmaine's quick research into the matter had resulted in finding an alternative and inexpensive source of similar ceramic tile. She showed some samples to Mary after coffee. Mary was pleased with the substitute and with finding a new material supplier.

A situational analysis of this event shows that the key to starting the resolution was (1) the prompt ventilation of Mary's concerns to Charmaine, and (2) Charmaine's quick attempt to moderate the problem and restore good relations for the long project ahead of both of them. There will be scars on Mary's ego for a little while yet. But fortunately, the healing process was started before Charmaine's poorly timed remark was magnified out of proportion. Further, it stopped disruptive elaboration of the event through the corporate grapevine.

Some points from the case study above to remember about prompt resolution of disagreements include:

1. Do not air problems outside the participant arena.
2. Try to settle individual and organizational disputes at the originating level wherever possible.
3. Do not allow a dispute to be pushed into the informal communication and conversation systems, in this case the corporate grapevine.
4. Go into a resolution process armed with solutions that might help neutralize the conflict.
5. Stop and reconsider before you talk loosely about someone else's mistakes and deficiencies. They may be neither.

6. The most ancient of good advice: if you can't say something nice about someone, don't say anything. Loose gossip has no place in the serious business world.
7. Resolve problems and conflicts promptly and as thoroughly as time permits.

Action 4: Become competent in proper application of the technical and management tools of our profession Quite often, particularly in the heat of verbal battles, we forget that our common objectives as construction professionals are to:

- Protect public health, welfare, and safety
- Fulfill our obligations to our clients and our employers
- Maintain professional, ethical, and moral behavior in our dealings with our professional associates.[1]

Our training and education have taught us to consider technical and professional excellence as measures of performance and behavior. Usually, competent professionals prefer to spend their time concentrating on the technical tools of their jobs. It is very important for those in this situation to realize that what they know and do technically is of little use if the information is not fully used or is ignored in achieving their professional obligations. It is easy and comfortable to get caught up in looking at the application of our technical skills in planning, design, and construction as useful data for someone else.

In a recent discussion of project management, one of the participants referred to this leadership style as *over-the-wall management.* He said he thought of the arbitrary acceptance or rejection of management obligations to take action as being similar to a game of catch in which the ball is thrown over a high wall to whomever, if anyone, might be on the other side of the wall. That mysterious figure was to catch the ball and carry on the game.

A very simple illustration of failing to apply the tools of our profession properly and consequently, to provoke over-the-wall management can often be seen in the procurement of materials and equipment. Let's look at the case of the fan without a home.

Case Study 3.6: We Versus They—A Case of Over-the-Wall Management

Leonard Pritchard was Traverse Construction Company's project manager on a new auditorium being built for the Clearstream Community College. The prime general contract had been awarded to Traverse on the basis of its lowest acceptable hard-money bid for all work except fixtures, furnishings, and equipment. These items were to be provided and installed by the

---

[1] Frequently called the hierarchy of professional obligations.

owner. Leonard had assisted Traverse's estimating department to figure the masonry and concrete on the job but had his hands full closing out two other projects at the time of bidding. So he had worked no further on the bid estimate and had not been involved in the job until it was almost ready to start in the field.

It soon became apparent to Leonard that exterior access to the auditorium mechanical equipment room located under the entrance lobby was restricted. Three large exhaust fans were to be installed in the equipment room but could only be set in place before the supported lobby deck was formed and poured. A possible alternative was to leave a section of the lobby basement wall exposed and open and the area adjoining unfilled until the fans could be brought into the building. Access from the inside of the building was out of the question.

Leonard found that Traverse's estimator had talked to the mechanical contractor's estimator as job proposals were being prepared. They had both decided that the access problem could be worked out when the job had been awarded and the fans had been ordered. Also, Traverse's estimator told Leonard that since Traverse intended to sub out the concrete work he wanted to decide what to do about the fans after the concrete contract was let. The estimator said that if the concrete contractor's bid had to be qualified by accommodating the access problems, it might adversely affect the concrete subcontractor's prices.

Dates set from a prebid schedule for various operations affecting various field and procurement actions were:

March 1	Start of project in field.
May 1	Start forming lobby basement wall.
May 15	Complete pour out and stripping of lobby basement wall.
May 17	Start forming lobby supported deck.
June 3	Pour out lobby supported deck.
August 10	Delivery of mechanical equipment room fan.

By the time the contract was awarded to Traverse and Traverse had awarded to the mechanical subcontractor, the access problem was still unresolved. Even the design team and the owner's representative were unaware of the difficulties because of the reluctance of the contractors to endanger their bid position by qualifying their proposals. So we now have four parties—the general contractor, the mechanical contractor, the owner, and the design team—that have, to a considerable extent, contributed to Leonard's project management problems. The solution, whatever it may be, will undoubtedly be more expensive to resolve now and will tend to drive the project participants into adversarial roles. The cause—inattention at the proper time to a problem that was quickly visible to a perceptive project manager at the wrong time—and tossing the problem management ball over a wall, hoping someone else would catch it and keep the game alive, is leading to disaster.

Taking the proper technical and professional management steps at the right time would have saved time, money, reputations, and frayed nerves on the Clearstream Community College. Timely and face-to-face managing of the interactions among the contractors, the owner, and the design team would undoubtedly have helped to get the project off and running on a more friendly and profitable basis for all.

Variations on the theme of abdicating professional and technical responsibilities, combined with over-the-wall management, are almost limitless and contribute immensely to the proliferation of destructive conflict and the resulting arbitration and litigation in the design and construction business. By not using information with the same care and attention that went into its production, we stand liable for any conflict or disruption caused by its poor translation. Competence in use is as important as competence in creation. Through this competence we help minimize conflict and make the high wall transparent.

Action 5: Don't lie—always tell the full truth   Few people have a good enough memory to lie consistently and effectively without contradiction. Those who do are seldom trustworthy enough to maintain relations with others without serious conflict—and, almost always, the liar is trapped in nets of his or her own setting. In the planning, design, and construction business, the products generated during each step in the process of building are defined by a very visible outcome. For example, an economic report or a traffic survey is a written, physically visible document that can easily be criticized if incorrect. Working documents are ink and graphite on paper. They have a specific shape and size and are visible and readable. Construction is the final expression of truth, and lies are easily discovered as the erection of a facility proceeds.

Lies told during any of these processes are either believed or disbelieved. If believed, they result in a loss of time, credibility, money, and confidence in the liar. Even worse, they result in similar losses for those affected by the domino effect of the lie. If disbelieved, the liar quickly disappears from the process. This may have a temporarily negative effect, create a critical vacancy and lost time, cause unplanned extra costs, disrupt the process, and destroy the credibility and reputations of those involved and affected. If you tell the truth, all the equipment you will need to remember what you said is your native intelligence and your learned knowledge, both communicated clearly to those affected by your statements and actions on the project.

Case Study 3.7: The Lie That Came Back to Haunt the Subcontractor

Paul Skokie was the president of Conskoinc, a firm that designed, fabricated, and erected precast concrete components for industrial buildings. He was a reasonably well known person in the local construction business. He was also recognized as someone who might occasionally stretch the truth about his products and his abilities, particularly when he was short of work.

Frank Clark is the owner of a small architectural and engineering firm, FCL Associates, that specializes in designing warehouses and industrial plants. He places a great deal of faith in the advice of his clients, his consultants, and the contractors with whom he works. His professional liability insurance premiums were the lowest in the local design community since he had had no failures or design glitches that would increase his premiums beyond those normally due to inflation and coverage policies.

Frank met Paul at the local builder's exchange golf outing one day and mentioned a new project upon which he was engaged for a client about 100 miles away in St. Mark's County. Paul, sensing a possible job and feeling that he could cover any small fibs, said that he had done some excellent competitive precast estimates for a project in that area a few months back. Unfortunately, he added, the project did not go ahead because the owner and the other at-risk partners could not obtain interim financing. He further said that the design and fabrication concepts used by his firm were so advanced that the prime general contractors bidding on the project did not fully understand how to manage precast erectors to obtain a trouble-free erection of the structure as he suggested designing and fabricating it.

Precast warehouse structures were somewhat new to Frank, who was accustomed to designing his storage buildings in steel. However, Paul's comments and promises appealed to Frank and he agreed to let Paul and Conskoinc do some preliminary structural designs and estimates to be incorporated into the presentation set to Mr. Montgomery, the client. Once Mr. Montgomery approved the presentation package, the project was to be released for final design and construction. Foundations were to start within a week after the release. When Frank Clark began to discuss the presentation package at the meeting with Mr. Montgomery and his partners, Mr. Montgomery asked about the relative merits of precast concrete and steel structures. He said a friend of his had considered using precast concrete in a similar building and had actually started construction using precast concrete components. He said that during erection, however, the structure failed, the precaster, Posten, Inc., went bankrupt, and the project literally had to be started over again. The time and cost loss put the job at high risk and it was finally aborted, with all participants being losers. Mr. Montgomery said the apparent reasons for the erection failure were inadequate design, poor fabrication procedures, and above all, inadequate temporary bracing during erection.

As the presentation scenario unfolded, the matter of the precast contractors available came into the discussion. Frank mentioned that he had been working with a preliminary design and estimate by Paul Conklin, the owner of Conskoinc. The names apparently rang a bell with Mr. Montgomery's plant manager, who slipped out of the room. He returned in a few minutes and when the conversation allowed, he mentioned to the group that Posten, the bankrupt contractor's company, had been headed up by none other than Paul Conklin. Hearing this piece of information, Mr. Montgomery closed the meeting and requested that FCL and Frank Clark go back to the drafting board and produce an alternative design in structural steel. Despite protests that at this stage of the project the design cost and the time delay would make this almost impossible without heavy penalties to the project, Mr. Montgomery insisted on the alternative design before giving any approvals to move ahead.

The apparent innocent fibs about the experience of Paul Conklin in St. Mark's County turned out to be lies about critical elements in any building system. The lies of Mr. Conklin were:

1. That the precast estimates he had done were excellent. In fact, one of the reasons for the structural collapse during erection was that inadequate temporary cross bracing had not been designed into the erection process.
2. That the project had not gone ahead, because the owner and the other at-risk partners could not obtain interim financing. The original financing to construct the building was in place. However, the collapse and the precaster's bankruptcy had made it financially unfeasible to proceed with the redesign and reconstruction of the collapsed structure. In effect, there was no financing available for proceeding a second time.
3. That the design and fabrication concepts used were so advanced that the general contractors could not manage the subcontractors using them. Actually, it was the inability of Mr. Conklin's previous firm, Posten, to erect its own precast elements that had caused the failure leading to subsequent abandonment of the project.
4. Then perhaps the biggest deception of all—that of Paul Conklin to himself in failing to recognize how the small fibs built a monumental lie that could cost many people their jobs and their reputations.

Some of the disastrous results of this set of lies and the resultant trouble may be seen by looking at the St. Mark's County project scenario three months later.

1. The project has still not been released for construction because Frank Clark and Mr. Montgomery just reached an agreement on the redesign costs last week. Frank is still two weeks from being willing to risk presenting a package.

2. The deterioration of Paul Conklin's reputation has lead to his closing his company at a serious financial loss to him and his backers.

3. The opportunity to try some unique and useful precast concrete concepts for storage buildings has probably been lost to the St. Mark's County area for as many as one or two generations of owners, designers, and builders.

4. The litigation process has already been started by the investors against Mr. Montgomery, and by Mr. Montgomery against Frank Clark. These will probably be settled since they deal mainly with delays to start of construction and not with actual construction penalties. However, the loss of three months' use of a warehouse could have become serious if the demand was extreme.

5. The professional liability carrier for Frank Clark is now investigating the charge by Paul Conklin that he has been harmed in that his design was not allowed to be carried to completion by FCL and Mr. Montgomery.

6. Chances are that Mr. Montgomery will not use FCL on his future projects.

Many other undesirable results can be pieced together in this brutal business scenario that started out as an apparent innocent fib or two at a social outing and a desire to give a good design firm's client an innovative and distinctive structure.

Action 6: Understand thoroughly the obligations you have to society and to your clients, your employer, and your peers. The obligation patterns built throughout the years contribute greatly to your ability to manage conflict well. Fulfillment of your obligations can be thought of as paying back a debt to the society that has generated a demand for and has bought and paid for your products and your services. There are also other obligations that you have to social groups. These other groups may include your clients, your employer, and your peers. All of these groups are made up of people who have contributed in some way to your success. They may also have contributed to some of the problems and obstacles that have prevented you from becoming as successful as you may have wished. But the fact remains that you must depend on them collectively to provide an ongoing source of income, gratification, encouragement, help, and assistance in good and bad times.

We find that when we fail to honor those obligations we have to others—individually and collectively—that the potential for conflict rises significantly. The case of Mike Carson exemplifies how loss of professional vision can cause destructive conflict and loss of positive direction.

Case Study 3.8: The Job-Hopping Engineer

Mike Carson was a good engineer. He had received high marks in college and he succeeded in receiving his Bachelor of Science degree in mechanical engineering. He had quickly learned to apply industrial mechanical engineering skills to the peculiar and often dissimilar demands of the construction industry. This quick learning skill tended to delude Mike into believing that his skills were self-taught and internally learned.

Mike was also a bit overbearing and self-centered, along with having a somewhat better opinion of himself than may have been justified. So when he found himself at an apparent roadblock to his advancement at Gibson and Associates, consulting mechanical and electrical engineers to the construction industry, he quit the firm.

This describes how the situation looks from the outside. Now, let's look at the inside picture. There is a bit more to this situation than just a career blip and an aggressive, impulsive young man resigning. Let's see what the problem really was.

Mike had become increasingly selfish during his early career and now, six years out of college, four firms later, married, the father of two children, he had yet to accept any active role in his professional and technical societies, had resisted volunteering for church assignments, and had developed a chip-on-the-shoulder attitude which said that "society, his job associates, his firm, and the world all owed him a living, for which he had already paid." This attitude was quickly noticed at Gibson and Associates. Despite the misgivings of Frank Gibson, the president, middle management kept Mike on, giving him increasingly important assignments. Finally, a mechanical contractor in town called Mr. Gibson and said he would like to discuss something with him over lunch. Mr. Gibson said fine and set a date. At lunch the mechanical contractor, Mel Trustin, said:

> We have a problem, Frank. One of your engineers has given us a hard time on several recent projects. We're pretty resilient against this kind of treatment, but in this case I believe it's time to review our position. Mike Carson, your project manager on one of our current large projects, has made several design errors and refuses to accept responsibility for them, claiming we are obliged to do what he asks, or in this case, demands. He's a pretty bright young man and we've helped him out of jams in the past to save him and us unnecessary problems. However, he is beginning to think this business is all take and no give.

> We've got to stop the erosion of our financial position on this project. I'm at a point where I may have to make a formal claim against the general contractor, the owner, and possibly against your firm. You and I are good friends and this is my last effort to remind your staff, and, I hope, Mike Carson, through you, that he has some obligations to his contractors, and clients, too. We need help. What can we do?

Frank Gibson was startled but not surprised. He returned to the office after lunch, confronted Mike Carson with the facts brought out at lunch, and asked him for a full explanation of his actions and position with Mel Trustin. Mike promptly said that he owed the contractor nothing and that in his opinion, the contractors on his projects had an obligation to pick up minor discrepancies in the construction documents that might cause problems. After all, said Mike, "we are all in this together and we have a mutual obligation to help each other. The contractor has all the money and can afford to take a slight hit to help the engineers, who carry the heavier load of professional liability."

Questioned further, Mike revealed that he had some design problems in the past that, he said, "could have been remedied by you sending me to technical seminars to learn some of the newest design techniques relative to building mechanical systems. It's the company's fault that I have to be so harsh on the staff and our contractors." Additional stormy dialogue ensued. The conversation came to an end when Mike stood up, said "I resign," and stalked out of the room.

Here we pick up the story where Mike's resignation has an external look of being just another good man leaving for a better job. In fact, the chances of Mike getting another job in the area have been considerably diminished by his actions and by impressions of Mike that have been left with Frank Carson, Mel Trustin, and the others who had actually been hurt by Mike's uncaring attitude toward them. The lack of concern about others and his obligation toward them has created conflict and has made Mike a man of suspicion. He has also caused potential legal trouble for his employer, the general contractor, the owner of the project, and the mechanical contractor on the job.

This is a serious problem and solutions will not come easily. However, the moral of Mike's story is that all of us have a major responsibility as designers, owners, and constructors to maintain a sense of obligation to society, our clients, our employers, and our peers. Without such a sense of obligation, society and others will soon ignore us and we will become a source of conflict that no one can afford to risk accepting.

Action 7: Understand and work to be effective in managing intersections of diverse interests   The generic construction industry has many different arenas of work, and consequently, many technical and management interactions that produce conflict and generate challenges for the competent manager. The complexity of the work arenas and their interrelationships was seen in the macro matrix in Figure 2.1. Mixing proper action (X axis), functional competence (Y axis), and responsible participation (Z axis) keeps the project team efforts focused on getting the job done. The leader in such a successful effort preempts destructive conflict by offering participants a multitude of chances to be successful.

George Lexington gives us some tips on methods of bringing together diverse viewpoints by an understanding of the issues and those involved in the next case study.

Case Study 3.9: The Case of the Potentially Successful Manager

George Lexington was a practical man. He was educated as an architectural engineer and had worked many years as a production architect in a 40- to 50-person design office. He had also written design programs for dozens of projects for the same firm. After 10 years with the organization he decided to change career direction. George joined List Construction, a general contractor construction firm, as an associate project manager on a new 32-story downtown office building for Third Bank Corporation. The project was fully designed and was due to start in the field in about three weeks when George arrived.

He was made responsible for managing structural and curtain-wall contract work under the overall direction of Tom Taylor, List's project director for the job. Mr. Taylor stressed at the beginning of George's assignment the need for constant and proper interfacing of structural and skin work with other elements of the building. Tom further pointed out that this interface might be made difficult by an intense territorial possessiveness within the design office of record, Larkin and Associates. The Larkin architectural production department head and its structural department head, both associates in the firm, did not like each other and did not respect the other's points of view. They often carried the rivalry into their work with contractors.

This interoffice management conflict was well known in the local construction market and was normally tolerated by contractors. Many merely added a small but consistently applied markup on projects designed by Larkin and Associates. This markup was known among contractors as "the Larkin component." On the Third Bank tower, List did not apply its usual Larkin syndrome markup of 0.10% of the total hard-money bid price to the Tower proposal. The difference between List and the next-lowest bidder was 0.09% of the total project bid cost. The job was bid from a short list of owner-selected contractors.

The challenge to George in his new job at List was made simple by Tom Taylor. He told George: "Erect the tower structure and skin within a price range that would justify not having added the usual Larkin markup." When George joined the firm, he was vaguely aware of the Larkin syndrome. He soon discovered much more about it as he was introduced to other members of the List project team, the Third Bank owner, and the principals in the Larkin firm. George promptly began to think through how he was going to accomplish his work that might be affected by excessive internal conflict among the design team. He quickly saw that one of the keys to successful action on this project, given that he and his new firm were technically competent, was to anticipate and minimize the potential for disruption by Larkin's internal rivalry.

Seven conflicts George felt highly probable were:

1. Structural or enclosure design revisions, given the inability of the Larkin department heads to get along with each other
2. Interference with a steady flow of submittal approvals on structural and dependent curtain wall elements of the building, due to design arguments
3. Possible claims from subcontractors for the structure and skin work due to slowing or interfering with their work continuity and procedures
4. Finger pointing and blame for disruptive actions caused by lack of clear-cut organization and communication patterns in the field
5. Erosion of List's profit margin by demands for excessive accommodation to work revisions
6. Loss of field attention by Larkin's structural and architectural department staff members caused by their desire to distance themselves from an internal management struggle
7. Potential for poor performance on his own part, caused by factors external to his responsibility and authority limits on the project

With these concerns on his mind, George used a technique that had been of great help to him in the past. He asked himself several open questions.[2] Five of the questions he asked were:

1. What is my mission on this project? (His mission was defined as the single most important thing for him to accomplish for List Construction on this project.)
2. What actions might I take that will produce the highest probability of success in achieving my project mission on the Third Bank tower for List?
3. What actions might I take that will reduce the probability of mission failure on the Third Bank tower?
4. What sectors of the macro matrix (see Figure 2.1) show the most probability of conflict within my area of responsibility?
5. Within the macro matrix, what actions can I take that will reduce or eliminate barriers to effective action in achieving my mission for List?

---

[2] Questions not able to be answered with a simple yes or no.

George's answer to the first question was that his mission on this project was:

- To produce a facility that pleases the Third Bank Corporation and adds value to its assets
- To build a facility that meets the quality, cost, and time standards required by List's contract, and returns the expected profit to List Construction
- To construct a building that meets the design mission and program of the programmers, planners, architects, and engineers for the project
- To manage his project work so that subcontractors to List Construction make a profit and enjoy the benefits of their good work

He addressed the other questions one by one, answering them in a manner demanded by the need to accomplish his mission. Once this process was completed, George wrote several objectives for himself. The seven objectives George set for himself included:

1. To develop and encourage a spirit of cooperation on the project through the use of partnering concepts.
2. To publish well-defined communication and responsibility guidelines for processing all administrative matters affecting the structure and the skin of the building.
3. To establish and maintain an appropriate documentation level throughout the project. Documentation is to be provided on an as-needed basis and must add value to the entire project effort.
4. To plan and schedule the work well and in conjunction with members of the project team that are affected by the project plan and schedule.
5. To reduce destructive conflict levels by encouraging settling job disputes early and quickly, at the originating level, and among those directly involved.
6. To prepare and publish a draft set of project close-out guidelines early so that all project participants under his management control would know what was required of them to move off the project properly and receive their final payments.
7. To adopt an active position as an advocate for those participants on the project that showed a desire to achieve his and his company's mission.

This starter list was to be kept open to additions and changes until it was working well and being implemented. George had developed a habit of looking at management systems as being totally open[3] (10 on a scale of 1 to 10), partially open (something less than 10 and more than 1), or totally closed (1 on a scale of 1 to 10). For example, a job meeting (considered as a system) might start at a degree of openness of 6. As the meeting proceeded, George would attempt to close down all old and new business to a rating of 1. Of course, it was almost impossible to close any meeting system totally since one of the purposes of a meeting was to uncover potential items that needed to be addressed after the meeting. However, the constant effort to produce decisions gave participants a feeling of accomplishment and direction.

Getting back to George's list of objectives—he considered this early list of management targets to have a degree of openness of 5. He knew that all the objectives defined in this list had to be achieved for success. George also realized that progressively closing the system would require defining essential targets that would emerge only as the project moved ahead. How do the steps taken by George relate to our guideline statement that a good manager of conflict must understand and work effectively in managing intersections of diverse interests?

First, notice that the case study involving George was entitled "The Case of the Potentially Successful Manager." The word *potentially* means that George was attempting to maximize the potential for success. There is no guarantee of success. Success results from improving the potential for success. Second, notice how George's five questions to himself, his mission statement, and his seven objectives were designed to minimize the impact of the seven destructive conflicts he saw as potentially affecting project success.

The advantages of addressing the total interaction needs of a job by a perceptive manager can be seen in the results of George's hard work on the Third Bank tower. Those results included:

1. Generating a set of questions and answers that can be discussed by George with his staff and his superiors to gain specific comments, improvements, and approval.
2. George's mission definition and objectives have set a pattern for development of a project-wide mission and objective statement.
3. The initial thinking may help show Larkin how some improvements in staff performance can bring them benefits. This is a somewhat vain hope at present, but mature thinking on such improvements often produces remarkable results.
4. George has set the stage for a full-blown partnering effort on the project at which his early thinking could very well form a basis for

---

[3] *Open* means open to the import or export of resouces and ideas.

establishing a healthy and open communication and work pattern for all on the project.

5. Discussions with List's structure and skin subcontractors can proceed with some confidence that George's efforts are sincere and meant to benefit all on the job. This expressed intent is often the key to unlocking the full benefit of good noncontractual relations.

Reducing destructive conflict and managing positive conflict well on a planning, design, and construction job are essential elements for success. They are also the key ingredients to understanding and using partnering tools successfully.

The roads to successful buildings are paved by hard workers and good managers all with good intentions.

—Don Templin, construction expert

4

RISK AND DISPUTE

SECTION A: RISK AND ITS ROLE IN CONFLICT

Risk is exposure to the possibility of harm, danger, loss or damage to people, property, or other interest.

The willingness to take risks is a measure of our faith.

intro

Risk and its acceptance, rejection, transfer, or avoidance have been recognized as factors in conflict since the beginning of time. Consider the risk that David took when he battled Goliath, the risk that General Eisenhower took when he invaded France in World War II, the risk that an ironworker takes when he is bolting steel held by a crane operated from 200 feet away, the risk that an owner takes when he executes a contract for a new construction project with a contractor with whom he has not worked previously, So, too, the willingness to take risks in the generic construction business is an act of faith. *summary* The behavior of many people today often makes the intelligent construction practitioner wary of misplaced faith, particularly faith in those with whom he or she has had little risk-sharing experience.

We see many examples of this in our day-to-day work in construction. A contractor fabricating and erecting a new close-tolerance curtain-wall configuration is not likely to trust the layout to a young surveyor who has just graduated from college and is on her first job. A mason working on a swing stage suspended from roof-mounted outriggers wants those installing the U-bolt clamps anchoring the outriggers to be people he can trust. The curtain-wall contractor and the mason in these examples are used to taking expected day-to day risks. They cannot afford to take risks that are out of balance with their expectations.

include

Steps to successful risk taking are simple. Boiled down to essentials they are:

1. Understand the nature of risk.
2. Understand the conditions under which you will be taking the risk. Draw a boundary line around the area of risk.
3. Identify the specific risks to be taken within the boundary.
4. Reach early agreement with those involved concerning how the risks and rewards are to be allocated.

Summary

Risk has become and will continue to be a very important element of the generic construction process. Risk has two faces—first, the face shown if risk is improperly understood, defined, and allocated. Here risk becomes a monster, devouring profit, escalating overhead, demoralizing participants, and providing grist for the mills of dispute. Risk's other face, shown when an understanding of risk is achieved, can be one of pleasure and enjoyment. In this case risk is understood and is allocated fairly and intelligently, the benefits are returned to all involved, and disputes are reduced to a minimum.

A simple example may illustrate this best.

Case Study 4.1: The Case of the Generous Boss

The Neville Law Center office building is done and has been closed out by the general contractor, Bonaventure Construction Company, and the owner/designer team. Bonaventure Construction's president has an external and internal history of rewarding successful performance by those contributing to over and above performance. Project construction on the Neville Law Center has been wildly successful from all standpoints. As a neutral outsider familiar with the project, you have been selected by Bonaventure's project team of five people to allocate and distribute a total $20,000 bonus promised before the project started to the team by Eric Platte, their company's president. Each of the five has contributed in his or her own way to earning the $20,000 for the group. Mr. Platte has given you an envelope containing forty $500 bills. How do you distribute the money to the five people on the team?

There are several ways of doing this:

1. Give an equal amount to each person.
2. Divide the money according to the company management rank of each person on the project team.
3. Divide the money according to the project-related management rank of each person on the project team.
4. Determine mathematically the percent of risk you thought each person assumed and allocate the money according to that mathematical risk.

5. Decide that your job as a recommending neutral entitled you to a share in the money and that whatever remains after your fee is deducted is to be divided according to one of the schemes above.

6. Allocate the money according to your perception of how much each of the five contributed to the project's success.

7. Select the person with the most apparent needs and give that person the most and grade the awards down according to need until the full amount has been distributed.

8. Divide the bonus with others on the project who are company staff but were not one of the five project team members.

9. Decline the request to act as a neutral.

You have gone over the possibilities with the team and they cannot agree which is fairest. Each plan elicits strong objections. The project team is divided and upset. A moment's consideration of any or all of the methods outlined will show that no matter which scheme you would recommend, it will meet with antagonism and dismay from some or all of the former project team. It appears that perhaps Mr. Platte violated several basic principles of risk analysis and management in respect to rewarding employees. What began as a well-intentioned reward must now be coaxed into being.

Let's examine the situation relative to the four steps to successful risk taking and see how what Mr. Platte thought was a good idea might turn into a minor nightmare.

1. *Understand the nature of risk.* Mr. Platte understood and properly managed the macro risks on this project. He saw the need to recognize the positive outcome of good hard work by his project teams. That was evident from the beginning. What he did not understand was the nature of the micro risk he took in promising to a temporary, and possibly changing team of people that if the project was successful by some yet undefined standard, he would reward them in some way. He missed the important fact that risk is inherent in almost any management decision. What his risk actually was in its smallest arena, that of the individual employee, totally escaped Mr. Platte's notice in managing the broader aspects of the project and the company.

2. *Understand the conditions under which you will be taking the risk. Draw a boundary line around the area of risk.*

- WHAT WAS MR. PLATTE'S PROMISE IN THIS MATTER? That he would reward, in some way, a group of people if the project they were engaged upon was successful, without specifying what contribution each person had to make.

- WHAT WAS MR. PLATTE'S RISK IN MAKING THIS COMMITMENT? That he would be able upon successful completion of the Neville project to divide an unspecified bonus fairly and satisfactorily in some manner to the project team. In this case Mr. Platte certainly did not expect that if the project was successful he would be criticized for being generous with the earnings. His basic mistake was this: He did not understand that a risk taken without any possible outcome accounted for and planned was a risk best not taken.

- WHAT BOUNDARIES WERE DRAWN AROUND THE AREAS OF RISK? None that would allow or encourage an intelligent allocation of the monies to be paid in bonuses at the end of the project.

3. *Identify the specific risks to be taken within the boundary.* Mr. Platte deferred specifying what he was risking until the moment that success had to be rewarded. At that point he saw dimly that his risk was far more than just paying out money from his company's profits. He quickly discerned that the potential for losing some, or perhaps all, of an excellent project team was a distinct possibility. This, of course, was due to the predictable individual feelings among the Neville five that they were not individually rewarded according to what they really felt they were worth.

Another characteristic of the reward is that it did not take into account the efforts of those not on the permanent project team. Many of these, such as the estimators, the project secretaries, the accountants, and others contributed valuable input as the project moved through various phases of work. Some of these may feel strongly enough about what they perceive as an inequitable bonus arrangement that they could consider leaving the firm. Mr. Platte has achieved almost the direct opposite of what he wished to achieve simply by not understanding the risk involved in this relatively small money-management matter.

4. *Reach early agreement with those involved concerning how the risks and rewards are to be allocated.* If Mr. Platte really feels extra rewards should be given to the project staff on successful projects, he should prepare a list of criteria by which these rewards will be determined. Of course, this process distills some of the spontaneity and excitement out of the process. On the other hand, the criteria lets the entire staff know what risks it is that Mr. Platte considers worthwhile rewarding. This may give direction to the efforts of the full project team. People who have the potential for success at their work usually understand the ground rules if they are formulated and communicated well. This allows them to work toward overachieving in a predictable manner consistent with their true abilities.

What did Mr. Platte get for his efforts? An angry team and no process to accomplish what he really wanted.

At the beginning of the job, Mr. Platte should have done the following: Upon award of the contract he should have said to himself that he would like to reward good performance on a job that probably appeared claim prone. He did do this but instead, made known his feelings of letting his past record and management performance lead the project employees to their own conclusions. His management style would have indicated to them a possible bonus for good performance. Don't broadcast good intentions unless there is a 100% chance of success.

What could Mr. Platte do now? Suggest to the team that they meet and discuss the matter in a closed conference. Then, from the solutions suggested, devise their own method of rewarding Platte's staff for the extra effort. Mr. Platte should further suggest that if they cannot solve the distribution problem without excessive destructive conflict, the money will be donated to a charity of their choice selected by their vote. This condition on the division of the money will force those primarily responsible for the success to reach a consensus or face losing the bonus to a good but external cause.

Summary The complex nature of risk is often misunderstood, particularly as construction tends toward greater centralization of management, authority, and responsibility. In this process risk is often shifted within the major parties to the contracts until it lands on someone willing to take the risk. Frequently, the risk is accepted because those taking it do not know or understand the implications of the risks they have assumed. Unfortunately, many risks that are innocently taken or are unfairly or ineptly assigned have high impacts on costs and profitability. All construction involves risks of some type, and these must be absorbed by some management method dictated by action guidelines. Once the general nature of risk is understood and the specific elements of risk become clear, you can take the action with a better chance of making it work well.

Specific elements of risk in construction are important to understand and to account for. They should be considered in every project irrespective of where on the line of action you or your firm is to participate. A comprehensive list of risk elements is given below. Note that many items in this list can be allocated to two or more of the major categories.

A. Constructibility of the project
    1. Maintaining integrity of the design and engineering program
    2. Uncertainty of new product development (construction products)
B. Contract
    1. Contract content
    2. Differing site conditions
    3. Late owner-furnished material and equipment
    4. Managing and resolving legal matters
    5. Unreasonable systems-performance guarantees
    6. Unsuitable owner-furnished material and equipment

C. Decision making
1. Availability of credible information
2. Changes in customer requirements
3. Delays in addressing problems
4. Delays in presenting problems
5. Delays in resolving problems

D. Financial
1. Adequacy of project funding
2. Bid errors
3. Changes in personnel managing funding processes
4. Establishing a profitable cost structure
5. Managing cost growth
6. Payments
7. Potential bankruptcy of parties to the contract who owe you money
8. Pro forma cost targets set for the design criteria
9. Shifts in budget priorities
10. Uncertain resource costs
11. Uncertainty of funding

E. *Force majeure* (an unexpected or uncontrollable event)
1. Acts of God
2. Community activism
3. Governmental acts
4. Local, regional, national, and international economy
5. Political climate
6. Political interference
7. Strikes
8. Subsurface conditions
9. Weather

F. Labor
1. Adequacy of labor force
2. Drooping morale
3. Jurisdictional restrictions and rules
4. Keeping job morale and attitudes healthy
5. Labor productivity
6. Productivity of personnel
7. Subcontractor capability
8. Underqualified people
9. Underskilled people

    10. Union strife
    11. Work rules
    12. Worker and site safety
G. Operational
    1. Adequacy of owner representation.
    2. Approval processes
    3. Availability of competent personnel
    4. Availability of tools and techniques
    5. Being a good neighbor
    6. Closing out the project
    7. Communicating with others.
    8. Delayed deliveries
    9. Delays and disruptions
    10. Heavy overlapping of similar resources
    11. Impact of claims
    12. Inaccurate resource costs
    13. Lack of technical knowledge
    14. Low-float tasks
    15. Maintaining a good work site
    16. Managing quality
    17. Material and equipment procurement
    18. Overlapping insurance coverage
    19. Paperwork and administration
    20. Permits and licenses
    21. Planning and scheduling the work
    22. Poor plan from which to derive a schedule
    23. Poor planning information
    24. Poor planning techniques
    25. Setting and maintaining policies and procedures
    26. Site access
    27. Turnover
    28. Unreliable resource information
    29. Value systems of others
H. Technical
    1. Appropriate designer involvement in construction
    2. Changes in needs or requirements of the finished product
    3. Changes in the owner program
    4. Changes to design

5. Changes to the design documents
6. Geotechnical characteristics
7. Innovative designs
8. Large number of defects
9. Maintaining construction document quality
10. Owner involvement in design
11. Processing revisions
12. Processing submittals
13. Project program
14. Sufficiency of plans and specifications

I. Time
1. Adequacy of performance time
2. Controlling time growth
3. Overoptimism in scope and duration of activities

All of these risk-taking elements act in varying ways to influence project success or lack of success. They create a wide variety of paths that can lead to success, dismay, failure, or even bankruptcy.

The partnering process is designed to pave the way to intelligent and effective risk assignment and acceptance. However, the responsible manager can take very early steps even before the formal partnering process begins to better identify, understand, and allocate risk. These steps include[1]:

- Identify potential threats to project success as early as possible.
- Identify what problems are likely to be caused by these potential threats.
- Determine the risks attendant on meeting the threats to project success.
- Evaluate the project team's ability to take the risks identified.
- Evaluate the project team members' abilities to take the risks identified.
- Identify the impact of legal contractual obligations on the risks identified.
- Evaluate and analyze existing project conditions and the forces they exert on the project.
- Determine the options you might have as a result of your evaluation and analysis of the risk potentials and nature, the team's abilities, the legal consequences, and the project conditions. Broad options include:
  1. Avoiding the risk.
  2. Pricing the risk.
  3. Managing the risk.

[1] Adapted from list by Thomas E. Papageorge, *Risk Management for Building Professionals* (Kingston, Mass.: R.S. Means, 1988).

- Ask the following questions for each risk you identify and then assign a value of 3, 2, or 1, representing a high, medium, or low risk.
  1. What is the probability that this risk will actually have to be faced?
  2. What are the circumstances that would trigger a risk-contingency plan into action?
  3. What would be the impact if this risk should have to be assumed?
- Establish and implement a systematic, continuous procedure to identify and adjust risk to acceptable levels to assure a high probability of project success.

Once potential risks are identified, management must make a decision about how to handle them. Disputes often are not settled promptly and fairly because risk has not been allocated correctly. A simple rule of thumb is to allocate risk to those best equipped to assume it.
Operational actions are critical to managing risk. Some helpful hints to assist in managing risks in planning, design, and construction are listed below. Remember, these apply to all phases of the line of action, to all functions, and to all participants expected to encounter and assume risk:

- Start the job at the right time.
- Profile the job before committing resources.
- Always remember—good management is risk control.
- Don't lose your personal intellectual grasp of risk on your job.
- Evaluate the quality of the total contract documents.
- Obtain and read all pertinent contract documents.
- Match your price to the project delivery system being used.
- Avoid being made a limited agent on a hard-money job.
- Avoid over-the-wall management.
- Keep abreast and aware of current industry trends, particularly organizational patterns.
- Be aware of your client's must, want, and wish list and respect it.
- Understand and account for other project participant's profit needs and desires.
- Don't hesitate to scrub your proposal if the risk is excessive relative to the rewards.
- Negotiate deadlines of high-risk tasks to accommodate potential slippage.
- Schedule tasks that can be postponed or canceled, if necessary, to later in the project.
- Be conservative in estimating task durations and costs.
- Insert contingencies as recognizable elements of the plan and schedule.
- Assign strong staff to high-risk jobs.

- Assign backup staff, however minimal, to any task where the loss of a team member would be damaging.
- Plan preventive actions that will be taken to reduce or remove risk.
- Plan contingency actions that can be implemented if a problem occurs.
- Identify circumstances that might trigger each contingency plan into action.
- Retain your optimism, solve problems, and keep morale strong despite setbacks caused by the winds of risk. This is your job as a manager.
- Consult good contract legal counsel *before* signing a contract.

SECTION B: ORIGINS OF PLANNING AND CONSTRUCTION DISPUTES

Disputes will occur in construction irrespective of how hard well-intentioned people try to prevent or minimize them. The construction business is replete with differences of opinion about thousands of subjects. Most of these disagreements are settled easily and at the level where they occur since they are about matters covered by explicit contract language or by good technical practices. If contract conditions are not explicit or are so ambiguous that a different interpretation held by one contract party and not held by another causes problems and cost overruns, agreement may not come so easily and the problem may have to be escalated to higher levels of management than where it originated.

Occasionally, another situation arises in which there is disagreement about what is or was meant by the contract language. In such soft noncontract matters, particularly where unexpected or unestimated costs may be involved, the resolution may become extremely difficult and outside advice might be needed. Another condition may be interwoven into one or more of the first three dispute situations: All parties may agree that there is a difference of interpretation or an error in interpretation, but fail to agree on the remedy to the problem. Failure to agree can affect cost, quality, performance, timeliness, and other task characteristics that make up the construction business.

This situation is illustrated in Case Study 4.2 with an actual project problem that was escalated through the inability of the participants to see the solution and disagreement over what was clearly a contract condition. The case study describes the situation surrounding Erickson Mechanical Services and its dispute with Giltland General Contractors over the payment of retention, a common cause of construction disputes.

Case Study 4.2: The Faulty Punch List Correction

Erickson Mechanical Services was a subcontractor to Giltland General Contractors, the prime contractor on a large industrial project. The dispute began when what had been a perfectly good industrial renovation job was near completion. The owner, Springbend Shock Absorber Company, was a reputable manufacturer of auto parts and accessories. Total project construction cost was about $12,000,000, with the mechanical contract being about $2,500,000. The mechanical punch list had been prepared by the engineer of record. Most of the items had been corrected by Erickson except for about $100,000 in outside piping work that could not be done until winter weather abated, in about another month.

Erickson submitted a final billing for the entire unpaid amount on the contract without deducting the $100,000 punch list item. Giltland and the owner deleted the amount from the current payment and, in addition, did not release any of the 5% retainage of $125,000 (as Erickson had hoped). Erickson complained to the general, Giltland, who said that all that was required to close out the job was for Erickson to come in when weather permitted and complete the punch list work, which at this point consisted solely of the $100,000 outside piping installation.

Erickson's president, Frank Kigil, took a hard-nosed stand that if Giltland and Springbend were not going to pay his bill he would not complete the punch list work. This stance, of course, totally missed the fact that Springbend not only still controlled the $100,000 punch list amount, but also the $125,000 of retainage. Mr. Kigil persisted and called in his attorney despite the warnings from his project team that Erickson, by its contract, still must complete the punch list work to collect the $100,000. The legal counselor and Erickson's staff also pointed out that if the positions hardened, Erickson might have a tough time collecting the retainage of $125,000. As experienced field professionals they knew how easy it was to keep recalcitrant contractors on the payment string for months if the owner and the general contractor felt like doing it.

The outcome was that Erickson and Mr. Kigil fought against doing the punch list work for several months. Finally, they were convinced by a friend of Mr. Kigil's that the situation was absurd, the reasoning was fallacious, and that Erickson might already be in breach of its contract for failure to complete the punch list work. The actual cost to Erickson for resisting the efforts of Giltland and Springbend, to complete the work, and to get out of the job was probably $4000 to $6000 at a conservative estimate, taking into account the loss of interest on the retainage, the attorney's fees, the delayed profit on punch list work, and the extra cost of remobilizing for the punch list work. Even more important, Erickson lost Springbend's interest in having Erickson on its projects in the future, Giltland's interest in working with Erickson as a sub in the future, and there was a lessening of confidence in Mr. Kigil's judgment by the total staff. Word quickly got around in the ever-present construction grapevine that Erickson had used poor judgment and had unnecessarily endangered the prime contractor and other subcontractors on the Springbend project.

Such disputes as described in this case study of an actual situation can be avoided when there is a mutual desire to solve problems within guidelines that have proven successful in the past. If we examine how differences of opinion have and are being settled fairly and promptly, we find some interesting ideas about conflict upon which to build resolution methods that have a high probability of succeeding.

1. Informal negotiation was the resolution technique before excessive legal constraints were imposed or accepted by the design and construction industry as a feasible resolution method.

2. Dispute intensity varies with economic times. In periods of exceptionally high economic activity, speculative money can be spent on costly resolution methods to gamble for a high return on the investment. In periods of low economic activity, money is usually not spent on expensive, high-risk, uncontrollable methods of resolution hoping for a favorable result.

3. In times of intense competition we cannot afford to spend our resources on high-risk gambles. Therefore, low-cost, nonbinding, mutually agreeable resolution processes have become increasingly popular.

4. The acrimonious atmosphere surrounding binding resolution methods has proven demeaning, negative, distasteful, and harmful to design and construction professionals who wish to practice honorably and effectively.

5. Short-term dispute resolution processes often encourage healthy ventilation of troubles and hurts. Properly controlled, this ventilation, can, in turn, help toward rapid healing of business and professional wounds.

6. A basic principle of conflict resolution is that the earlier in a construction project the participants seek to resolve disputes by the use of an intelligent resolution process, the more this process will contribute to project success.

7. Even when problems turn into nasty disputes, litigation or other binding settlement processes should not be the initial method used to resolve them.

8. Advance commitment to nonbinding resolution methods contributes to solving problems effectively and fairly as they arise.

9. On-site dispute resolution often helps dispose of problems as they arise and before they multiply.

10. Dispute resolution proceedings should be conducted expertly and effectively by experienced design and construction practitioners.

5

THE PROBLEM JOB AND HOW TO IDENTIFY IT EARLY

Project success is very much an intangible when we are just beginning the process of nursing a design and construction project through its beginning stages. Even seasoned design and construction practitioners err occasionally by taking early design construction activities too lightly. I believe that a project does not grow into success—it is pushed and coaxed into success by intelligent management and excellent leadership. But there are other factors contributing to project success. Intelligent management and excellent leadership are often provided intuitively by a competent project manager. What is often not provided is the contingency plan—the alternatives available when the predictable elements of dispute make their appearance, even before the designer has a pencil line laid on paper—or even before field actions have begun.

It is apparent that with potentially troubled jobs the problem resolution process must start before the problem appears. In Chapter 6 we look at the determinants of project success, but let us turn here for a moment to determinants of project failure. What are the telltale marks and scars that tell us that we may have a problem job on our hands? These jobs are sometimes said to be claim prone. When a claim by one party on the project against another party on the project cannot be resolved by mutual agreement, the product is a contested claim.

Long and sometimes bitter experience over many years has demonstrated to many design and construction professionals that most problem, or claim-prone, jobs are predictable. The problem job is usually marked most visibly by increased costs of design, construction, and operations; it is almost always a job that takes longer than had been planned to design and construct; and it is invariably an effort that produces sharp disagreements and lingering hard feelings among the participants. During the very early stages of a project, it is often possible to gain a good insight into the expected nature of the work we are about to do. In fact, it is good policy for the perceptive and concerned owner, designer, and contractor to become familiar with those characteristics that early identify a job as having claim-prone features. Twenty-three of the more visible telltale features are listed and described below.

1. A wide spread in proposal prices on hard-money projects
2. Issuance of a large number of prebid addenda and clarifications
3. For subcontractors, a poor general contractor, if the job is being built by one prime contractor
4. For projects with separate primes, poor reputations of other prime contractors
5. More than four to six prime contractors involved on project (for normal building work only)
6. Poor reputation of architect or engineer of record
7. Several prime design team organizations
8. Excessive how-to-do-it emphasis in contract documents
9. Nonliable or not-at-risk parties involved in responsible positions
10. Large numbers of allowance items
11. Zero or excessively small tolerance specifications
12. Poorly defined authority and responsibility structures in the offices of the design team, the owner, and the general contractor or other prime contractors
13. Inexperienced specialty contractors
14. Excessive number of owner and designer preselected suppliers for key material and equipment
15. Large dollar amount or numbers of owner-purchased equipment or materials
16. Location in strike-prone or jurisdictionally sensitive areas
17. Heavy use specified for untried products and equipment
18. Nonliable party involvement in establishing delivery commitments (i.e., not-at-risk construction managers, planners, architects, engineers, or owner representatives)
19. Excessively long time periods to award contracts after a proposal is submitted (*Note:* This often occurs in public work where many nonoperational approvals and agencies are involved.)
20. Poor owner reputation
21. Extensive declaimers on purchase orders, meeting minutes, and other official or semiofficial documentation
22. Extensive early documentation inconsistent with the complexity of the project

23. Unqualified, mismatched, or dysfunctional project management staff assigned to the job by owner, user, designer, or contractors

The list above is by no means complete, nor is it meant to imply that a job having these characteristics will necessarily be a problem project. The list is a well-founded, honest, conscientious effort to state certain unique construction job features that have been identified in projects that have actually ended in dispute, mediation, arbitration, or litigation. In the examples of problem jobs given below, I have stated three of the important elements to consider when determining its claim-prone potential: the project characteristics, the root of the potential problem, and the reason why the project may be prone to generating contested claims.

1. A Wide Spread in Proposal Prices on Hard-Money Projects

The Project. A hard-money, publicly owned waste treatment plant is being built, on which no prequalifications were placed on the prime contract bidders. Liquidated damages are to be assessed for time overruns.

Root of Potential Problems. The wide spread in proposal prices appears to stem from premature issue of the construction documents before they were properly checked by the design team and the owner, and excessive numbers of prebid addenda and clarifications issued during the bidding period.

Reason for Claim-Prone Potential. Many of the elements of the project requiring prebid change or clarification are probably going to be questioned as the project proceeds in the field. The shifting of risk required by the hard-money delivery system from the owner to the prime contractor will undoubtedly cause conflict and disputes among the designer, owner, and contractors. In essence, the true scope of work is probably not yet fully defined and will be determined by arguments over what is actually included in the project and what was intended to be included.

2. Issuance of a Large Number of Prebid Addenda and Clarifications

The Project. A local public library is to be remodeled and an addition is to built. The job is being bid on a total fixed-price basis. The owner's library staff has managed and directed the design of the facility and is going to manage the construction through the prime contractor. No liquidated damages are to be assessed for time overruns.

Root of Potential Problems. The large numbers of prebid addenda and clarifications were caused by the architect and engineers trying to design a facility from an inadequately written project program. Extensive revisions to the design concepts and details caused the design team to exhaust its fees without actually completing an adequate set of bidding documents. Subsequent to issuing the bid package, the changes and errors had to be corrected by prebid addenda and clarifications.

Reason for Claim-Prone Potential. The design of this library will probably never be complete. Yet the architect, the engineer, the contractor, and the subcontractors will all be held accountable for providing what the owner felt was his intent when the job started. The obscure risk assignment, the inexperienced owner's job management, and the single-criterion project delivery system have combined to make this job nearly impossible to build within a predictable cost limit. In addition, the unbalanced fee expenditure by the designers has precluded the probability of an adequate fee being available for high-grade field administration.

3. For Subcontractors, a Poor General Contractor, if the Job is Being Built by One Prime Contractor

The Project. The job is an addition to a large private residence that is being constructed by a small general contractor acting as the prime. The general has a reputation for shopping subcontractors for low hard-money pricing, not paying his bills promptly, negotiating down his subcontractors on their retention, back-charging subcontractors for many items normally paid for by most good primes, and seldom getting through a job without legal disagreements with his owners and clients. In addition, the general contractor's field management is known to be made up of unknowledgeable friends and others to whom the general contractor owes favors.

Root of Potential Problems. What makes this a potentially claim-prone job is the need for a high level of documentation by subcontractors, the careful attention that must be paid to all payment and collection practices, and the lack of subcontractor-oriented prime contractor management. Also, the lack of good management at the client contact level leaves the subcontractors to fend for themselves in planning and scheduling work. Any subcontractor who takes on such a job as this is assuming an at-risk position for his payments. He also stands to have to endure a long close-out process and difficulty in collecting his final payment on the project.

Reason for Claim-Prone Potential. The biggest threat in this situation is to the financial integrity of the subcontractors. As a side observation, the general contractor encounters a very high potential for not being able to get competitive sub prices on other projects. This means that the likelihood of the general contractor remaining in business is reduced considerably by not being price competitive. The potential for bankruptcy puts all involved in the project at excessively high risk.

4. For Projects with Separate Primes, Poor Reputations of Other Prime Contractors

The Project. Grover Construction Company is 60 years old. It was founded by the current president's great-grandfather. Grover has prided itself on being an ethical, up-front builder that negotiates well, pays its bills promptly, treats its subs well, and has never been involved in any litigation or binding arbitration. The company has had disputes and contested claims in the past but has resolved the problems most of the time to the complete satisfaction of all concerned. Grover has recently, and against the better judgment of the president, bid and obtained a hard-money water treatment plant, in which it is the excavation, concrete, masonry, and carpentry prime contractor. The owner for the water authority, Transton Township, is acting as its own general contractor on the project and will manage its work through the present water plant superintendent.

Other prime contractors include Lochan, the heating, ventilating, and air-conditioning contractor, Carlsbad, the plumbing and piping contractor, Davidoff, the fire protection contractor, Rasmuth, the electrical contractor, and Linus, the controls contractor. All have good reputations except for Rasmuth and Linus, the electrical and controls contractors, respectively. These firms are known for their sharp shopping expeditions among vendors, their tendency to claim for extra costs based on work installed before them not being acceptable to them, and being slow to complete setting embeds and sleeves in concrete and masonry correctly and in a timely manner. It is June and because of the size of the job, much concrete and masonry work must be complete before heavy winter weather starts in mid-November.

Root of Potential Problems. The work of Grover and the other primes with good performance records is heavily dependent on the timely performance of the electrical and controls contractors. The potential problems revolve around the need to complete as much work as possible before the harsh winter weather severely limits outside work and close-in of the buildings. Trades installing in-wall work must be available and work closely with the other primes to set their work. This progressively frees up concrete pours and masonry erection. These are critical early elements of the work and must be readied for the long winter period of construction. Full cooperation between contractors is essential. The management of the project is under the owner's control. The water plant superintendent, who will be in charge, although a good process operator, has little experience in new construction of the size anticipated.

Reason for Claim-Prone Potential. The need to cast large amounts of concrete and close-in critical buildings for inside trades is paramount as the Transton plant gets under way. The construction timing lends itself to negative leverage from contractors who follow, and then lead other contractors in the installation of their work. Total confidence by the prime contractors in the ability and trustworthiness of their fellow prime contractors is essential. If Grover and the other reputable primes are to be subjected to delays to in-wall electrical and control work, this is a time of year when it has the greatest negative effect. A day of work lost during good weather may require two or three days to make up when the weather is bad.

If the electrical and controls contractors perform according to their past reputations, the neophyte construction manager for Transton Township will have difficulty keeping them working. They will probably seek all excuses they can for delays to their work and will impose claims for their delays on other primes and subcontractors. This invariably results in contested claims, hard feelings, cost overruns, and delays to the work. Grover knows it must be careful when forced to work with other contractors having doubtful reputations. This was the reason for the reluctance of Grover's president to bid the job.

5. More Than Four to Six Prime Contractors Involved on Project (for normal building work only)

The Project. A large urban area school is to be built utilizing as much local contracting ability as possible. The total cost of the school is $15 million. The school board is committed to awarding a prime contract for construction work groupings with a value over $20,000. Considerable latitude is allowed in grouping the packages so as to be able to temper the number of packages by the management abilities available to run the project. The contract is to be managed by a full-time board of education facilities employee and a small facilities department on-site staff. Fifty-two prime contract packages have been bid, all to be under the direction of the board of education project manager.

Root of Potential Problems

1. The number of separate contracts is considerably greater than is usually encountered in contracts using separate prime contracts.
2. The large number of prime contracts is being managed by the owner (the school board facility staff), who has a traditionally adversarial position with contractors.
3. There is a high potential for gaps in the contract work, all of which is being managed by the owner with limited abilities and authority to authorize needed corrections to both the design and construction work.

Reason for Claim-Prone Potential. The resolution of conflicts, errors, disputes, and other constructive or destructive influences on the job is left in the owner's hands. The owner's representative (as a field executive) may not be able to rule in a firm, fair manner on all matters in which parties other than his own are involved. In addition, there is a high probability that with the large number of prime contracts awarded, some of the weaker firms will experience business difficulties such as bankruptcy, inadequate financing, and low levels of staffing. All these seriously affect the entire team's performance and profitability potential.

6. Poor Reputation of Architect or Engineer of Record

The Project. An industrial plant is being remodeled which involves large amounts of field measuring, on-the-job decision making, and considerable redesign as the project proceeds. The prime contract is being let to a single general contractor, Eagle Construction, on a time and material basis with a guaranteed maximum price (GMP).

Root of Potential Problems. The architect/engineer has a local reputation for shopping contractors' prices, insisting on shifting risk to the contractor, and processing revisions and submittals slowly and with large numbers of corrections.

Reason for Claim-Prone Potential. The project delivery system may not accommodate sizable job changes necessitated by field discoveries as the remodeling proceeds. The design team is not accustomed to working on a job where the risk of change must be shared with all project team members. These differences in attitude coupled with the design team's inexperience and proven incompetence on projects of this nature make destructive conflict nearly inevitable.

7. Several Prime Design Team Organizations

The Project. A proposed 300-acre combined industrial and office park is being erected in the heart of the old Iowa city of Frostange, a community of about 150,000 people with a strong manufacturing base and a high level of community spirit. The Economic Development Division of the city, and the Industrial Development Corporation, a nonprofit land-holding entity, are entering into a contract with the design team for the park. It is the intent of the city and the corporation to prepare and issue a full set of site-preparation contract documents. The document package will contain working drawings for full utility and road construction of the entire site.

Development of the Frostange Vista Business Community is to proceed in stages. Three 100-acre parcels are to be brought on line over the next four years. It will require careful land planning to maintain the desired staged-use occupancy schedule and to attain minimal disruption of new facilities built earlier. The total site development program is to be brought on line over the next six years. This has all been defined in a well-written staging and performance program prepared by volunteer members of the Industrial Corporation.

All of the program committee members have an interest, driven by their financial investment and their community interest.

Several prospective planning and design offices have expressed interest in the project. Preliminary qualification screening has reduced the number of firms able to do the job to six: two from San Francisco, one from Chicago, one from St. Paul, and two from the Frostange area. After much interviewing and discussion, the selection task force recommended the following:

- That one of the California firms be retained to do the land-use planning
- That the Chicago firm be retained to design the site utilities
- That the St. Paul firm be retained to design all paving and site structures
- That one of the local Frostange firms be retained to design the site topography and landscaping
- That a project manager from the Frostange department of public works be put in charge of the design project and report to the city's Economic Development Division

The economic market study has been completed, the preliminary site development program has been written, the design work has been funded from industrial development bonds, the selection recommendations have been accepted by the department of community economic development and the Industrial Corporation board of directors, and the prime site design contracts have been awarded to the firms selected.

Root of Potential Problems. A complex design program generally requires a single point of responsibility resting in an organization, an individual, or other unit entity. This could be done in the Frostange program if the project manager is vested with the authority needed to direct the work. His authority and responsibility must be clearly defined. If it is not clearly defined, the distance and the regional differences in the prime design contractors will make good management virtually impossible. The client holding the design consultant's contracts must be clearly identified and be a legally authorized agent of the financing and ownership sources. This is essential in order that the project manager be able to manage the entire program of work successfully.

Long distances between the consultant's operational bases and the regional differences among participants always influence conflict potential. Great distances between related design operations slow down communications, increase the possibility of misunderstandings, and force the use of remote transmission devices, such as telephones, E-mail, modems, faxes, and other such indirect contact. These still do not adequately replace the quality and quantity of work that can be produced when those involved are physically near and communicating with each other in person.

Regional differences influence the design approach, particularly of such elements as pavement, landscaping, utilities, and the impact of winter weather on the facility. Differences in design procedures often cause disputes and varying opinions that are literally impossible to resolve. Such differences create conflict and turf building, and encourage a clique mentality among those in favor, or those not in favor of a particular design element.

Reason for Claim-Prone Potential. A major problem in the Frostange program is provision of a single-point responsibility management. Because of the geographic and time differences in the design firm's operational bases, only a strong, competent, understanding owner or project manager can direct this program successfully. Such a person is always in high demand, and the chances of finding one who could be spared from the department of public works (DPW) may be remote. The city probably has a very ambitious capital improvement program in work. This is indicated by the decision to embark on the Frostange Business Park venture. The DPW director is not likely to look kindly on shifting his or her best people to another assignment at this time. So the quality of management has a smaller probability of being adequate than is desirable. Weak management usually initiates a struggle for position and power in design firms having strong ideas.

8. Excessive How-To-Do-It Emphasis in Contract Documents

The Project. A large chemical fluid waste storage tank farm is to be constructed at a regional airport just outside a densely populated urban area. The tank farm will be designed and built by an at-risk construction manager who will be responsible for constructing the facility to the owner's specifications. The contractor is working on a time-and-material arrangement with a guaranteed maximum. The owner's design specifications, from which the contract was established, define each component in detail and in some cases even give the chemical composition of products to be used. One of the products specified is an exterior tank foundation coating designed to resist stains and discoloration from petroleum products while maintaining its original color.

Root of Potential Problems. The detailed specification for the coating had been rescinded by the manufacturer since the product had not met the stain-resistant qualities claimed for it. The owner was not aware of this change and had double specified the product. One specification, the prescriptive description, was for the chemical composition, which could be met. The other specification, the performance description, was now for a performance that could not be met.

Reason for Claim-Prone Potential. The conflict in composition and performance of the product shows the danger of overspecifying materials and equipment. The problem becomes especially acute for products that are evolving. In this case the coating used began deteriorating under gasoline spills and overfills. It discolored and began to peel soon after application. This required that the coating on all tank foundations be removed, and a new coating meeting the performance specification applied. Deciding where the responsibility lies and who is to pay for the correction will be difficult and time consuming.

9. Nonliable or Not-At-Risk Parties Involved in Responsible Positions

The Project. The city of Cranmore is about to start construction on a new historical museum estimated to cost about $7,000,000. The funding for the museum is to come from monies allocated in the capital improvement budget of the previous city administration. Hope Trudy, a small, competent local architect, has been retained by the Cultural Center Authority to prepare the architectural and engineering construction drawings for the project. The Cultural Center Authority reports directly to the mayor. The new museum program has been prepared by a consultant hired by the museum board of directors, an independent group elected by museum donors.

At the request of the museum board, a nonliable construction manager has been retained to work with the design team in estimating costs and selecting the building systems and construction materials. The construction manager, Douglas and Associates, reports to the city engineer, who will pay their fees and costs out of the city's capital programs budget. Douglas and Associates will ultimately assume management responsibility but not financial liability for each construction contract awarded by the city engineer through them. Architectural and engineering design is under way.

Root of Potential Problems. This complicated organizational structure and diffused responsibility pattern is a possible source of job trouble. There are several contributing factors. First, the funds available must be channeled through the agency or organizational sources from which they came. This makes the payment process erratic and inconsistent. Second, control of the project program was left with the museum director, to whom the programmer reports. This division of the early work may separate the program from the design of the facility and cause gaps in the design. Third, control of the working drawing phases are under the Cultural Center Authority, whose existence is independent of the city administration. Fourth, the construction manager has been assigned to the city engineer as a consultant during design and as a nonliable manager for construction. Responsibility for its actions is completely separated from the project programming and design team management. Communications, responsibility, and organizational lines are poorly defined. Above all, notice that the responsibility for cost control is merely implied in the various assignments and that none of those directly involved as yet have a well-defined financial liability.

Reason for Claim-Prone Potential. The architect, the programmer, and the construction manager, each of whom exerts considerable influence on the cost structure of the project, all report to different agencies and individuals. This will ultimately cause confusion as to who is to control construction costs, quality, and scheduled performance. The ultimate contractors at-risk on this project will be working for managers not at risk. The potential for bad communications, poor business practices, design and construction cost overruns, and delays to the project by an obscure responsibility and authority pattern make this project subject to claims and political dishonesty.

10. Large Numbers of Allowance Items

The Project. A designer and manufacturer of large luxurious yachts, Catalpa Boat Company, has had a very successful three-year sales record. The demand for its products has forced the owners to build a new branch manufacturing plant and sales office in Florida. Working drawings for the plant are nearly complete, and design of the boat engineering studios and offices is in the final stages. The architect/engineer, Maloney and Associates, is to print and issue the final construction documents in three weeks. Both Maloney and Catalpa are lining up potential general contractors to bid the project on a hard-money proposal.

In the flurry of work to complete the drawings and because of a lack of color and material decisions by the owner, Maloney has recommended that the carpet, the furnishings, the show room fixtures and furnishings, and the office mahogany paneling be specified by allowances. He has estimated the total cost of these items at $300,000. This is based on current discussions with the owner and his wife about the quality desired.

Root of Potential Problems. The cost of the allowance items according to current discussions, and as a best guess of the design team, will probably change. Interior item costs tend to increase as more deliberate design attention is given the selection of materials and furnishings. Thus, by the time the final interior design documents are issued and the contractors price the work, changes in final costs to be incurred can be expected. On the Catalpa project, the final cost for items included in the allowance schedules came in at $450,000, considerably more than the original allowance.

Mr. Catalpa was unhappy and blamed the Maloney design team. Mr. Maloney was unhappy and blamed Mr. and Mrs. Catalpa. The contractors and subcontractors were unhappy because they were delayed by lack of information and had to reschedule several critical interior work items. Nobody was happy with the job.

Reason for Claim-Prone-Potential. If allowance item costs increase over the original estimates of the architect, the owner may become very upset and fail to understand the real causes of the increase. These may be increases in costs due to inflation, inaccurate estimates by the design team, changes to the scope of work by the owner, or poorly specified materials and equipment by subcontractors. Whatever the causes, the results are unhappiness and erosion of confidence in the entire project team. In addition, delays in deliveries and revisions to work already in place could well cause claims against the owner and the contractors. All these possibilities make the extensive use of allowances a cause for concern on a project.

11. Zero or Excessively Small Tolerance Specifications

The Project. Two large, adjacent, reinforced concrete buildings are to be extensively renovated into a new interconnected national headquarters building for a large electronics manufacturing company. The buildings were built in 1919, and on-site photos taken during construction show that rough manual labor was used in great quantities for forming, setting reinforcing steel, and pouring and finishing concrete. The tools used for setting grades and elevations appear to be of a most primitive type. The original building use was to be as a secure warehouse for heavy office machinery. You are the general contractor on the remodeling project. Your contract is as a liable or at-risk construction manager with a time-and-material agreement specifying a cap on the costs. You have been given a notice to proceed and are trying to execute the contract for all the most troublesome long-lead-time items.

Root of Potential Problems. A special difficult-to-get floating computer floor is being used throughout the building in all data processing areas, and you have been made aware of a peculiar difficulty. The computer floor contractor called and said that one of his competitors called after the selection and said that the computer floor had to be designed, fabricated, and installed so that all joints in the floor between panels adjacent columns were to be exactly on column centerlines. No deviations are to be allowed. He said that the general requirements gave the architect the right to do this. You were aware of this condition but did not realize that the architect or owner intended to enforce the provision to the letter in an old concrete building. You also interpreted the column centerline to mean the centerline of the new plaster column covering. Your floor contractor now tells you, after a conversation with the architect, that the architect says that it means the center of the structural columns. Currently, because of poor construction documents, and some serious differences between interpretations of the specifications by you, your bidders, and the architect, relations between you, the owner, and the architect have become quite strained.

Reason for Claim-Prone Potential. We can work backwards in this brief potential problem statement to see how a current minor annoyance has become a potentially expensive and disruptive problem. Relations between you, the owner, and the architect have become strained for reasons that have little, if anything, to do with the computer floor. However, now that an interpretation is needed of the documents in the matter of the floor, the architect is saying,

"The specs are clear. The general requirements give me the right to demand zero tolerances if I feel it is in the best interests of the owner. I do."

He may be right unless some common ground for working out this over-blown problem is found quickly. If unresolved, the problem may force you to take another subcontractor at a much higher price. Even more serious, you might find it is still impossible to build to a zero tolerance as demanded, with the higher subcontractor price. This could lead to all kinds of contested claims that will have to be settled in some manner.

12. Poorly Defined Authority and Responsibility Structures in the Offices of the Design Team, the Owner, and the General Contractor or Other Prime Contractors

The Project. The project is a large multimedia production studio for the use of several participating organizations. There are 10 of these organizations that have agreed to contribute equally to the total cost of the facility, estimated at $15,000,000. The project must be on line within 14 months from the start of design. Budget estimates must not be exceeded, and construction is to start long before all construction contract documents are completed. Tom Placet is an employee of the largest of the 10 members of the conglomerate and the person appointed to be the interim representative of the group. He is not certain of the limits of his authority but tends to assume more authority than he may actually have.

Laura Thompson, the design principal of Thompson and Associates, has begun her design work and is now settling the final details of her contract with the conglomerate. She has asked Tom Placet, the representative of the 10 firms building the project, with whom it is she is to execute her contract. He replies that the corporate entity, Ten Strike, Inc., has not yet been formed and that the largest of the organizations involved will pay the design bills until the corporate formation is complete. Miss Thompson has also asked who is to represent the client in her staff's work on the project. Mr. Placet says that although he is not knowledgeable about design and construction, he will act as Ten Strike's voice during the early program and design phases. He quickly adds, though, that four or five of the principals from the conglomerate also want to be involved with the program and early design. In fact, they have formed a design task force to manage the facility work for Ten Strike.

Root of Potential Problems. The tardy incorporation of the conglomerate and the confused and inexperienced management representation by Mr. Placet is a strong signal that the remainder of the project might be overdirected and micromanaged by the 10 owners. Another potential problem is that the appointment of a task force to oversee program and design work is liable to produce unhealthy and unresolved conflict among the parties. This is especially dangerous in the early and highly subjective phases of the work. Here, many decisions are more a matter of opinion than of technical discretion. They require single-point responsibility to be exerted in the decisions to be made.

Appointment of a program and design task force indicate an unwillingness on the part of the owners to turn the decision making over to a single knowledgeable representative. This will slow down the design process and will probably carry over into construction. Most important, it produces an obscure organizational structure in which the true ultimate decision maker is difficult or impossible to identify.

Reason for Claim-Prone Potential. An obscure organizational structure complicated by lack of single-point contact and doubtful representation is a caution signal. Such a combination will usually cause decision-making delays, create crossed communication paths, produce inferior architecture and engineering design, and create tensions and conflicts on the job. All are counterproductive to the practice of good design and construction. This is particularly so when the project is time and cost sensitive, as is the Ten Strike job.

13. Inexperienced Specialty Contractors

The Project. Hockey is a passion with the the residents of Secord City, the location of Tobin College. They, the Tobin alumni, the student body, and the faculty have vigorously supported a proposed new ice arena. The Tobin College All Stars have completely satisfied their fans by winning the conference championship for five straight years. Finally, the contributions and other financial supports have been counted and the design and construction of the project can begin. The size of the arena is set at 4000 seats, a moderate-size facility well within the financial capacity of the college.

Frank Trigger, a Tobin alumnus and volunteer project representative, has been appointed the owner's project manager on the new job. Frank knows many small business men and women in Secord, a rural community of 80,000 people. Many of them have contributed to the financing of the new hockey arena and are delighted that the job is about to begin. Allenton and Bayer, architects and engineers from a larger neighboring community, Princeton, have been selected as the designers of record. They have a good reputation for functional, attractive, workable, economical designs for small sports facilities. Lehigh Constructors, a reputable general contractor, also from Princeton, has been selected to advise during the design period and gradually to assume the role of an at-risk general contractor. All design and construction personnel have been advised by the Tobin College president and board of regents that strong consideration in selecting subcontractors is to be given to local supporters of the Tobin All Stars.

Root of Potential Problems. In evaluating the project, Lehigh Contractors knows that even though the arena is only of moderate size for hockey arenas, the potential for serious difficulties is great. One of the most difficult challenges faced by Lehigh and the owner is the requirement that strong consideration be given local subcontractors who are hockey supporters. The subcontractors available in Tobin have built many facilities, including farm structures, small shopping centers, and some moderately sized pole barns and industrial structures. The largest job ever built by Secord contractors was a new 100-room dormitory. It was occupied about five years ago.

Reason for Claim-Prone Potential. Frank Trigger, John Lehigh, and Tom Bayer, the principal architect on the job, are meeting to assess risk and its assignment on the project. Mr. Lehigh points out that there probably are not enough experienced subcontractors in Secord to fulfill the wishes of the Tobin College president. For complicated trades such as the structural frame, the scoreboard work, the ice system, and even much of the routine mechanical and electrical systems, out-of-town contractors may have to be selected. The reason is simply that these trades require skills and equipment not available in Secord.

Frank Trigger says that he certainly does not wish to offend financial supporters of the arena, particularly those local builders and construction-related vendors who have been vocal and enthusiastic about the arena. John Lehigh says that the potential for delays, cost overruns, and even claims could severely damage the euphoria now pervading the Tobin College campus. He suggests that the college sports staff and the management meet with local supporters to find a way to allocate the work fairly among local and nonlocal business men and women.

Tom Bayer agrees. However, Frank Trigger vetoes the idea and says "let's go ahead under the present conditions and see how the whole matter evolves." "The future home of the All Stars may be heading for construction problems," says Mr. Lehigh to Tom Bayer as they leave the meeting.

14. Excessive Number of Owner and Designer Preselected Suppliers for Key Material and Equipment

The Project. A group of retail trade associations has decided to form a development corporation and to build a 25,000-square-foot office building to suit the particular needs of its members. The group has retained an architect and engineer on a fixed fee for design of the total project. The corporation has located and gained control of a desirable site and has acquired financing commitments adequate to proceed with writing the design program and preparing the schematics and final construction documents. Group members have agreed that they want to award a hard-money construction contract. Because of their extensive business contacts with others in the construction business, they have told the architect that they want to preselect or tightly control the award of subcontracts by the prime general.

Root of Potential Problems. Where the owners or design team members exert a tight rein on the contractors and subcontractors desired on the hard-money job, they often preselect who they want and dictate that these contractors are to be used. If the list of preselected subs is large or if those preselected are critical or control a large part of the cost, the prime general is limited in the selection of subcontractors he wants to use to optimize his profitability. In this case the project is being built by a large group of price-oriented clients, the trade association executives. The conditions placed on preparation of the construction drawings and specifications will undoubtedly result in communication confusion during the bidding period. This, in turn, will probably cause short circuiting of the usual control and management channels used by good general contractors.

Reason for Claim-Prone Potential. The prime general contractor in this case, whoever it is, will have to compete on a project where an important share of the project costs will be incurred on trades in which he had little, if any, part in selecting the contractor to do the work. However, he is obligated to take the price submitted by the preselected contractor and to include it in the bid. This takes away many of the advantages of retaining a good prime general on the project. It increases the cost of the project to the owner because of the lack of competition at the sub contract level. It also puts an unbalanced risk on the prime general because his contract is to be let on a low bid basis. In addition, the general will undoubtedly have trouble managing some of the subs because they have been preselected by the owners or the architect and engineer.

15. *Large Dollar Amount or Numbers of Owner-Purchased Equipment or Materials*

The Project. Sylvester, Inc., is a small centrally managed chain of retail hardware stores that generally build their own outlets in various cities under a partially negotiated time-and-material contract using contractors with whom they have previous experience. The chain normally provides the light fixtures, floor tile, wall coverings, and roof-top units for each store. The prime contractor is responsible for their installation. The chain has had some management changes in its facility group and is awarding this project by hard-money bidding. However, they are using the same arrangement for owner-furnished items as they normally would in their time-and-material contracts. Sylvester, Inc. has also opened the bidding to other than their normal short list of prime contractors.

Root of Potential Problems. The change in procedure from a short-listed group of prime contractors to an open list, and from a time-and-material contract to a hard-money contract, can signal a stiffer competitive price stance by the chain. The adoption of that new stance may also be the result of changes in the facilities' department's management structure. This new management seems to indicate that its attitude is to start lowering costs by accepting bids from contractors that might not be qualified for the short-list type of work.

Delivery of owner-furnished materials and equipment, insuring and storing them prior to installation, and then getting the subcontractors to install them are critical actions that may adversely affect the project. The installers might have made a higher profit by furnishing these materials as on a conventional job. In this case, Sylvester, Inc. has shifted potential profit away from the installers. The practice could also distort the bidding process for inexperienced subcontractors. These conditions may actually work to Sylvester's disadvantage.

Reason for Claim-Prone Potential. Delivery delays outside the control of the contractor, damage to materials shipped and stored improperly, and the reluctance of some subcontractors to install materials furnished by others are three major elements that can easily cause a contested claim. These are common occurrences on jobs where owner-furnished equipment is being installed by on-job contractors. With experienced owners, primes used to working with the owner, and subcontractors who know the conditions under which the materials are usually supplied and handled, the probability of dispute might be relatively small. In fact, several large national firms presently build under the premise that they will provide several materials to the project, and they do it successfully. For the neophyte owner or contractor, however, this project delivery system increases the potential for destructive conflict.

16. Location in Strike-Prone or Jurisdictionally Sensitive Areas

The Project. A large power plant is being built in a strong multiunion area, where disputes are frequent, not only between contractor management and union management, but between unions and often between trades. You are a cooling tower subcontractor to the prime general. Your purchase order is for a fixed fee with all general conditions and general requirements of the prime contract being a part of the purchase order. The general contractor's agreement with the owner is as a turnkey, design-build, and 10-year-operate contract.

Root of Potential Problems. In this jurisdictional region, the power plant owner, a regional public utility, has agreed to be a signatory to the current labor agreement with the plumbers, pipefitters, cement finishers, and laborers. In this agreement, the work rules spell out field operational rules considerably different than those which govern the trades in areas where you normally work. Although you may have been aware of the restrictions before you proposed on the project, in such cases the potential for jurisdictional disputes might catch you in the middle. For instance, the rules may not allow your foremen on the job to work with their tools. Yet your agreement with the general specifically allows you to employ working foremen. The unilateral agreement of the owner with the unions may force you into expensive management changes and extra overhead not anticipated.

Reason for Claim-Prone Potential. If you proceed in accordance with the agreement you have with your client, the general contractor, and use working trade managers, the owner could very well dictate a change in your work arrangement on the job. If you, in turn, attempt to use your contract with the general as a rationale for proceeding, you could well face an argument from all concerned that the general requirements and general conditions binding the general are also binding on you. This produces a serious problem for you on three fronts:

1. You may now have to file a claim of some type against the general or even the utility.
2. You may have to proceed using nonworking field management in your key cost-sensitive activities.
3. You have been marked as a troublemaker by the utility, the general, and the union hierarchy in the area and can count on considerable trouble settling the matter amiably and to everyone's satisfaction.

17. Heavy Use Specified for Untried Products and Equipment

The Project. A 58-story office building in a densely populated urban downtown area is being built by an at-risk general contractor holding an agreement for all work, including base building work, landlord–tenant work, and tenant–tenant work. The project is being built under a guaranteed maximum agreement with all savings going to the owner. The owner has required his architect to exert every effort to be innovative and forward thinking about selections of new materials, systems, and equipment in the building. This, the owner hopes, will give the building high visibility and increase the probability of better-than-average rents and a high occupancy rate.

Root of Potential Problems. One of the elements that was seen as a potential showpiece is the curtain wall. The architect and engineer have selected a new method of laminating a panel of thin stone veneer to a plywood backing. The plywood backing is fastened to a prefabricated structural steel frame. The unit frames are erected from the outside. The entire assembly seems to be inexpensive and the limited tests that have been run indicate the individual panels stand up well to most weather elements.

Reason for Claim-Prone Potential. The exterior panels had been tested on small assemblies only and there was no experience with the panels on total projects of any size. When 90% of the panels were in place, the first of the panels that were erected began to delaminate. The problem was found to be in the exposure of the laminating adhesives to the large range of outside extremes in temperature and the rapid change of temperature as the sun moved around the building from east to south to west. This daily change had caused movement of the stone facing against the plywood backing and ultimately caused the stone to crack and work loose.

An accurate early risk analysis of the material combination was not possible because no small or large expanse of the material had been exposed for any sizable time to the elements in this particular city. It is estimated that replacing the curtain wall will probably take two years and cost about the same amount of money expected to be spent on the entire original building.

18. Nonliable Party Involvement in Establishing Delivery Commitments (i.e., Not-At-Risk Construction Managers, Planners, Architects, Engineers, or Owner Representatives)

The Project. A small high-fashion shopping center is to be built, featuring 20 exclusive women's and men's shops and 10 personal service businesses at the high-priced end of the line. It has a small, very highly styled, enclosed mall that is a showplace for the rest of the shops and services surrounding the mall area. The facility contains about 90,000 gross square feet. The mall will be in a high-income suburb of a wealthy urban community, and because of the lack of surrounding land, and, consequently, competitive facilities, nearly 80% of the center has already been leased or committed to by tenants.

The owner has several other such retail complexes in various parts of the country. She has set a policy for all of them to have her specialty interior design team select and set delivery dates for nearly all landlord-furnished tenant work. She also encourages her tenants to retain the shopping center architect to design the tenants' interior work. This is store work at the tenant's expense. In this system, delivery dates for practically all major items of interior finish work furnished by the landlord are established concurrently with the award of the landlord-work construction contract.

The successful landlord-work general contractor must phase the construction program so the job is ready to receive both owner and tenant material and equipment being delivered in accordance with the early commitments by the owner and by the tenant architect. The prime contract will be a time-and-material agreement, with a guaranteed maximum price (GMP), and liquidated damages of $1000 per day for time overruns. This is the first time this owner has imposed liquidated damages, and she is not certain that it is a good decision.

Root of Potential Problems. The prime general contractor selected has not worked with this owner–architect–tenant team previously and is only vaguely aware of the commitments the owner and designers have already made to certain suppliers. He also does not intend, at present, to use several of the subs from whom the design team has already obtained delivery promises. Now, as the contracts are let, the commitments of the owner and the architect are being challenged by the general contractor and a serious rift is developing that threatens to hold up award of contracts.

Reason for Claim-Prone Potential. As soon as a time-sensitive retail project contract is awarded, the work must proceed promptly with no delays. If delays are imposed on the contractor by conditions over which the contractor has no control, contested claims are inevitable. Even more critical is a constantly eroding relation between the contractor and the owner caused by something they have seen work well in the past. The difference here is that this contract imposes a severe penalty for late performance—earlier contracts with this owner did not have a liquidated damages clause. In this case, problems, cost overruns, and deterioration of job relations at all levels are very likely. This, in turn, affects all parties and decreases their potential for profitable work.

19. Excessively Long Time Periods to Award Contracts after a Proposal is Submitted   (*Note:* This situation often occurs in public work where many nonoperational approvals and agencies are involved.)

The Project. The department of public works (DPW) and the Grand City water department accepted proposals on March 1 for an addition to their new water filtration plant. The low bid is $2.50 million, with the second and third bids standing at $2.55 million and $2.62 million, respectively. The low bidder has been recommended for award by the DPW staff and the selection has been relayed to the city council for placement on the agenda.

In the past, the DPW director has been able to move construction contracts through the review and recommendation period within a week of receiving bids from contractors. Recommendations are sent to council, where the normal time from submission of DPW recommendations to approval of award has been no longer than two weeks. From the council the documents go back to the DPW for administrative work on the contract and issuance of a notice to proceed. This has been taking from one to three weeks, depending on the ability of the contractor to obtain bonding and to meet insurance requirements of the contract.

On this project, the DPW recommendations have been made and the council has them available for placement on its next agenda. However, a recently appointed city manager and a newly elected mayor have decided that the process has been too loose in the past and are insisting on more evidence that the contractor being recommended has the legal and financial horsepower to do the job. This is the first time that such challenges have been made to the usual DPW prequalification process.

Unexpected, but now required project paperwork causes the DPW to miss the first weekly council meeting, and a technical agenda deadline glitch keeps it off the second week. The recommendation for approval is made by the council the third week but conditional upon bond, insurance, and subcontractor selection criteria being met before full approval. The award is now one to two weeks beyond the normal award time and shows signs of slipping another two to four weeks before a notice to proceed can be issued. Already the DPW director is hearing from the apparent low bidder that it can hold its original price only for another week—long before the final council approval is now expected. In addition, some prices for smaller work, which have been bid during the past three weeks, show increases over estimated costs by the DPW. This has not occurred in the past.

Root of Potential Problems. The award process, which has worked very well in the past, is now undergoing some bureaucratic changes that have not yet been tested for their validity. The purpose of improving the choice of contractors by tightening the award process has not been yet proved necessary. Delays to the water project threaten to move the job into at least three to six additional cold weather weeks on early trades by delaying the start of construction. The need to recoup the potential costs of cold weather construction is of sizable concern to the low bidder.

Reason for Claim-Prone Potential. An increase in the time it takes to award a contract can boost the difficulty of accurately predicting conditions under which the project will actually be built. This is particularly so if the delay to contract award is a change from the usual pattern. In the case of the water plant addition, contractors in Grand City have been used to having awards made within four to six weeks of the proposal submission date. This process is now expected to take at least nine to eleven weeks and will probably change the weather conditions under which much of the job is to be built. This will undoubtedly increase the cost of general requirements, general conditions, and possibly raise the price of some weather-sensitive subcontract work. This increased time of award is watched carefully and is undoubtedly one of the main reasons for the uniformly higher prices being submitted on some of the other current capital improvements work.

20. Poor Owner Reputation

The Project. You are a contractor engaged to build a strip office park of about 80,000 square feet for Leaton Delaney, a local developer. The space is for commercial and professional offices of varying sizes according to tenant needs. The base building is under construction with the structure closed in and a standard office front in place including doors and glazing. Underground rough-in for standard toilet locations in the tenant spaces is part of the base building work, as is an acoustic ceiling grid and modular lighting connected to electrical distribution boxes above the ceiling. Acoustic ceiling panels are to be provided but not installed. They will be stored on site. The site is to be fully paved, striped, and lighted under your contract.

You have built several smaller projects for Mr. Delaney before. However, you have generally had problems collecting your monthly and final payments.

You have not gotten into financial troubles because of the small sizes of the jobs, your relatively heavy workload, and the good potential for larger work that seems now to have arrived. Your jobs have all been on a time-and-material basis with no upset price (guaranteed maximum) on your work. You notice that your client is getting later and later paying you and the other prime contractors. You now have money coming to you from three small earlier Delaney jobs. The other contractors working for Mr. Delaney have warned you of pending financial troubles for you if you do not clear your current debt load due from him.

Root of Potential Problems. An increasing use of monies due you and others to finance other ongoing development undoubtedly accounts for the slow payment process. That it is not caused by current problems of Mr. Delaney that might be resolved soon is abundantly clear. The slow payment is a deliberate attempt to hold accounts payable for Mr. Delaney's use. The practice has serious potential for trouble, particularly since this project is nearly five times the size of any other project upon which you have been engaged for this client.

Reason for Claim-Prone Potential. Not only you but also your fellow contractors have encountered serious payment problems with Mr. Delaney. You have determined that it is probably a studied attempt to keep his creditors financing other development. Work for him has you locked into his construction team. Now you cannot risk losing all your monies by refusing to take on this project. If you refuse to do the job and try to get what is due you, you will probably spend a large amount of time in arbitration or in court trying to collect your money from Mr. Delaney. He has all the cards in his hand at this time.

21. Extensive Declaimers on Purchase Orders, Meeting Minutes, and Other Official or Semiofficial Documentation

The Project. A new $800,000 water pipeline is to be laid from the Stoneridge City water plant extending east to a recently built subdivision, Huntington Woods. The line has been a center of controversy. Most Huntington Woods residents are very happy with their individual wells, supplied from a good aquifer. Most do not feel that they need a central system. The water tastes good, is free of contamination, and is moderately soft. Concerns have been expressed that the treated water from the plant, which draws on wells over the same aquifer as the Huntington Woods wells, will offer no benefits not presently being enjoyed by the subdivision residents.

The expansion of the distribution system will also require a $700,000 improvement of the present wells and treatment facilities. The financing is to be provided by Utilities Engineering, Inc., a private firm that will also begin operating the plant when the expansion comes on line next year. They will gradually assume full ownership of the plant over the next 10 years. The design is complete, the construction contracts have been let, and the preconstruction meeting was held yesterday. The contractors are to mobilize and move on site tomorrow. The contract is a single prime with a fixed price.

Root of Potential Problems. At the preconstruction meeting, Union Mechanical, the mechanical contractor, asked who was to take notes at the meetings. Utility Engineering's on-site representative said that the specification required the architect/engineer, Loquer Engineering, to fulfill that requirement. Union agreed but said that it will also take meeting notes and publish these to the city of Stoneridge, Utilities Engineering, the engineer of record, and the general contractor. Union said that it wants their understanding of the proceeding in their files for future reference. The project is one that has conflict potential from the community and Union wants to protect its financial position on the job. Union further said that the ownership arrangement was different from that normally expected, and this has also led the company to want its own set of records.

There was a great deal of discussion, agreement, and dissent with publishing a second set of minutes. Some of the other subs sided with Union, while the owner and the engineer thought that it was a waste of time and effort. The general contractor, with whom Union has its contract, was noncommittal about the additional minutes. It is obvious that unless better rapport between the contractors and the owner is reached, the documentation, declaimers, disclaimers, and claims on the job are going to be almost unmanageable.

Reason for Claim-Prone Potential. Extensive documentation imposes a cost burden on any job, particularly when it is designed to cover any situation with potential for conflict and possible claims. The conditions imposed on a project by such strongly expressed objections and because of factors that should have been apparent before the project was bid will lower the probability of project success. In addition, the parties are starting a new project in one of the worst ways—with an admission that Union and the subs that agreed with Union knew the specified requirements before bidding. Yet they proposed on the job, were awarded the contract by the general, and now want to impose an extra documentation requirement on the job.

The question is why? The main reasons are, in this case, distrust of others on the project team, suspicion of the financing source and the payment process, concern that claims will be ignored, and fear that negative community pressures could create job disruptions and escalating costs to members of the construction team. Few jobs can be successful starting out with such feelings being expressed so strongly so early in the project.

22. Extensive Early Documentation Inconsistent with the Complexity of the Project

The Project. A new 300,000-square-foot light industrial plant of structural steel is being constructed with conventional exterior concrete sill walls and industrial steel sash above the sill wall. It will have a medium-duty interior slab on grade and conventional light industrial interior rough and finish systems. There will be an attached office of 60,000 square feet. The offices are designed to be simple and utilitarian. Site work consists of parking for 500 automobiles, some trailer storage, and well-landscaped walking areas around the visitor parking and employees outdoor walking and lunch area.

Root of Potential Problems. The prime contractor has been retained on a fixed-cost basis and was selected as low bidder in a price-competitive analysis of five proposals from a set of short-listed contractors. You are the owner of the plant and have been assured by your prime general contractor that the plant, which is now under construction, can be completed in 14 months and will be available for substantial occupancy about March of next year. However, you are somewhat concerned about the amount of correspondence being generated between the prime general contractor and the paving contractor, and the prime general contractor and the precast sill wall contractor. The paver and the precaster are subcontractors to the prime general.

You are receiving copies of letters to both of these contractors from the prime general. You are also getting correspondence from the prime general about the importance of minimizing approval delays by your architect/engineer of record. To the best of your knowledge, the submittal logs you have seen indicate that to date, the shop drawing turnaround is normal and as agreed on in the early planning meetings.

Reason for Claim-Prone Potential. It appears that the prime general contractor has some information that you, the owner, do not have. In this case, the missing information is that the paving price was shopped very hard by your general and he got a low bid based on the site being ready for paving by late June this year. This is just before the big rush to complete educational units for fall start of school. If the prime general misses the early paving window, he faces a sizable cost increase in the paving contract or a substantial claim for additional costs from the paver.

The sill wall precaster's price was also shopped heavily and was provided by the precaster only because he had an April opening in his production schedule. However, if the prime general is not ready with foundations to receive the precast, or if there have been delays to detailing, approval, fabrication, and delivery of precast panels, the wall panel contractor will claim sizable extras. With this missing information in hand, you, the owner, can perhaps understand the prime general's concern with documenting his relations and the conditions of the project and their impact on site paving and sill wall fabrication. An increased understanding will allow you to respond quickly and appropriately if friction between the general and its subcontractors escalate. In this situation a good issue resolution system should be in effect.

23. Unqualified, Mismatched, or Dysfunctional Project Management Staff Assigned to the Job by Owner, User, Designer, or Contractors

The Project. A new high-visibility urban-sited college library building is to be built on the campus. The facility has been planned and programmed by the college as a forerunner facility characterized by special finishes and advanced electronic operations.

Root of Potential Problems. The design firm's project manager is a production-wise, design-sensitive, architectural professional; the owner's field representative is a pragmatist who looks at this job as one he wants to finish so he can move on to his next project; the contractor's project manager and superintendent are both inexperienced on fussy, complex, detailed, and high-design jobs.

Reason for Claim-Prone Potential. Each responsible manager from the three key organizations has different attitudes, different levels of skill, and different perceptions of what this project is and what the intent of each participant should be. The disparity in the background, beliefs, and desires of these major players could easily produce continuous conflict and disagreement about how each of the building components should look and should be constructed. The potential for project difficulties is high.

In the discussions above we have outlined some of the factors that can cause a job to fail and the participants to quarrel. The extent and variety of situations that make a project claim-prone are indicative of the immense difficulty we encounter in attempting to make the profession of design and construction free of unresolved destructive conflict.

6

PROJECT SUCCESS AND DISPUTE RESOLUTION

SECTION A: HOW TO MEASURE PROJECT SUCCESS

To see how dispute resolution becomes an operative system for problem solution, we must have a clear view of what is meant by success in the design and construction industry. Many factors influence the degree of success achieved on a project. They include everything from how the project originates to how well the completed facility is managed, operated, and maintained. An analysis of many good and not-so-good projects over the years has lead me to believe that five fundamental influences should be given careful attention by the managers of any project:

1. Validity of the defined project goals and objectives
2. Type and amount of profit desired
3. Excellence of the job plan and sequence
4. Competence and disposition of the participants
5. Kinds and intensity of problems that are encountered

On any construction project, from concept to full use, at least six major goals must be achieved. The degree of achievement of these goals determines the degree of project success. Very simply, for a project to be successful, the client, owner, and user must be assured upon completion of the project that:

1. The facility program and the facility design have met the needs, desires, and wishes of the owner, user, designers, and the constructors.

87

2. The planning, design, and construction work on the project has been accomplished within the time and cost structure required and desired.

3. All relationships on the project have been maintained at a high technical and professional level and have proven rewarding for those involved.

4. The people involved at all levels of work on the job have realized a financial, professional, and technical profit for themselves and their associates by being on the project.

5. The project has been closed out with little or no residual potential for major operational or maintenance problems.

6. The entire design and construction process has been free of unresolved contested claims for additional money, additional time, damage payments, and of the potential for future financial demands after the job has been closed out.

I have frequently referred to the importance of the facility program.

> A program is a narrative describing the needs and character of the proposed operation, the requirements of the user and owner, the nature of the environment to be planned, designed, and built, and the corresponding characteristics of the space that will satisfy these needs and requirements.

Sometimes a program is called the brief.

Lack of a good statement of the owner's and users' needs is one of the most frequent causes of flawed and troublesome construction efforts. The analogy can be made that a building design might have a potential for significant excellence, but if built on the wrong site, to a size greater than needed, and at a price beyond means, no degree of design or construction excellence will atone for the program errors. A good program statement clearly defines measurement yardsticks for the entire project. It describes:

1. The character and needs of the proposed user operation
2. The requirements of the user and owner
3. The nature of the environment to be planned, designed, and built
4. The characteristics of the space that will satisfy the users' and owner's needs and requirements
5. A pro forma analysis and project budget that properly accommodates three levels of user and owner needs:
   a. *Must list:* those items that must be included in the scope of work to make the project a go. If any of the items in the must list cannot be included, the project is a no-go.
   b. *Want list:* those items that are wanted and might be possible to include in the scope of work, over and above the must-list items,

since they provide a definable and acceptable rate of return on their cost.

   c. *Wish list:* those items that the owner and users wish they could include but might not be able to, for budgetary or other reasons. Note that affordable wish-list items are best added, not deleted, as the project moves into construction.

6. An analysis and preliminary recommendation of the project delivery system best suited to the project

Any valid evaluation of progress success, or lack of it, must include a method of measuring profit. Experience with many different people, companies, organizations, committees, social structures, and other goal- and objective-oriented groups indicate that there are at least seven kinds of profit that those participating can expect to receive. Not all are a part of any single person's expectations. The participants can, however, expect to gain all the profits through their combined efforts.

The Seven Types of Profit

1. *Financial:* an improvement in a money position
2. *Social:* a gratifying experience contributing to society's well-being
3. *Self-actualization:* a gain in personal nonfinancial satisfaction by contributive work
4. *Value system:* a reward gained by application of values in which one believes
5. *Technical:* acquisition of technical skill or technical data of value
6. *Enjoyment:* personal enjoyment of a situation gained from involvement in it
7. *Educational:* learning made possible by efforts exerted in any given situation

A simple example of various profit types and how they can contribute to the potential for project success may clarify the importance of understanding profit motives.

Case Study 6.1: The Rest Stop Project Profit Potential

The Kentucky Department of Transportation (KDOT) is constructing a state park rest stop. It is to be a high-visibility tourist center and a showplace for the beauties of the state and nearby areas. The prime general contract has been awarded to the low bidder on a competitively bid hard-money proposal. Referring to Table 2.2, this project delivery system can be described as follows:

- *Agreement premises:* A3, totally qualified, low price wins
- *Authority limits:* B3, as contractor, no agency
- *Payment methods:* C1, fixed total for all costs

Some of those involved in the project include:

- *Franklin Construction, the general contractor.* Franklin's goal is to do a high-quality job within the contract scope and make a financial profit. The company also wishes to use this small project to train a new superintendent and project manager to work profitably.
- *Carl Jones, KDOT's resident inspector.* Carl Jones wants to add a good project to his professional record. The inspector's goal is to get the best quality work, in the time specified, with no additional cost. Then Carl will feel that both he and KDOT will have made a profit.
- *Leslie Carstairs, the Department of Natural Resources technician monitoring known contamination abatement.* Miss Carstairs is concerned primarily with protecting the health, safety, and welfare of the public as affected by the environment. When this is achieved, she and the KDNR have made a profit.
- *Longston & Associates, the architect and engineer of record.* They want a good-looking, smoothly functioning facility that they have designed within their fee and abilities. Their profit is in the fee and in the satisfaction of a good design.
- *The project subcontractors and vendors.* The members of this group look to the general contractor to help them make a financial profit by its skills and leadership. They also want to construct a project they are proud of and can show off to their friends and family as an example of their workmanship and construction abilities. When this happens they are profitable.
- *State Bureau of Tourist Affairs.* This is the operator who occupies the rest stop and hopes that its excellent design and safe, clean, attractive spaces will convince the public that Kentucky is the place to visit. When the tourist public likes what the Bureau of Tourist Affairs is selling, the Tourist Affairs staff and the public have both made a profit.
- *The U.S. and Kentucky taxpayers, the financiers and users of the facility.* After all is said and done, the real profit goal to consider is that desired by the individual members of the public. What they want is a facility that meets their needs. They want a rest stop that was worth their hard-earned tax payments. When this is achieved, the public has realized a profit on its hard work.

You can see that the profit drive ranges from a pure and simple desire to make money, to having fun on a vacation. Your job, whatever role you play, is to ensure that all involved make the profit they desire and deserve.

When this happens you will probably have built a successful rest stop and have a profitable project to your credit by whatever standards you measure profit. So far we have considered the end results by which a successful project may be identified. This is not totally adequate in a how-to discussion of success. The process of achieving success itself must be clearly defined so that guidelines to success are visible each step of the way.

Some well-recognized authorities say that there are nine major steps to follow in successful design and construction. These are basically the elements in the line of action shown in Figure 2.1. How well you do these tasks will determine how successful you will be. The steps are:

1. *Conceive the basic project.* Visualize and state the fundamental nature of the proposed project, its purposes, and its base characteristics. Those responsible include the KDOT, the designers, and the Department of Tourist Affairs.

2. *Prepare the project program.* Set down the physical characteristics of the total project in written and graphic form to define clearly the facility conceived. Those responsible include the KDOT, the designers, and the Department of Tourist Affairs.

3. *Articulate the program for approval.* Merge the concept and the program into written and graphic construction language that can be understood and approved by the ultimate decision makers for full design. Those responsible include the KDOT, the DNR, and the designers.

4. *Approve the project.* Release the concept and the program so that the full design and construction process can be started. Those responsible include the KDOT, the DNR, and the designers.

5. *Design the project.* Prepare full contract documents for construction use. Those responsible include the KDOT and the designers.

6. *Construct the project.* Award and build the project ready for turnover to the owner or end user. Those responsible include the contractors, the KDOT inspector, the DNR, and the designers.

7. *Turn over the project.* Release the constructed project to the owner or end user with full documentation needed to operate and maintain the completed environment. Those responsible include the contractors, the KDOT inspector, and the designers.

8. *Operate the project.* Take over, run in, and make the new facility fully operational. Those responsible include the KDOT and the Department of Tourist Affairs.

9. *Maintain the project.* Keep the new facility in proper operating condition. Those responsible include the KDOT and the Department of Tourist Affairs.

Note that the responsibility for acting in each of these nine steps to a successful project crosses back and forth between all the parties who have a role in planning, designing, constructing, regulating, operating, and maintaining the rest stop. It is the effective passing of the relay baton to each party in turn that keeps the process working well. The process should be functionally seamless.

We have examined end measures of success and have looked at the generic process of achieving project success. Now let us consider some of the disruptive forces that diminish the probability of project success. There are many types and kinds of design and construction problems. However, the ones that have been most disruptive to project success are those that attract extensive involvement of outside nonliable third parties. These are the people who profit through the design and construction professionals' seeming inability to resolve damaging issues among themselves.

The most troublesome issues are those that involve the common contested claim. Dozens of things can go wrong in design and construction that might generate serious conflict and result in a contested claim. However, 10 of these have proven to be most troublesome in our profession. They are:

1. *Constructive acceleration:* an action by a party to the contract that forces more work to be done with no time extension, or the same amount of work to be done in a shorter period of time

2. *Constructive change:* an action or inaction by a party to the contract that has the same effect as a written order

3. *Defective or deficient contract documents:* contract documents that do not adequately portray the true contract scope

4. *Delay:* a situation, beyond the control and not the fault of a contract party, that causes a delay to a project

5. *Differing site condition:* a situation in which the actual conditions at the site of a project differs from those represented on the contract documents or from reasonable expectations of a site in that area

6. *Directed change:* a legitimate change within the contract scope for which the owner is obligated to pay

7. *Impossibility of performance:* a situation in which it is impossible to carry out the work within the contract requirements

8. *Maladministration:* the interference of one contract party with another contract party's rights that prevents the latter party from enjoying the benefits of least-cost performance within the contract provisions

9. *Superior knowledge:* the withholding of knowledge during the precontract period and the time subsequent to contract execution which affects adversely a second party's construction operations in matters of importance

10. *Termination:* dismissal of a party to a project contract for convenience or default

A successful project is also one that has been kept free, not necessarily of conflict, but of damaging conflict. It is well to keep in mind four basic premises about project success:

1. Project success starts with honest people behaving ethically with a high degree of competence.
2. Project success further requires these people to understand the profit needs and demands of all those involved in the project and to help achieve a profit for all.
3. Project success next requires these honest, competent, ethical leaders to understand fully the best sequence to be followed in managing a project competently.
4. Finally, project success demands that we reduce meaningless conflict to a minimum and spend the time wasted in such contests to improve resolution skills among ourselves.

SECTION B: PROJECT SUCCESS AND DISPUTE RESOLUTION

Project success requires resolution of both constructive and destructive conflict. Unresolved conflicts and disputes often require that we consider a neutral nonbinding view where we desire positive solutions. Equally critical to quick, fair resolution is that we avoid forced third-party solutions such as may result from binding arbitration or litigation. Alternative dispute resolution methods are being called into play more and more to settle early conflicts and disagreements that resist resolution by the usual dialogue and negotiating tools used daily at most construction sites.

In broadest terms, alternative dispute resolution (ADR) is a method of resolving disputed design and construction claims outside the formal arbitration hall or the courtroom. It is a method resulting in an opinion or decision that often is not legally binding on the participants. There are many definitions of ADR; a few are given below. Some are paraphrased from references; for others, the origins are more difficult to trace.

Selected Alternative Dispute Resolution Definitions

- A method more appropriate for settling a dispute than either direct negotiations or traditional litigation.
- A process of dispute resolution that is an alternative to the traditional judicial process.
- A process that is designed to facilitate achieving a private settlement or resolution of a dispute by the parties to the dispute themselves with the aid of a person or process which assists the parties in that solution.

- A process of dispute resolution that is an alternative to the traditional judicial process.
- A group of formal and informal procedures that serve as alternatives to litigation.
- An approach to resolving a construction claim outside a courtroom.
- The use of informal dispute resolution techniques to resolve contract disputes.
- A method of avoiding the use of litigation to resolve contested claims where the parties affected recognize that such claims might occur. The parties then provide a mechanism or alternative dispute resolution method to achieve timely solutions to these potential problems.

The number of ADR techniques currently in use or being considered in the planning, design, and construction business is staggering. This is not necessarily because of the number of techniques but because of the complexity of many of the processes. Some of the ADR methods proposed or in current use include:

- Advisory arbitration
- Advisory opinion
- Arbitration
- Architect/engineer rulings
- Binding arbitration
- Court-annexed arbitration
- Court-appointed masters
- Direct negotiations
- Dispute resolution board
- Early neutral evaluation
- Expedited binding arbitration
- Fact-finding
- Incentives for cooperation
- Independent advisory opinion
- Informal spontaneous negotiation
- Intelligent and proper risk allocation
- Internal negotiation methods
- Mandatory binding arbitration
- Mandatory nonbinding arbitration
- Mandatory pretrial negotiation
- Mediation
- Minitrial

- Neutral fact-finding
- Nonbinding minitrial
- Partnering
- Private judging
- Private litigation
- Resolution through experts
- Special tribunals
- Standing neutral
- Step negotiations
- Voluntary binding arbitration
- Voluntary nonbinding arbitration
- Voluntary prehearing negotiation

Many of these methods involve the use of formal legal processes to guide the participants through a maze of procedures, people, and regulations. For the most part, the legal orientation when applied to ADR methods may not properly fit preventive methods of dispute resolution. It is that preventive feature that distinguishes the set of negotiating and resolution tools that seems to have made partnering a forerunner technique in providing the vehicle for a smooth sail through troubled dispute waters. However, a summary knowledge of certain ADR methods will lead to an intelligent use of those best suited to prevent disputes or to nip the dispute in the bud while it can be resolved constructively. In partnering, it is these prevention techniques with which we are most concerned.

What are the advantages of using a conflict resolution system that helps settle disputes early and keeps them from escalating into a full-blown binding arbitration or court battle? Some advantages that have already been realized on very real and very difficult projects are listed below.

- The process gives the participants a privacy that encourages informal, morally agreed on resolution—no public records such as court files are generated for inspection by those outside the dispute.
- Little, or very limited discovery is permitted or required in the resolution process.
- Hearings, as compared to litigation, are usually conducted in informal settings.
- Participant control is high concerning the time and length of the hearings.
- There is little need and no requirement that a party be represented by counsel.
- Costs for fees, expenses, and attorneys are less than for binding arbitration or litigation.

- Persons who are industry peers and knowledgeable about the design and construction profession conduct the hearings and make the decisions.
- The informal process, quickly mobilized to help resolve problem situations, helps defuse difficulties before adversarial attitudes can grow and harden.
- Resolution participants with specific qualities, credentials, and abilities can be preselected by mutual agreement.
- Disputants retain major control of the resolution timetable.
- Minority opinions can be expressed and considered if there is more than one person refereeing the dispute resolution.
- ADR forces the parties to identify their problems and try to solve them as they occur.
- Use of ADR processes seem to have lowered construction costs as parties to future projects see benefits accrue to them by prespecified alternative settlement methods.
- Monitoring project conditions before the problems occur, as is done in some alternative resolution methods, helps moderate stubborn stands by the parties.
- Use of ADR encourages dispute resolution by the parties to the dispute itself.
- ADR tends to eliminate the constant adversarial atmosphere that often surrounds a claim-prone job.
- Informal resolution techniques help the parties better understand and interrelate the issues in dispute.
- ADR assists the parties to find solutions that allow them to walk away from the process with a settlement in hand.
- Moderated conflict resolution often allows the parties to do business with their former adversaries once the dust of conflict settles.
- The system of professional internal resolution provides approaches that often encourage reason at the negotiating table.
- ADR helps to resolve past and current problems quickly so that the participants can focus on future business and technical and professional opportunities.

The ingredients needed to succeed in resolving disputes are not complex to understand. Nor do they make difficult management demands on a project team to assemble and shape effective action. Prerequisites for implementing successful and effective ADR systems are simple:

- Intelligent and realistic performance commitments
- A strong desire for a fair resolution, equitable for all involved
- People in charge who want a fair resolution

- A resolution technique that is acceptable to those involved
- The knowledge of how to arrive at a resolution system that can produce a fair decision
- A proper allocation of risk where it is best accepted and managed
- An understanding of, and agreement with, the belief that if you are not entitled to it, don't try to get it
- To work and act from a position of fairness and intelligent self-interest rather than solely from a position of power
- To suppress greed
- To try to establish an honest, well-expressed feeling of trust among participants
- To gain support from the participants and stakeholders
- To assign experienced, competent people to responsible management positions
- To use ADR systems in a manner that has proven effective and successful
- To establish a method of measuring results of using the ADR system
- To have empathy
- To recognize and celebrate success

SECTION C: HOW TO APPLY AND USE ALTERNATIVE DISPUTE RESOLUTION

Discourage litigation. Persuade your neighbors to compromise whenever you can. Point out to them how the nominal winner is often a real loser: in fees, expenses and waste of time.
                                        —Abraham Lincoln, July 1, 1850

The bewildering array of ADR information and systems make it essential to narrow the choices of an ADR system to those most useful and understandable. These must be techniques that the design and construction professional can use, understand, and apply among his or her peers without excessive legal risk. My experiences and study indicate that the graphic depiction of a step-by-step application of ADR processes that address a graduated need for appropriate resolution is of great help in explaining and using ADR. Such a picture is painted with very broad strokes in Figure 6.1, where the vertical line XX at the far left represents the point where a mutually acceptable contract or agreement is in effect for the project. XX is a high point where all the people on the job are friends and are looking forward to a pleasant, successful, and profitable project. At point XX, there are usually few recognizable conflicts that are of major concern to the design team, the owner, or the contractors.

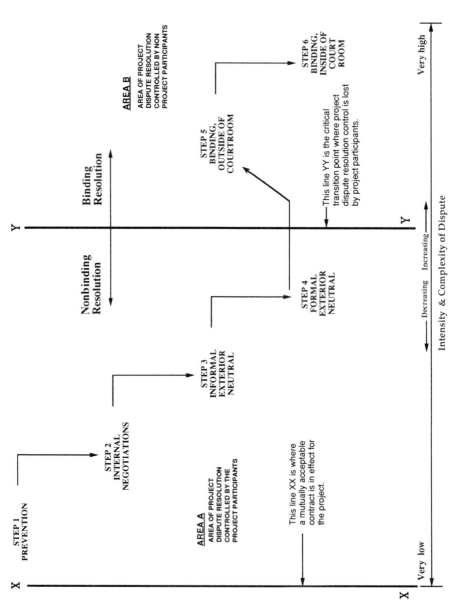

Figure 6.1 Steps in issue and dispute resolution, detail level 1.

STEP 1 PREVENTION

STEP 2 INTERNAL NEGOTIATIONS

STEP 3 INFORMAL EXTERIOR NEUTRAL

STEP 4 FORMAL EXTERIOR NEUTRAL

STEP 5 BINDING, OUTSIDE OF COURTROOM

STEP 6 BINDING, INSIDE OF COURT ROOM

Nonbinding Resolution

Binding Resolution

AREA A
AREA OF PROJECT DISPUTE RESOLUTION CONTROLLED BY THE PROJECT PARTICIPANTS

AREA B
AREA OF PROJECT DISPUTE RESOLUTION CONTROLLED BY NON PROJECT PARTICIPANTS

This line XX is where a mutually acceptable contract is in effect for the project.

This line YY is the critical transition point where project dispute resolution control is lost by project participants.

Intensity & Complexity of Dispute

Decreasing   Increasing

Very low

Very high

The horizontal line from left to right depicts an increasing intensity and complexity of a dispute. A low level of conflict is indicated at the contract execution point. If the level remains low, the project continues just as it was started at vertical line XX and is completed without ever moving too far to the right. This is one of the few cases where a totally static system situation may actually be the best set of conditions surrounding the project. As the job moves through the phase of the work along the line of action it will probably begin to accumulate disagreements that most of the time can be prevented from escalating or be resolved by internal negotiating among the parties to the various contracts. As long as the conflicts can be settled by the disputants with minimal outside help the project is considered to be in area A, the area of project resolution controlled by the project participants. This is likely to be the highest-probability arena in which to successfully resolve the problems causing the dispute.

Where disputes are of moderately high complexity and do not lend themselves to such internal settlement processes, the intensity of the dispute might escalate, and the problems may be discovered to be more difficult to resolve than originally thought. Here, the movement from left to right in our ADR picture will usually require that an opinion or suggestion be provided through some informally conducted negotiations, headed by a neutral, nonlegal, and respected design and construction practitioner or practitioners. If the dispute still remains unresolved, the rules of negotiation and settlement may have to be more rigidly imposed and adhered to than in earlier efforts. This generally means that formally prepared statements of the problem areas and the other characteristics of the dispute must be prepared and presented in a structured manner.

To summarize, the four levels of alternative dispute resolution can be described in terms of intensity and complexity as shown in Figure 6.1. Movement to the right indicates a technique applied to a more intense and complex situation than that to the left. The ADR methods to the right of XX and to the left of YY are designated as steps one, two, three, and four.

1. *Prevention actions.* Prevention literally means establishing and managing the project so that the means to resolve contested disputes are defined and applied before the dispute occurs. Prevention is often the technique that produces the maximum harmony at the least cost.

2. *Internal negotiations.* Internal negotiation techniques require that the parties involved conduct negotiations once the dispute has arisen and a contested claim appears. Usually, internal negotiation methods require consensus among the participants. Conducting internal negotiations is a relatively low-cost method of settling arguments, and when well-intentioned parties to the dispute are parties to the negotiations, the method works remarkably well, if applied intelligently.

3. *Informal exterior neutral methods.* External neutral methods entail the use of an industry-knowledgeable, unbiased, disinterested party to serve

as a referee, a dispute resolver. The setting for hearing the facts in the conflict is usually an informal job office setting. The process can normally be conducted at a relatively low cost. There may, however, be a need for some preparation and a resulting cost expenditure by the neutral as he or she becomes familiar with the project and the conflict.

4. *Formal exterior neutral methods.* The use of formal exterior neutral techniques requires that an external unbiased outsider(s) serves in a relatively formal setting to provide direction and advice for best resolution methods. The process can still be undertaken at a relatively low cost. However, the use of more structured procedures may require considerable preparation and may also require legal input from attorneys and other advisors. The lines of conflict are usually sharper and more carefully drawn than in the methods described in 1, 2, and 3.

If steps 1 to 4 have been attempted and no resolution is in sight, movement across vertical line YY must be considered. Line YY is a major transition point along the way where control of the resolution process is transferred to nonproject participants. It is a move which signals that the participants have lost project dispute resolution control. It is usually the point where the resolution process must be submitted to a binding decision rendered outside or inside the courtroom.

Taking a project dispute across line YY should require that an inordinate intensity and a very complex set of conditions exist. And oddly enough, it is usually up to the disputants to decide whether or not to move across the line. Once line YY is crossed, the project is now in area B of the Figure 6.1. Area B is the area of project dispute resolution controlled by nonproject participants. It is here where all the undesirable features of any contested claim and intense conflict appear. Financial loss, excessive unprofitable time spent on the job, increasing risk of failure of the business, and the draining of resources into the hands of those who often most profit from a lack of resolution. These and many other debilitating side effects begin to appear. The participants are no longer in the design and construction business: they are in the legal business.

The characteristics of each situation will dictate the choice of a system. Keeping the definition of the content of steps 1 to 4 in mind from our previous discussion may provide the classification definition needed. If we add boxes and the names of specific ADR techniques to the Figure 6.1 picture of the dispute resolution process, we can see and evaluate the techniques to be considered. Figure 6.2 shows some of the actual systems that might be appropriate to use and that meet the criteria outlined in the step descriptions. I have selected and shown a set of methods that have gained general acceptance in the planning, design, and construction profession in recent years.

It is important to understand the characteristics of each of the techniques and to see clearly where each fits into the progression from the most simple to the most complex. In Figure 6.2 are described some of the more common

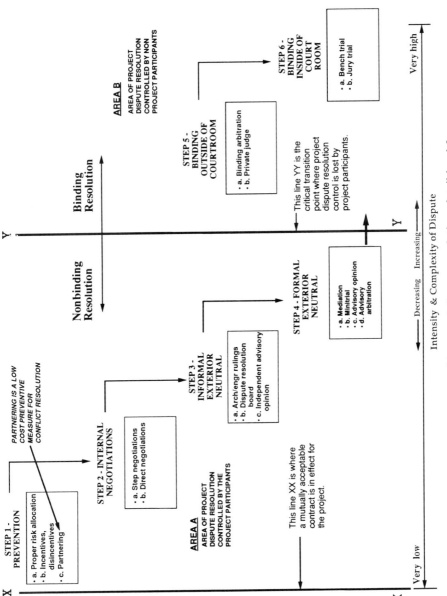

Figure 6.2 Steps of issue and dispute resolution, detail level 2.

101

and most popular of the resolution systems. Descriptions may vary slightly depending on the nature of the conflict, the location of the job, the characteristics of the organizations involved, and the conditions imposed on dispute resolution by the contract documents. However, they offer a sound and reasonable starting point to consider. The numbers of the step designations below correspond to those Figure 6.2. The techniques available within the method are lettered and correspond to the techniques named in the step boxes in Figure 6.2.

Step 1: Prevention methods   Prevention methods concentrate on putting the burden of dispute resolution in the hands of the project management. This is done by attempting to predict what is characteristic of the project that might cause destructive conflict. The causes of problem jobs as outlined in Chapter 5 present some of the clues to conflict and exposure risks encountered when a project action poses a threat to successful completion of the job. Prevention methods include:

a. Proper Risk Allocation. Risk control and allocation is a technique that all owners and all planning, design, and construction firms should practice whether or not there is a potential for contested claims. Risk should be assigned routinely to the parties who can best manage or control the risk—for example:

- To the architect if the owner has prepared a well-conceived and clearly stated program from which to begin design development
- To the owner if the architect/engineer is expected to assemble and write the program
- To the contractor where full, well-prepared, and quality-checked construction documents are available
- To the owner where construction begins before construction documents are complete

Attempts to shift risks to owners, architects, engineers, contractors, or others not able to absorb these risks is not cost-effective. Such efforts may:

- Reduce competition
- Increase costs due to greater contingency allowances
- Increase costs and reduce effectiveness because of the potential for greater numbers and intensity of design and construction project disputes

b. Incentives and Disincentives. Incentives are rewards for above-average performance. To be most effective, incentives should be built into the contract provisions so they are guaranteed automatically for outstanding work. Incentives can produce savings in time and, in some cases, costs where time savings can be translated directly into costs, as in a theater or sports stadium (see Chapter 13, Section A, for more detail on incentives–disincentives). Rewards are usually paid out in the form of bonuses. In the current construction business the use of incentives has proven of great value in highway and related work (usually let by a governmental agency).

In an incentive-driven job, the incentive date is the date upon which a significant critical point in the project must be reached. These dates must be explicitly defined. Most of time they are specified by the performance expected of the facility. For instance, in a highway program a completion point may be set to the minute by a performance specification that establishes the start of incentives as "when the northbound lanes are open to full traffic from station 10 to station 20."

The finish date is usually defined in the contract documents by the client or owner. If the incentive date is met exactly, there is no incentive or disincentive. If the incentive conditions are reached before the incentive date, a reward, usually in dollars per day, is paid for each day by which the incentive date is bettered. If the incentive date is not met, the disincentive arrangement is in effect. Then an amount, often the same as the incentive amount, is levied as a penalty against the contractor for every day the date is exceeded.

c. Partnering. The partnering concept has proven a powerful participant-driven method of anticipating problems that might be encountered on a project. It suggests early-on building solution models that work. Partnering stresses good-faith agreements, emphasizes teamwork, and encourages good communications among the stakeholders and others at risk on the job. We discuss partnering in great detail throughout this book.

Step 2: Internal negotiation methods   Internal negotiation methods require that the parties to the dispute try to reach a resolution by consensus within the disputant group. The parties involved with internal negotiation techniques usually conduct their own bargaining sessions. An effective way of using internal negotiations is to include a requirement in the issue resolution policy that any conflict will first be submitted to a selected group of project stakeholders for advice as to settlement methods and possible resolution. In this method, the parties involved conduct their own internal negotiations once the conflict is recognized as a potential threat to project success. The system usually requires consensus for a solution. Resolution is reached through informal discussions—first at the originating level and then at any higher level needed to reach full accord. The technique is relatively cost free since the negotiations are between job personnel and are conducted within the scope of daily managerial duties of those involved.

Internal negotiation methods include:

a. Step Negotiations. These usually begin at the level where the dispute originates. Internal step-negotiation methods are sometimes specified in a partnering charter as an early attempt to resolve issues. In step negotiation, the conflicts are first considered by those directly involved. If no resolution is reached, the conflict may then be submitted to a prespecified group of vocationally related stakeholders for advice from them as to settlement methods and a possible resolution. If no resolution is reached at the advisory level, the issue is gradually moved up to the next-higher management level until a management plateau is found where solution and agreement of the dispute are reached.

b. Direct Negotiations. These may start at the ultimate-decision-maker level. They are those negotiations in which the matter in dispute is first taken to those who have the authority to make a final binding decision in any project-related matter. These are called the ultimate decision makers (UDMs). Occasionally, it is difficult to locate the true ultimate decision makers in project work because of the matrix nature of responsibility and authority patterns. In these cases, there may be some intermediate management consideration of the matter prior to it reaching the ultimate decision-making level.

This direct approach over the heads of the immediate participants is risky and its selection as an ADR method may tend to lower chances for future lower-echelon settlements. On-site managers may tend to move any serious dispute immediately to a higher level rather than be responsible for a settlement with which their superiors might not agree. If the direct negotiation technique fails to resolve the matter, the conflict is often moved to the informal external neutral technique for consideration.

Step 3: Informal external neutral methods In an external neutral system, an outside neutral serves as an informal resource to help resolve the conflict by guiding the discussions toward the discovery of a solution satisfactory to all parties. The method, because of its informality, entails a relatively low cost. Using neutrals usually requires nominal preparation so as to provide the neutral with adequate information by which to decide intelligently. This is a low-cost, but effective system of resolution when it seems that consensus and settlement can be arrived at if the issues are looked at fairly. The process does require higher-than-normal abilities and experience of the party providing the advisory opinion.

Informal methods of settling disputes are those in which the ground rules for the resolution discussions are kept simple, and the meetings are held at or near the job site. In construction, the preferred participants in the external neutral process are technically trained and educated professionals active in the planning, design, and construction disciplines. These neutrals must be capable of listening, analyzing, and evaluating objectively construction-related demands or claims that are in dispute. The product of the neutral's work is usually a recommendation to follow a course of action based on the neutral's opinion as to the outcome of the action. Recommendations are generally not considered binding.

If the neutral is a member of a pool of experts and is available at the call of parties in need of a neutral, he or she is called a standing neutral. Professional and technical societies and associations sometimes form an on-call pool of standing neutrals for the convenience of those seeking advice, counsel, and expert opinion about projected outcomes in a construction related dispute. This method involves relatively little cost but may require some nominal preparation time in preparing background material for the neutral's study and analysis.

Informal external neutral methods include:

a. Architect/Engineer Rulings. The architect or engineer of record is the person who prepares the contract documents for a part or all of the project. Usually, this person is responsible for the correctness of the documents and is expected to place a registration seal on the documents to be used. In times past there was a moderate or good chance that the architect and engineer of record could provide a fair decision that would be accepted as unbiased and expert. However, in today's litigious climate, the maintenance of an unbiased, neutral position taken by any member of the design team of record has all but disappeared.

The danger of being accused, and perhaps projected into a legal quarrel, casts doubt on the ability or desire of the responsible design professional to render an objective opinion in conflict situations. This is particularly so where the designer or the designer's clients may be in the wrong. Therefore, the submittal for resolution of other than routine problems to the designer of record occurs infrequently in the construction business.

b. Dispute Resolution Board. A dispute resolution board is a group of neutrals selected from outside the project, preferably at the onset of the project. These neutrals should have good experience, excellent technical construction knowledge, good professional standing, and unbiased positions in matters affecting the project. Various numbers of people might be chosen for the board by different selection methods. The major requirements are possession of unequivocal fairness and a selection process that provides a good balance of experience and professional background.

One of the most commonly reported selection systems is used when the project organization is limited to an owner and a single prime contractor. One board member is selected by the owner and approved by the contractor. One is selected by the contractor and approved by the owner. A third is selected by the first two board members. The third usually acts as board chair. As with other neutrals used in alternative dispute resolution, the board members must have no conflicts of interest relative to the project and should be very familiar with the planning, design, and construction industry. The board's findings in a dispute are not binding on the parties. It is a reasonably low cost method of settling disputes and tends to produce recommendations that deter further submission of the issue to binding techniques.

c. Independent Advisory Opinion. In this method an outside neutral is usually selected by mutual agreement among the parties to the dispute. The neutral expert meets informally with the parties involved, preferably at the job site, listens to informal discussions from all concerned, and renders a prediction as to the ultimate outcome of the dispute if not resolved at this level and by those in this meeting. This system has the benefit, as do most neutral-party systems, of bringing a fresh outside viewpoint to the project and its difficulties. The objectivity of the neutral coupled with his or her opinion rendered can go a long way toward providing a sober, realistic look at the situation when only small differences stand in the way of resolution.

Step 4: Formal external neutral method   In these systems of resolution the neutral is used in a manner very similar to the step 3 techniques described above. However, step 4 resolution methods are likely to be for greatly escalated conflict situations or for more complex and higher-risk disputes than those encountered in the step 3 methods. In formal external neutral resolution the selected neutrals are guided by more highly structured guidelines than dispute neutrals in step 3. Several parties may be involved, preparation for the discussions or hearings may take longer that for step 3 methods, and the work may require some formal legal advice and assistance. Relative costs are still modest compared to the binding resolution techniques that may have to be used if the alternative dispute resolution system does not bring settlement.

Four techniques are worth mentioning within the formal external neutral system.

a. Mediation. This method can be either formal or informal and has a high potential for quick, equitable settlements in planning, design, and construction matters that may be fairly complex. The mediator is selected so that his or her talents specifically fit the demands of the dispute. Selection is usually determined with a minimum of dissent since the appointment must be a mutually satisfactory choice from what can be a small, limited list. The choice of a mediator rests totally with the participants directly involved; the selection is made for technical and professional reasons.

The possibility of the mediator adopting an adversarial or advocacy position is kept low by the responsibilities of the position. Informal mediation requires the neutral to set the ground rules and guidelines for conduct of the hearings and whatever investigation must be made. Since the mediation decision is not binding on the parties to the dispute, the recommendations of the mediator should be accompanied by a detailed description of how the decision was made and the key elements of the resolution. The criteria used by the mediator are usually outlined clearly in the mediator's report, and the resolution recommendations are described in detail.

b. Minitrial. A minitrial is a private settlement method usually initiated by an agreement between the parties to a dispute. The minitrial is voluntary, nonbinding, and nonjudicial. It consists of an adversarial presentation of each side of the dispute to a small panel of executives representing each side. The executives selected should have full authority to settle. The panel might include a neutral with a construction background.

The hearings of each side's presentation are presided over by a third-party neutral who helps the participants set guidelines for the process and may be asked to provide an advisory opinion on the matters in dispute. The length of hearings should be short, a matter of one or two days, and be followed by a short period of negotiation by the executive panel. The minitrial is concluded when the parties negotiate a satisfactory agreement or agree that another route is appropriate to follow.

c. Advisory Opinion. In the formal advisory opinion procedure the neutral expert meets formally with both parties, obtains information from both, and renders a prediction as to the ultimate outcome if adjudicated. The major difference between the informal independent advisory opinion technique outlined above and the formal advisory opinion is the degree of structure. In the informal process little if any preparation for the meetings is needed by the parties, and they present their own cases.

In the formal process each party's position is outlined in a paper submitted to the neutral before the hearing and evaluation. The neutral expert meets in a formal setting at a neutral site away from the location of the dispute. Each side makes a short presentation, addressing the issues in dispute. The presentations are in a setting that allows questions and requests for clarification by the neutral as the presentations are being made.

Once the presentations have been made, the neutral works with the opposing parties to help identify items on which they agree or are close to agreement. There may be an effort to reach full agreement on these at the meeting with the neutral. The next items to be discussed are dispute questions whose answers are unknown and other issues about which the parties still disagree. After assessing the attitudes of the parties to the disagreement, the neutral provides an appraisal of liability, damages, professional harm, job disruption, and other elements of damage to the participants and the project that may be produced by failing to settle remaining undecided points now.

d. Advisory Arbitration (Voluntary Nonbinding Arbitration). There is an abbreviated hearing before neutral expert(s). The arbitrator(s) issue an advisory award and render a prediction of the ultimate outcome if adjudicated. This process is described by some as a catalytic procedure to stimulate agreement. The process of binding arbitration is followed except that the decisions and awards by the neutral arbitrator are nonbinding and rendered as an opinion.

To understand what is meant by voluntary nonbinding arbitration, it is necessary to understand what mandatory binding arbitration is—usually a well-defined process marked by formal hearings, the presence of legal counsel, and run by guidelines that govern how decisions are to be rendered. Binding arbitration is considered a direct alternative to litigation. It is supposed to trade the possibility of appeal from judicial decisions for the speed and economy of a final decision made outside the courtroom.

In advisory arbitration the dispute issues are modeled and tested before a neutral expert. The opinion of the neutral may assist to obtain a nonbinding agreement since the neutral's decision will probably mirror any decision that might be obtained if binding arbitration were used. This may show the disputants what the actual result might be if the dispute were continued into the binding resolution area.

The techniques outlined above are called alternative dispute resolution techniques since they attempt to resolve problems, disputes, and conflicts among the parties directly affected. They also avoid the use of binding exterior decision making as is used in formal legal processes. If the application of these techniques does not bring about a mutually agreeable solution to the problem, the parties may be forced to move across line YY into area B in Figure 6.2. This is the area of project dispute resolution controlled by nonproject participants. The methods available in area B are controlled primarily by those outside the planning, design, and construction industry, and the decisions rendered are usually binding.

An example of the scale of expense encountered when a decision is made to cross the YY line best illustrates the price of long settlement processes that move from nonbinding solution attempts to binding solution attempts.[1]

Case Study 6.2: The Escalating Office Building Claim

The Rayell Corporation had contracted with the Hartford Construction Company to construct a large office building in Alabama. The job was a hard-money project that had been bid competitively (project delivery system A2/B3/C1/D1–a1,3,4; see Table 2.2). About midway, the project turned into a problem job and several difficulties caused Hartford to consider how best to recoup the office and field monies that it claimed it had cost them to accommodate the problems. The Hartford project team estimated that it had cost the firm about $400,000 to remedy the problems they had encountered. This estimate had some negotiating margin built in. However, it was kept low because the construction team members felt they would far prefer a negotiated partial settlement than to get into an extended legal battle with Rayell.

Informal negotiations dragged on for about a year with no settlement yet in sight. In fact, the claim amount was escalated to $700,000 because of the constantly increasing complexity and time demands of the claim

---

[1] The details and names in all case studies have been changed. However, the circumstances surrounding the cases illustrate situations actually encountered.

under discussion. However, to this point, there were few, if any, legal costs incurred since negotiations had been kept at the project levels.

After about 18 months of haggling, Hartford and Rayell instituted litigation against each other. When reestimated with the projected expense of legal action, time delays, executive time, and the higher risk now to be taken, the claim cost jumped immediately to $3,000,000. The case dragged on for five more years, with the damage claim escalating to nearly $5,000,000 exclusive of legal fees. Finally, an out-of-court settlement was reached seven years after the original claim was submitted by Hartford. The settlement agreed upon was imposed on both parties and totaled almost $10 million, with $4 million being spent on legal costs and related expenses to pursue the legal route of settlement.

Case Study 6.3: The Difficult Precipitator[2]

A reputable steel and heavy sheet metal erector, Strong Steel, was selected by price-competitive bidding to erect owner-furnished structural and plate steel for a new precipitator in southern Ohio. The hard-money bid was for $7,500,000. The precipitator was to be located next to a large fossil fuel power plant belonging to the Midwestern Power Group. The project was supposed to begin in March, but various design and site access problems caused the work to be delayed. In early July, the Strong Company was allowed on the site and immediately began erecting the steel on foundations built by Midwestern. Structural steel had been delivered to the job site earlier and was available immediately.

The project proceeded well to about August, when Strong began to encounter serious trouble plumbing and trimming the structural steel and plate erected to date. After about one month of checking, measuring, and reviewing records, Strong discovered that the center column pier supporting the steel column, from which all work was dimensioned and all elevations set, was $1\frac{1}{2}$ inches below the elevation indicated on the erection drawings.

Correcting the error took almost one month and the weather began to turn wet and nasty. Steel and plate erection continued until the onset of freezing weather in mid-November. From then on field work proceeded only as weather permitted. Work was also slowed by misfabrication and mismarking of the structural steel and plate furnished by Midwestern. The winter weather continued harsh and bitter through December, January, and February. Strong lost nearly four times the number of weather days normally expected in this location. Trade disputes over weather-related working conditions and site congestion also interfered with planned erection of steel.

Repeated efforts to process changes and increased costs failed for a variety of reasons: some political, some regulatory, some just due to hard-

---

[2] A precipitator collects flue gas solids and is usually installed just ahead of the stack entrance from a boiler.

ened positions and stubbornness on the part of both Strong and their client. In late February Midwestern dismissed Strong from the job for being behind schedule. Strong's equipment was seized by Midwestern and a successor contractor was hired to complete the work.

Little effort was made to resolve the issues remaining and Strong sued Midwestern. After a long and tedious bench trial, during which Strong went out of business, the circuit court judge produced a finding of fact in the case. He ruled that Strong was entitled to $5 million in compensatory damages plus $13 million in punitive damages. Strong had already been paid $5.7 million on their $7.5 million contract when they were dismissed for cause by Midwestern. The successor contractor to Strong was paid $5.7 million to complete the project. The cost of the precipitator to Midwestern was not $7.5 million, but $29.4 million, nearly four times the original bid price.

Strong went bankrupt, Midwestern's chief operating office was fired, and the awarded money was tied up for about 10 years by repeated legal appeals. A careful analysis of this case indicates that Midwestern and Strong could probably have settled all their differences before Strong was dismissed from the job for about $4 million, a saving of $19.7 million over the amount awarded.

Almost any of the alternative dispute resolution systems shown in Figure 6.2 could have been applied to the project described in the precipitator case study above. Probably the most effective would have been the use of a dispute resolution board, since the problems actually encountered were predictable. However, the participants' inability to perceive merit in any viewpoint but their own, their concern over being blamed for environmental problems, and just simply, their unabashed greed and arrogance made a relatively routine construction project a traumatic experience for all involved.

Settlement methods available under area B techniques are shown in Figure 6.1 as steps 5 and 6. These methods for the most part are binding upon the participants, and decisions are usually backed by judicial authority. The methods of working to a binding resolution are usually defined by a set of legally dictated processes established by laws enacted by a legislature. The process of settlement by binding methods is engaged in by the participants knowing that whatever the decision is, they are legally bound by it.

The preparation time and expense in binding arbitration or litigation can be considerable. The parties are usually represented by attorneys who, in turn, select and retain witnesses, consultants, and other experts familiar with the processes of planning, design, and construction and who have a working knowledge of the law. The process of preparing a case, submitting to the discovery of evidence, and trying the case in the arbitration hall or the court-room is expensive, of long duration, and usually destroys good relations the parties might have had before the hearings.

There are many subsystems of binding resolution. However, I have limited the brief discussion here to three of the main methods used: binding arbitration, private judging, and litigation.

Step 5: Binding resolution outside the courtroom

a. Binding Arbitration. In binding arbitration the parties to a dispute agree to submit their differences to the judgment of an impartial person or group appointed by mutual consent. The parties argue their case to the arbitrators under somewhat structured rules. The arbitrators, after hearing the case, decide on the resolution and submit it to the parties. It is then binding.[3] Binding arbitration saw an increase in use beginning in the 1920s, most notably because a number of states enacted legislation favorable to its use. The system served the planning, design, and construction industry well for many years. However, in recent years, the rapid increase in expense, the volume of cases, and the increased complexity of design and construction, have tended to make binding arbitration less attractive than nonbinding forms of resolution.

b. Private Judge Method. Private judging for resolution of disputes is accomplished by the employment of a private person, qualified and having credentials to perform in an adjudicative role. The private judge is often a retired judge, or possibly an attorney, or a qualified expert who is not an attorney. The private judge conducts the hearings in private, and the proceedings can be as formal or informal as desired as long as they conform to the judicial standards under which the private judge is operating.

Often, the private judge is used as an alternative to the public court system when the public system is unable to deal in a fair and expert manner with a dispute. This may occur when the case load is excessively heavy and delays may hold up formal public proceedings. It may also be encountered when the available court personnel lack the specialized skill needed to decide properly in a case. Like all binding processes, the private judge system requires careful preparation of the case and usually, representation by legal counsel. The expense for this may be considerable, but it is still less than might be encountered in the public court system.

Step 6: Binding resolution inside the courtroom

Formal litigation in the public courts is probably the highest-cost settlement system the disputants can use to settle their claims, conflicts, and disputes. Litigation is defined as being subjected to or engaging in legal proceedings within a public court system. Formal litigation is conducted under very tight procedures spelled out in voluminous books of laws and regulations. Public court proceedings are

---

[3] This is the situation provided that state law does not require submission to the legislature for payment approval of a settlement. Such restrictions are occasionally found on state-owned public work projects.

liable to be very public, very expensive, very slow, and often expose the disputants to being judged by unknown people of unknown qualities when they serve as judges and juries.

For instance, litigation allows and encourages an early period of discovery that is characterized by demands for what are called interrogatories and depositions. These are very expensive processes conducted by the parties to discover evidence to use during the trial itself. All things considered, it is easy to understand why litigation is usually taken as a final step after all other methods have been tried and exhausted. There are two types of court hearings: the bench trial and the jury trial. Each has characteristics that influence the presentation and the results expected.

a. Bench Trial. A bench trial is conducted by a judge and without a jury. It has several advantages over jury trials in design and construction disputes. As a legal professional, the judge has knowledge of and access to resources and technical advice that may help him or her considerably in understanding the points at issue. The judge has the flexibility of being able to have points and issues clarified rapidly as the trial proceeds. The relatively small number of people involved allows the clarifications to be presented in a dialogue between those directly involved in the decision process.

Much of the evidence presented in a design and construction lawsuit may be highly technical and beyond the capabilities of a lay jury to understand. Time available will not usually permit an explanation that will satisfy six to 12 people on the jury. The probability is far higher of a single professional (such as the judge in a bench trial) better understanding the explanation. If a design and construction case is to be tried in a court trial, it is almost always better for all parties to have a bench trial.

b. Jury Trial. In a jury trial the case is heard before 12 or fewer laypersons selected from the community. These people are charged with the responsibility to decide on how to resolve the disputed matters. A jury trial in design and construction often takes longer and costs more than a bench trial. The reasons are simply a matter of the time and the cost of longer and more complete explanations needed of technical matters—matters that are generally foreign to the laypeople usually found on a jury.

In addition, the nature of the material needed to make decisions is not easily transferred by most lawyers and expert witnesses into understandable arguments easily comprehended by a layperson. An argument about bonding or about latent underground defects may lack adequate emotional and subjective reference points for one not familiar with the reason it stands as a point of conflict. Usually, a jury trial is the least desirable of all methods of resolving disputes in a planning, design, and construction matter.

7

PARTNERING CONCEPTS AND STRUCTURE

SECTION A: INTRODUCTION TO PARTNERING

The first phase of a complex 100,000-square-foot office addition and remodeling project is approaching completion. The project manager for the contractor is enjoying an early morning cup of coffee as he opens a letter he just received from the executive vice president of the firm who owns the facility. The letter addresses the use of partnering on the project.[1] He reads: "This letter is our testimonial to the success of the project charter developed during our day long meeting at the start of the project field work. As you know, representatives of your firm, along with the architects, engineers, system suppliers, and all subcontractors, participated along with the owner." The executive continued: "When the subject of partnering was first suggested by your company, we had some reservations as to whether or not the time, effort, and expense would prove beneficial for all those involved."

The letter concludes: "We are no longer skeptics but firm supporters of the partnering concept. We are well into the project with no lost time, no accidents or injuries, and are on schedule. Everyone on the project is working in harmony and we are well on our way to completing a high-quality facility on time and within budget. Thank you for suggesting the partnering concept to us. It has proven invaluable."

---

[1] Paraphrased from an actual letter.

What is partnering? It is:

- A design and construction standard for human behavior in noncontract matters
- A guidebook to etiquette in any enterprise, particularly where formal rules may not provide a guide to courtesy, common sense, empathy, and responsible action
- A moral agreement in principle
- A process to break down barriers that would otherwise hinder good working relations
- A system of conducting business that maximizes the potential for:
  - Achieving project intent
  - Attaining specified quality
  - Encouraging healthy, ethical customer/supplier relationships
  - Adding value
  - Improving communication
  - Providing methods of project condition measurement and feedback
  - Resolving conflicts by nondestructive techniques
- A high-value operating system
- A business system to improve potential profitability
- A marketing tool to assist competent planning, design, and construction firms reduce the potential for debilitating competition
- A predesign management technique to set noncontract operating ground rules that add value
- A predesign program system to help set concepts, ideas, intent, and direction for the internal staff of the planning, design, and construction owner and client through improved communications
- A preconstruction management system to set operating ground rules not covered by the contract
- An agreement among the project participants designed to help reduce overhead
- A preventive action to reduce destructive and expensive conflict
- A revisiting and updating action to validate, confirm, reinforce, or revise original operating ground rules that need review
- A process to remove barriers to good working relations and improved project morale
- An aid to help reach agreement on common goals and objectives leading to successful projects
- Part of a larger effort to seek effective, alternative methods of resolving destructive conflict

The ingredients of a good partnering system are:

- Top management support of team concepts
- Knowledgeable participants
- A well-written set of guidelines to developing and using the partnering system
- Regular and authentic project performance evaluation against contract and noncontract standards
- A method of resolving disputes early and at their origin

Expectations of a good partnering effort are:

- Improved job working relations
- Improved project communications
- Heightened awareness of project mission
- Lowered insurance costs
- Quick, fair resolution of project disputes
- Higher respect for the needs of others on the job
- A reduction in the number and intensity of contested claims
- Increased profitability

SECTION B: PARTNERING DEFINITIONS AND CHARACTERISTICS

We encounter partnering concepts in all aspects of our business, professional, technical, and social life. They surround us, engulf us, are visible from every window! In Section A of this chapter, I attempted to convey a macro sense of what partnering is and should be. However, a professional practitioner of generic construction should be prepared to explain in detail to serious readers and listeners what he or she really means by partnering and which definitions of partnering are acceptable or authentic.

Various organizations and individuals have proposed many different definitions of partnering. Each has strong points and they all have a place in the planning, design, and construction industry. I include below some of these definitions.[2] They provide a well-based conceptual reference from which we can begin our journey to visit the methods of reducing destructive conflict in the construction business and profession through the use of this old–new technique.

---

[2] Paraphrasing is by this author; where paraphrasing has been used, an effort has been made to identify the source, where possible.

1. Associated General Contractors

   Partnering is a way of achieving an optimum relationship between a customer and a supplier. It is a method of doing business in which a person's word is his or her bond and where people accept responsibility for their actions.

   Partnering is not a business contract but a recognition that every business contract includes an implied covenant of good faith.[3]

2. Construction Industry Institute

   The partnering concept centers around a long-term commitment between two or more organizations for the purpose of achieving specific business objectives by maximizing the effectiveness of each participant's resources. The relationship is based on trust, dedication to common goals, and an understanding of each other's expectations and values. Expected benefits include improved efficiency and cost-effectiveness, increased opportunity for innovation, and the continuous improvement of quality products and services.[4]

3. American Society of Civil Engineers

   Partnering is an effort that attempts to merge the contractor's, the owner's, and the engineer's interests into a single project goal. Partnering involves cooperative project management among the contractor, the owner, and the engineer.[5]

4. U.S. Army Corps of Engineers definition A

   Partnering is the creation of an owner–contractor relationship that promotes the achievement of mutually beneficial goals. It involves an agreement in principle to share the risks involved in completing the project and to establish and promote a nurturing partnership environment.[6]

5. U.S. Army Corps of Engineers definition B

   Construction partnering means developing a cooperative management team with key players from the organizations involved in a construction contract. The team focuses on common goals and benefits to be achieved through contract execution and develops processes to keep the team working towards those goals. Partnering means exercising leadership for the entire engineering team.[7]

6. American Arbitration Association

   Partnering is a synergy—a cooperative, collaborative management effort among contracting and related parties to complete a project in the most

---

[3] Paraphrased from *Partnering: A Concept for Success* (Washington, D.C.: Associated General Contractors of America, 1991).

[4] *Partnering: Meeting the Challenges of the Future,* Partnering Task Force Interim Report (Austin, Texas: Construction Industry Institute, August 1989).

[5] Paraphrased. The full text from which the paraphrased material was derived can be found on page 62 of *Dispute Avoidance and Resolution for Consulting Engineers* by Richard K. Allen (New York: ASCE, 1993).

[6] From City and County of Denver, *Practical, Profitable Partnering, Denver's Team Approach to Urban Reconstruction* (Denver: Department of Public Works, 1993).

[7] From *Construction Partnering: The Joint Pursuit of Common Goals to Enhance Engineering Quality* (Omaha, Nebr.: U.S. Army Corps of Engineers, 1991).

efficient, cost-effective method possible, by setting common goals, keeping lines of communication open, and solving problems together as they arise.[8]

7. Conglomerate definition A

Partnering is a system of conducting business that maximizes the potential for:

a. Achieving project intent.

b. Obtaining specified quality

c. Encouraging healthy, ethical customer–supplier relationships

d. Adding value

e. Improving communication

f. Providing methods of project condition measurement and feedback

g. Providing methods of resolving conflicts quickly by nondestructive means at optimal levels of management

8. Conglomerate definition B

Partnering is:

a. A preventive action to reduce destructive conflict

b. A predesign management system to set operating ground rules not covered in the professional services contract

c. A preconstruction management system to set operating ground rules not covered by the contract

d. A marketing tool to assist competent planning, design, and construction firms in reducing the potential for debilitating competition

e. A preprogram system to set concept, ideas, intent, and direction for the internal staff of the owner and client

f. A revisiting and updating action to validate, confirm, reinforce, or revise original operating ground rules that need review

g. A planning, design, construction, and turnover guide for the unspecified, non-contract conduct of the project team

Conglomerate definitions 7 and 8 are those that I have derived from the others. They are an attempt to define partnering by the manner in which the system is used in active implementation. Overall, the variety of definitions set down several ways of using basic partnering concepts, while giving each system a distinct identity. Notice how the definitions gradually move from broad-based general ideas into specifics as the conceptual definitions are applied to explicit design and construction practices, such as programming, marketing, and turnover.

An interesting feature of partnering is the difference between the Associated General Contractors definition (1), and the Construction Industry Institute definition (2). The Associated General Contractor statement describes what is considered project or tactical partnering. The Construction Industry Institute statement describes what some design and construction professionals

---

[8] From *Construction Industry Dispute Avoidance: The Partnering Process* (American Arbitration Association, NCDRC: New York, NY, 1993).

call strategic partnering. The purpose of each of the two kinds of partnering is somewhat different. In partnering the words *tactical* and *strategic* also have slightly different applied meanings than traditional definitions give them. In a partnering sense, the word *tactical* applies to the management skills required to achieve micro and current end results. Project partnering is for a specific project only. The word *strategic* in a partnering context applies to the management skills required to attain a macro end. Strategy is sometimes considered the action required to plan and direct large, long-range programs important to the health of an organization. In the design and construction profession, strategic actions encompass multiproject techniques that help ensure long-term healthy growth.

Definitions of partnering in light of tactical and strategic differences are as follows:

- *Project partnering:* a method of applying project-specific management in the planning, design, and construction profession without the need for unnecessary, excessive, and/or debilitating external party involvement.
- *Strategic partnering:* a formal partnering relationship that is designed to enhance the success of multiproject experiences on a long-term basis. Just as each individual project must be maintained, a strategic partnership must also be maintained by periodic review of all projects currently being performed.[9]

The practice of strategic partnering covers an enormous range of planning, organizational, staffing, directing, controlling, management, marketing, and sales functions. Since entire books can be and have been written on these topics, it would be presumptuous of me to try to rewrite or paraphrase them in this book. Therefore, I have maintained the primary focus on project or tactical partnering for the generic construction profession.

Defining and using partnering methods must ultimately lead to formulating and implementing explicit goals and objectives. These then provide a foundation for the operating system from which we effectively manage and construct the work before us. The conglomerate definitions of partnering are part of a world-of-words system that allows conversion of ideas from a philosophic base to a workable office and field action base. In this sense partnering is no more than honest philosophic and pragmatic action at work.

Proven benefits of using a partnering system from which to manage planning, design, and construction are many. Good partnering systems can be expected to:

- Form a basis for the use of relatively inexpensive, quickly formulated preventive methods of dispute resolution

---

[9] From a definition of strategic partnering by Ida B. Brooker in University of Wisconsin Project Partnering Seminar, 1994.

- Encourage conducting business with minimal destructive conflict
- Help break down obstacles to good working relations
- Provide a marketing tool to assist competent planning, design, and construction firms reduce the potential for debilitating competition
- Reduce destructive conflict
- Assist to reach agreement on common goals and objectives
- Help achieve project intent, and specified quality
- Encourage healthy, ethical customer and supplier relationships
- Add value to all elements of the process and the product
- Improve communication
- Provide project condition measurement and feedback

Conceptual definitions of partnering are essential to convey the relatively full significance of the system. The meaning of partnering can be stated most clearly and directly by using phrases describing how people get along with each other when honor, honesty, and integrity are characteristics of the contract arrangements. As our profession of construction has become more confrontational and litigious, the simple nonlegal phrases used in the past to describe issue resolution seem to have disappeared. Some of the written and spoken phrases by which partnering concepts are conveyed when a person's word is considered as morally binding as a written contract included:

"Let's shake on it."
"What can I do so you'll come out whole in this mess?"
"Is there any way we could get this done without an act of Congress?"
"Do we have a gentleman's agreement?"
"Let's look at the drawings a bit more closely."
"Let's tally up the favor score."
"Let's settle this over a beer."
"Would you do me a favor?"
"How about some free hoisting for a back-charge reduction?"

SECTION C: THE INGREDIENTS OF A PARTNERING SYSTEM

I was touring Ireland when I found a perfect beginning for thinking about this book. An inscription dating back to 1492 and now in St. Patrick's Cathedral in Dublin, Ireland, tells a hard-headed tale that has partnering significance today.

The Door of Reconciliation

In 1492, two prominent families , the Ormonds and Kildares, were in the midst of a bitter feud. Besieged by Gerald Fitzgerald, Earl of Kildare, Sir James Butler, Earl of Ormond, and his followers took refuge in the chapter house of St. Patrick's Cathedral, bolting themselves in. As the siege wore on the Earl of Kildare concluded that the feuding was foolish. Here were two families worshiping the same God, in the same church, living in the same country, trying to kill each other. So he called out to Sir James and, as the inscription in St. Patrick's says today "undertook on his honour that he should receive no villanie."

"Wary of some further treacherie." Ormond did not respond. So Kildare seized his spear, cut away a hole in the door and thrust his hand through. It was grasped by another hand inside the church, the door was opened and the two men embraced, thus ending the family feud.

The expression "chancing one's arm" originated with Kildare's noble gesture. There is a lesson here for all of us who are engaged in "family feuds," whether brother to brother, language to language, nation to nation. If one of us would dare to "chance his arm" perhaps that would be the first crucial step to the reconciliation we all unconsciously seek.[10]

As we begin to discuss and analyze the ingredients that make up a successful partnering system, we can put the little bit of Irish history related in the Door of Reconciliation to work in reducing the philosophy of partnering to practical and tough-minded applications. This is our purpose as we begin detailed considerations of how partnering actually fits into our design and construction business world today. Partnering is often thought of as a subsystem of alternative dispute resolution. In Figure 6.2, partnering is shown as one type of preventive ADR and only one of 12 ADR systems. Yet partnering is an overarching way of viewing the entire process of managing projects and organizations.

The definitions of partnering given earlier in the chapter employ the same words that often are used to describe the qualities of good project management and competent project managers. Expectations of good project management frequently are the same as expectations of the successful use of partnering systems outlined above. In this way we may use partnering in an organizational sense when the spirit of partnering is incorporated into the the total operating mode of the organization. Of course we call this organizational partnering. This is shown graphically in Figure 7.1 where the total organization (A) is shown encompassing individual projects such as projects X, Y, and Z. The total organization contains an internal alternative dispute resolution (B) system by which daily internal conflicts are worked out and resolved. It also contains an almost invisible organizational partnering system (C), not usually stated,

[10] From a postcard entitled "The Door of Reconciliation" published at St Patrick's Cathedral, Dublin, Ireland.

(A) total organization

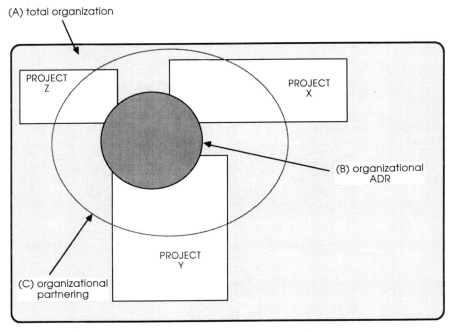

Figure 7.1 Organizational partnering.

but nevertheless continuously acting as a structure for preventing destructive internal disputes.

Then, within each project there is a partnering and ADR system that operates similar to organizational partnering and ADR systems. This partnering system is shown in Figure 7.2, where each single project W, X, Y, and Z contains a project partnering system A and an alternative dispute resolution system B. These are shaped and influenced by the organizational partnering and ADR systems of each entity involved in any given project as a stakeholder. However, the partnering and ADR systems shown in Figure 7.2 are concerned only with the project work by the project team members or stakeholder.

Project systems must consider sizable numbers of participants from outside each organization. Therefore, project partnering and ADR systems are very complex and are usually cast into a form that gives consistent guidance across organizational boundaries to all project team members. An early requirement of moving effectively from partnering philosophy to partnering application is to define the major ingredients of a system that can use partnering effectively. Each of the ingredients listed below is critical to partnering success. More may be required than the six listed. However, without these six the probability of partnering success is notably diminished.

Ingredient 1: A project or business plan to which partnering can be applied A project to which partnering can be applied is easy to acquire. A

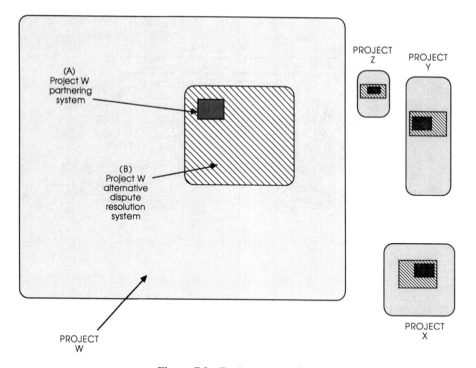

Figure 7.2   Project partnering.

project is defined as a set of work actions having identifiable objectives, and a beginning and an end. Usually, it is the spark that starts the entire process. To acquire and support a workable partnering system, a project must merely exist. Practically all organizations and projects qualify for the use of partnering.

For project-specific partnering the project must be clearly defined and its boundaries delineated carefully. Stakeholders in a project partnering arrangement are together only for that project. The project management structure is a one-time-only organization. A good beginning is critical to success. Its definition in the minds of the stakeholders must be crystal clear. The scope of the work to be done must be explicit and the legal contract conditions fully understood.

Ingredient 2: An intelligent and competent project team made up of people who want the project to be successful (see Chapter 6 for a discussion on how to identify a successful project)   Most planning, design, and construction participants fill this bill. In our business we seldom encounter a person who actually wants a project to fail. A common characteristic of design and construction professionals is that they have personally selected this profession. Often they entered it when they were young and inexperienced. Once involved, the participants seldom leave. Design and construction are dynamic, and challenging opportunities abound. The road to success can be followed in any direction and to any heights you want. The trip itself is often dangerous, but usually exciting.

Effective partnering requires that the people involved be competent and want their projects to succeed. It requires that these people understand the underlying purpose of partnering—*to add value.* This basic premise drives all technical and management construction systems that have earned a place in the professionals' management portfolio. Those professionals must understand how the tools are used to build potential for adding value that leads to success.

It is interesting to note that the requirements of effective project management and the essential qualities of good project managers overlap and often include the very elements that create a good partnering system. After watching hundreds of managers on a multitude of design and construction projects, the common elements of importance to effective project management that I have observed include:

- Technical competence
- Proper project planning
- Good project staff morale
- Clearly defined authority lines
- Clearly defined responsibility lines
- Respected leadership
- Clear understanding of the project mission
- A sensitive monitoring system
- Prompt and effective resolution of problems
- Discerning points of view that focus on the important issues
- Effective modes of action that accomplish project missions and objectives
- A project-wide desire for excellence
- Inquisitive minds
- A collective sense of humor
- A well-prepared set of contract documents
- Collective patience
- Collective endurance

Essential qualities of project managers who must achieve success through their efforts include:

- An ability to move from perceptions of a micro situation to perceptions of a macro situation and back again at will
- An ability to work well with people
- A desire for excellence

- An inquisitive mind
- An ability to manage conflict
- A sense of humor
- Good mental peripheral vision[11]
- Education in related fields
- Training in related fields
- Leadership ability
- Related technical and professional credentials
- An understanding of the true role of profit in our society
- A potential for being creative
- Good communication ability
- Intelligent consistency
- Honesty and integrity

It is not unreasonable to conclude from this list of project-related elements and qualities that good managers help produce good projects. In fact, if the elements and qualities of mutual cooperation, good management, sound design, and financial stability are there and being used well, it may be immaterial whether a partnering system be formally adopted or not. This ideal arrangement is seldom encountered, however, and here is where project partnering becomes a system to be considered.

Ingredient 3: Knowledgeable technical and management leaders within each of the organizations involved   Excellent partnering requires outstanding leadership and well-applied abilities by those in responsible stakeholder positions. The planning, design, and construction profession requires that people in the business have a knowledge that can withstand constant testing and challenge. The major intellectual, managerial, and leader qualities needed for a successful partnering system include:

- Good thinking patterns
- Well-thought-out approach patterns
- High standards of ethical and moral behavior
- A strong desire to achieve project success
- Ability to win the trust of other participants
- A desire to allocate risk properly
- Sound experience in construction techniques and practices
- A knowledge of effective methods to resolve destructive disputes

---

[11] The ability to perceive and apply, through conscious or unconscious sensing, the meaning and importance of images, thoughts, noises, and spoken sounds surrounding us without losing the ability to think through the main subject being considered at the moment.

- An ability to obtain and keep upper management support and participation
- An ability to generate support from all participants
- Realistic and accurate performance standards
- A self-imposed method of accurately measuring performance

In the construction profession, the participant—whether planner, architect, engineer, constructor, or manufacturer, from interns and apprentices through to the highest-placed executives—are continually challenged to improve, to acquire credentials, and to teach. This continuing process of learning and educating has helped maintain design and construction professionals at a consistently high level of competence.

Good leadership is considered critical to effective partnering. Unfortunately, it is not nearly as readily available as is knowledge of the business. Leadership quality at all levels in our private and public organizations is relatively low, especially when measured against the expectations of those who are led. This deficiency seems to run in cycles: low when the value of leadership is not recognized as needed or desirable; and high in emergency situations where instant and forceful action is demanded for survival. In most construction-related organizations the internal leadership needs are great and often filled well. However, the leadership of joint project efforts in construction-related projects is often poor and badly applied by those who are assigned leadership roles. Good leadership, applied intelligently, is critical to partnering success.

Ingredient 4: An effective project organization that melds the desires, the needs, and the leaders into a working team that has a high success potential Several identifiable relations exist on most projects today. The perceptive manager will recognize and use these relations to strengthen the ability of the team and its members. The strengthening is designed to improve the project and add value for the participants, the client, the user, and others who will benefit from the project being well implemented.

The five most important interrelations are also the most fundamental. Their definitions and scope include:

1. *Reporting.* These are the official channels through which individuals give, or receive raises, appraisals, and evaluations; are fired or assigned; or are provided professional, vocational, and personal identity in the organization. The true organizational superior of an employee is usually that person with whom he or she maintains a reporting relation. The line expressing reporting relations has an arrowhead at one end pointing to the superior.

2. *Formal functional.* Formal functional relations are the organizational connections that show the distribution of data, information, and decisions

that flow along formally defined transmission lines. Formal functional communications are usually written and are normally both from and to individuals and groups. Formal relations are defined precisely and most day-to-day business is accomplished within the formal relation framework. The line expressing a formal functional relation usually has an arrowhead at each end to show a mutual exchange of responsibility and authority. If there is a higher authority to be implied, a single arrowhead can be used, pointing to the superior party.

3. *Informal.* These are the connections that are formally invisible but exist nevertheless and are crucial to communication health. Informal relations are the natural channels along which organizationally related material is most easily and comfortably transmitted. The informal relation exists by mutual consent of the parties to the relation and is stimulated to maximum effectiveness by a mutual profit gained from the relation. Little, if any, authority is normally expressed in informal relations. Communications are usually oral and one-to-one. Often, informal relations define the hidden organization structure. A line defining an informal relation is usually shown dashed with an arrowhead at each end.

4. *Staff.* Staff interrelationships are the business patterns through which individuals or groups provide consulting services necessary to achieve goals and objectives. Staff personnel usually have little or no authority over those outside the staff group. The line expressing staff relations has an arrowhead at each end.

5. *Temporary.* Temporary relations are created when extraordinary or unusual management demands must be met. The temporary relation is usually unstable and should be kept active for only short periods of time. The line expressing a temporary relation can have an arrowhead at one or both ends, depending on the nature of the relations. Extensive use of temporary relations creates business dysfunctions, breaks down morale, and causes internal tensions.

Recognizing these five basic relationships and what they mean to the project and its direction is critical to partnering success. It is the unusual person who can see the full impact of a certain set of organizational relations and how they affect the probability of project success. The partnering system, through its charter, will help sharpen the team's focus on the way the members work together and with others outside the team.

Ingredient 5: A willingness to "chance one's arm" as did the Earl of Kildare at the Door of Reconciliation Essential to effective partnering work is a willingness to take the risk that the system will work. There is no guarantee of success in anything that we do. Life is still an art and not a science. Therefore, we must make assumptions about people, muster our management and leadership skills, plan our course of action, and select the correct time and place to apply these elements. Then we act! We may still fail but should not allow failure to overpower us. A sustaining force is always provided by the assurance that others who have followed similar paths have succeeded. So, assembling the tools that have worked well for those who have already been successful, and plunging into the effort, is the next step in ensuring partnering success.

Ingredient 6: A set of tools by which the partnering program can be structured and implemented This ingredient is what this part of the book is all about—creating or acquiring a set of tools by which a partnering program can be structured and implemented. In the project partnering effort, four fundamental tools are needed to prepare and implement a successful effort:

1. A short, well-written, easily understood mission statement.
2. A document to which the stakeholders[12] on the project are willing to become signatories. This document is commonly called the charter.
3. A method by which the stakeholders can evaluate the effectiveness of their partnering efforts on a periodic basis.
4. A conflict and dispute resolution system that is acceptable to the stakeholders. Usually, this system contains one or more of the several alternative dispute resolution methods available.

As you can imagine, the proper combination of these physical, intellectual, and management components does not come automatically or easily. The workable partnering system must be tailormade to fit the job, the stakeholders, and the particular design and contract circumstances that surround a project at the beginning of each phase of operations. The partnering system helps articulate, communicate, and reach moral agreement on noncontract matters of importance. It is geared specifically to helping the project managers, by the application of their skills, to bring a project to a successful completion.

The components of partnering are relatively simple in concept and easily defined. They are the mission, the charter, the evaluation system, and the dispute resolution system. The project mission can be expressed as being *a statement of the most important result to be achieved by completing the project successfully.* Many companies adopt a mission statement somewhere along with the rise of the company's fortunes. Other companies consider their missions to be self-evident and make no formal statement.

Mission statements are of importance in preparing the partnering charter. The mission statements for an individual, for a project, and for a company on the project are all built on each other. They all reflect the attitudes of people who are important to the success of any planning, design, and construc-

---

[12] Stakeholders are the parties at risk financially and legally, or in an extended sense, those affected and potentially put at risk during the execution of a planning, design, or construction contract.

tion program. The company executive staff members prepare a mission statement of what is most important to their company so as to derive from this statement the philosophy, the goals, the objectives, and the staffing needed to achieve that mission. They also adopt a mission so as to inform their customers and the general public what the company wants to achieve to earn its keep in society.

A project management staff establishes a mission statement so that the many diverse groups that may work together on the job can merge their company missions into one that is best for the project. The closer the company mission statements match the project mission statement, the higher the probability, if the mission statements are truthful, that the project will be successful for the owner and for the organizations involved in planning, designing, and building the facility.

A person probably will not have mission statements for each of the many components of life upon which he or she is constantly engaged. In particular, the vocational mission statement is likely to be merged with other work-related desires. This may render the individual vocational mission definition literally invisible for a specific project. An analysis of many individual mission statements written by request for specific projects shows that the most important results for the person have a close relation to the most important results desired by the project team and the organizations involved. In effect, the mission statements at the three levels of project action—individual, project team, and organization—indicate that each shapes the other.

The mission statement forms the basis for combining potential project problems with solutions to form objectives. These objectives are combined into a set of behavioral rules for stakeholders. The resulting document is called the project charter. A partnering charter is essentially a document prepared by the project stakeholders and containing a project mission and a set of informal objectives and guidelines to successful performance in the execution of noncontract project matters.

Experience has shown that a good method of writing an effective project charter is to preface the main body of the charter by the project mission statement derived from organizational and individual mission statements. As the mission statement is being refined, the project team prepares background material for defining objectives on the job. The problems that might occur on a similar job or on this specific job are defined by the participants by discussing the problem statements with other stakeholders on the project who have similar design or construction responsibilities.

For instance, the participants in a charter being written for the design phases of a project might divide into discussion teams consisting of an architectural group, an owner group, a consulting engineer group, and a construction manager group. On a building project the discussion teams usually consist of the owner, the architect, the engineers, the site work contractors, the substructure and structural contractors, the mechanical, electrical, and controls contractors, and the interior finish contractors. There may be further detailed func tional divisions of the discussion groups if the project complexity warrants.

The purpose of the discussion groups is to keep problem and solution conversations within a common agreement on basic causes and correction techniques. Within these reasonably friendly arenas, discussions of what problems are caused by the various people on the project can usually be conducted effectively and with reasonable candor. As the discussions proceed, the results are presented by each discussion group to each of the other discussion groups. These presentations are limited strictly to a noncritical review of the material produced by the discussion group.

After the mission writing and the potential problem identification is complete, the problem items resulting from group discussions are merged, printed, and provided to the entire stakeholder staff attending. Working in the same functional groups as before, the individual discussion groups now are charged with the task of defining as many solutions to these problems as time allows.

The solutions are usually cast in the form of recommendations answering a basic question: "What can we do to help promote good relations and excellent performance on this project in light of our mission statement and the problem-potential statements written earlier?" The answers are molded into a one- or two-page document called the charter. This rough statement is reviewed by the stakeholders as a body, refined by them, and ultimately printed in a final signatory charter. The stakeholders sign the charter if they wish, and the signed, informal moral agreement becomes the standard of behavior in noncontract matters for the project.

SECTION D: OVERVIEW OF THE EVALUATION SYSTEM

The third partnering system component, an evaluation system, is prepared and implemented once the mission statement and the partnering charter are signed and in place as an accepted working tool. The periodic partnering performance evaluation is, in a nutshell, a project report card, grading the project team on how well it achieved the mission and the recommendations contained in the charter over the length of the evaluation period. The purpose of periodic evaluation is to provide a measurement yardstick by which stakeholders can determine, analyze, articulate, and communicate project performance trends. They can then, with intelligent and competent management, convert the analysis to management actions that improve the project.

Notice in our discussions the constant emphasis on stakeholder preparation of partnering components. The entire partnering system is like a chain whose links must be equally strong to be effective. The perfect chain is an elegant tension mechanism. Unbalancing the chain system by using links of unequal strength erodes the utility of the system. Poor, inadequate, or delegated partnering component work may weaken the strength of the partnering system.

take responsibility and assume the authority for the full
' partnering techniques.

n evaluation system can be seen by extracting a specific
ual charter. The example extract used is from the Magne-
ᴊ America die-cast facility building project charter. This charter
ᴩ epared by the owner, the at-risk construction manager, the design team,
and key subcontractors on the project. This constituted the stakeholder group.
The charter is reproduced below for ease of reference.[13]

Charter for Magnesium Products of America Project

Mission

As a team, to design and safely construct a world-class magnesium die-casting
facility on time, within budget, while developing positive and profitable
relationships among all team members.

Objectives

1. Alternative dispute resolution
   a. Resolve problems in an effective, timely, and fair manner at the lowest level
   possible.
   b. Address all issues in a professional, nonpersonal, timely manner.
   c. Maintain an active, open issue log.
2. Close-out and final payment
   a. Implement a rolling punch list technique.
   b. Prepare and publish acceptable close-out guidelines.
3. Communicate effectively at all levels
   a. Attend meetings and be accessible.
   b. Maintain flow and tracking of documents and information.
   c. Maintain effective written communication and documentation.
4. Control cost growth
   a. Document approved changes and forecast potential changes.
   b. Define and resolve outstanding design issues quickly.
   c. Recognize the owner's desire to control operations and maintenance costs.
5. Expedite project payments
   a. Maintain invoicing procedures so as to facilitate prompt payment.
6. Quality Control
   a. Define quality objectives.
   b. Conduct preinstallation meetings.
   c. Keep all field testing current.
   d. Perform quality review throughout design and construction.
7. Job morale, attitudes, good neighbor, good work site
   a. Maintain a clean and safe work environment.

---

[13] From a charter prepared by stakeholders on the Magnesium Products of America die-casting
facility in Eaton Rapids, Michigan. The at-risk construction project manager was the Christman
Company, Inc., of Lansing, Michigan.

   b. Maintain a teamwork attitude with openness and respect for other
      contractors' work.
 8. Leadership, responsibility and authority definition
   a. Identify and communicate organizational responsibilities to all team
      members.
   b. Issue timely goals/objectives for weekly activities progress meetings and
      field activity.
   c. Educate team to specific needs/design of all process equipment.
 9. Maintain partnering effectiveness
   a. Prepare and publish partnering effectiveness procedures.
10. Planning and scheduling
   a. Publish an exception report highlighting critical path, behind schedule, and
      changes.
   b. Monitor, update, and issue project schedules.
   c. Communicate schedule changes and requirements.
   d. Issue timely goals and objectives for weekly activities, progress meetings,
      and field activity.
11. Revisions and submittals
   a. Interface shop drawings between trades.
   b. Define submittal priorities, turnaround commitments, and lead times.

In the charter, the subject of cost growth as influenced and managed by
the project stakeholders was addressed in objective 4. An evaluation task
force of stakeholders made a judgment early in the job as to what weight
would be given the objective to determine its contribution to project success.
A weight of 5 indicated that the objective was absolutely critical for full
mission achievement. A weight of 1 indicated that the objective was of very
low importance in achieving the mission.

The task force assigned objective 4 a weight of 3.59. This was their perceived
importance of the objective to achieving design and construction project success in accordance with the mission and the charter. The evaluation task
force then made a decision as to the value to be assigned to a satisfactory
performance of the project team in achieving the mission and objectives. Par
performance was defined as the degree of success that would have to be
achieved if the mission and objectives were met satisfactorily during the evaluation month. A value of 5 indicated that the objective was being met as well
as it could. A value of 1 indicated that the project performance was totally
inadequate to meet charter objectives essential to achieving the project mission. For objective 4 the par importance was set at 3.44.

Once the weight and par value of each objective were selected, the weight
was multiplied by the par value to give a rating of the team's efforts that
represented a satisfactory standard against which to measure project performance at each evaluation point. A month-by-month evaluation of the project
encouraged the stakeholders to carefully review all performances, including
their own, to determine the good and the bad about job conditions. The

constant review was a powerful force to achieve project excellence. It was also a wonderful diagnostic tool when things began to go wrong.

Evaluation methodologies may vary from project to project. In most situations the system of evaluation is subjective, soft, and often personal. It should be kept relatively simple and be easily understood by all involved. Evaluation is a critical element of a good partnering system, and an evaluation system should be prepared by the stakeholders as soon as possible after the charter is adopted (see Chapter 12 for a detailed discussion of the partnering evaluation system).

SECTION E: OVERVIEW OF THE ALTERNATIVE DISPUTE RESOLUTION SYSTEM

The fourth element of a good partnering system is a conflict and dispute resolution system that is acceptable to the stakeholders. The conflict and dispute resolution system is a process designed to achieve a private settlement or resolution of a dispute by the parties to the dispute themselves, without using binding judicial procedures. Alternative dispute resolution techniques to be used in the application of a partnering system are defined, as is the evaluation system, by a task force of project stakeholders. The task force will usually recommend the use of one or more of the dispute resolution techniques described in Chapter 6, Section C.

A brief review summary is given below of the main methods of nonbinding, nonjudicial issue resolution. Note that the specification of a dispute resolution technique within a partnering agreement implies that partnering is already in use as a preventive technique.

1. *Prevention methods:* usually produce maximum harmony at the least cost and include:
   a. Intelligent and proper risk allocation
   b. Incentives
   c. Partnering
2. *Internal negotiation methods:* relatively low cost, requires consensus for success. There are two scales of negotiations:
   a. Step negotiations, which usually start at dispute originating level
   b. Direct negotiations, which often start at the ultimate decision-maker level
3. *Informal external neutral methods:* relatively low cost but effective resolution system. Types of neutral participants and their rulings include:
   a. Architect/engineer-of-record rulings
   b. Dispute resolution board rulings
   c. Independent advisory opinion

4. *Formal external neutral method:* a relatively low-cost system but may require greater preparation time than other less formal methods. The techniques used by formal external neutrals include:

   a. *Mediation:* formal hearings by a neutral third party

   b. *Minitrial:* private settlement initiated by agreement between the parties

   c. *Advisory opinion:* neutral meets with both parties and renders prediction of ultimate outcome if adjudicated

   d. *Advisory Arbitration:* neutral conducts formal hearing and issues advisory award; may also render prediction as to ultimate outcome if adjudicated

I recommend that the dispute resolution system be prepared by a task force after the partnering charter is written and signed by the stakeholders. At that time the character and overall nature of the project team have become visible through the intense discussions that mark a partnering charter meeting. It is also best to allow the stakeholders time to digest and consider how they—as individuals and organizations—are going to fulfill their obligations under the charter. This may shape their ideas about what approach the issue resolution task force should take to tailor and establish a workable settlement method best suited to the project and its implementation team.

8

WHERE PARTNERING APPLIES

Before a partnering effort is implemented, the partnering steps that will be taken should be systematically identified and planned to ensure that the effort will produce maximum value results. Time spent in properly planning the production of the partnering components will pay handsome dividends. The partnering components—mission, charter, evaluation, issue resolution—are each complex subsystems of the whole partnering method. Care in their preparation through attention to the steps in partnering planning will garner support from and give confidence to those essential to project success.

SECTION A: DETERMINING THE NEED FOR PARTNERING (SHOULD WE USE IT?)

Design and construction partnering systems are closely tied to the desires and capacity of industry's management. Partnering and management must be considered together when evaluating the probability of partnering success. In most cases, and on most projects, conditions exist that provide a reasonable assurance of successful partnering. The fact that partnering has already survived many acid-test years indicates that there is an underlying strength in the system.

My favorite partnering story is told in a picture of a barn raising dating back to the late nineteenth century. It shows about 80 laughing and celebrating men, women, children and even a serious horse, celebrating a barn topping-out near Massillon Ohio. The description tells the tale: "The people of America's

Partnering at work. (Courtesy of Massillon, Ohio Museum.)

heartland have always been neighborly about barns. In this nineteenth-century photo, a proud crew has just completed erection of a massive new structure for hay and livestock."

The words *neighborly* and *proud* seem to be good starting points in determining a need for partnering. If the elements of neighborliness and pride in the project are actually or potentially present, partnering can probably succeed. Of course, it is always reassuring to have as many reasons present as possible to ensure that the partnering effort has a high probability of succeeding. Besides neighborliness and pride, we might include these other ingredients of a successful partnering effort:

- A strong desire to achieve project success for all
- Making and accepting intelligent commitments
- Refusing to accept or impose unreasonable risks
- Working and acting together ethically, morally, and with integrity
- Working and acting from a position of fairness rather than power
- Suppressing greed
- Using ability and performance to establish an honest atmosphere of trust among participants
- Gaining support from the participants and stakeholders
- Assigning experienced, competent people to responsible management positions
- Having empathy and using it to understand others' points of view
- Preparing a good charter, a good partnering evaluation system, and a good issue-resolution process
- Allowing time to make the partnering system work
- Recognizing and celebrating success
- Gaining the support and participation of higher management
- Developing and using guidelines and evaluation systems for measuring performance quality
- Employing good thinking patterns
- Setting and using achievable standards of ethical behavior
- Having and using technical competence that is needed for project success
- Having and using good planning, design, and construction experience
- Measuring project team performance accurately

 're are some instances where partnering may have a low potential to
    . In these cases, forcing the application of partnering could be coun-
     ve, make the system difficult to use, and lower the chance of
       ect success. Projects such as these may occur if the following
          resent:

- The parties intend to pay lip service only to the partnering effort.
- People in key technical or management positions choose to resist intelligent issue resolution discussion and fair decision making.
- Early commitments by the owner have made healthy, easily understood contract relationships difficult or impossible to maintain.
- Several parties to the contract prefer to resolve disputes by contested claiming and binding resolution.
- Poor contract documents form the basis of the partnering effort.
- Excessively one-sided conditions are placed on subcontractors by prime contractors or on the prime contractors by the owner.
- Unfair or obscure payment-processing systems are specified and enforced.
- Risk has been poorly defined and allocated unfairly.
- Political influences driving the project are used to override sound design and construction practices.
- Dishonest accounting and payment practices are prevalent at various levels of the project contract structure.
- Contract general requirements and conditions vary from contractor to contractor on the project.
- Technically incompetent people are in charge of the project work.

Most of the ingredients and circumstances causing partnering difficulties stem from poor management attitudes and practices. Bad management action adversely affects outcomes desired in making a partnering effort. To use partnering effectively, management must understand the value to be gained by its use. Rigorous and extensive office and field experience on partnered projects indicates its worth. Partnering properly used and applied has proven overwhelmingly beneficial. Some of the more notable results show us that partnering:

- Lowers costs to resolve conflicts
- Stimulates quick settlement of conflicts
- Allows knowledgeable professionals to make resolution decisions
- Puts the decision makers close to the resolution process
- Lessens the probability of appeal for the nature of decisions rendered
- Allows participants to maintain privacy in the resolution process
- Increases the probability of fair resolution by timely consideration of disputes
- Helps to cross critical transition points[1] by setting the ground rules for the crossing

---

[1] The point in a project delivery system at which the responsibility and authority for the work passes from a supportive group to the ex'e'cutive group.

It is interesting to look at a few of the documented partnering experiences of the U.S. Army Corps of Engineers. The Corps was among the first owners and users to realize and benefit from the power of partnering to help achieve higher levels of project success. Comments from the Albuquerque District Corps of Engineers in a "Guide to Partnering" dated 1991 reports: "Our experience is positive based on six contracts, with four of them substantially complete." Benefits include:

- Disputes reduced—no formal claims
- Common objectives achieved (schedule, safety, etc.)
- Increased responsiveness
- Higher trust levels
- Improved communication
- Excellent cooperation and teamwork
- Increased-value engineering proposals
- Developed expedited process for tracking and resolving open items

Another set of partnering comments by Colonel Charles E. Cowen—then Commander of the Portland District Corps of Engineers—in the paper "A Strategy for Partnering in the Public Sector," dated April 1991, indicates that on projects of which he had knowledge, the benefits of partnering included:

- An 80 to 100% reduction in cost growth over the life of major contracts
- Time growth in schedules virtually eliminated
- Paperwork reduced by 66%
- All project engineering goals met or exceeded
- Projects completed with no outstanding claims or litigation
- Safety records improved significantly
- Pleasure put back in the process for all participants

The examples above show what is needed to make partnering work, the project characteristics that might make partnering difficult to use, the value that might be added by partnering, and what the experience can mean for a competent and responsible owner/user. This track record should provide the serious professional practitioner with the confidence that he or she needs to begin to formulate a strategy and tactics for successful partnering.

A key element of success in partnering is to have a reasonably clear picture of what the management structure of your organization[2] looks like. Sound, well-thought-out internal management relations usually help to ensure that your participation in partnering will succeed in improving your total profitabil-

[2] The management and operational structure through which individuals cooperate systematically to conduct their business.

ity on the project. A brief analysis of the management interrelations as described in Chapter 7 that should and do exist is of help in understanding and improving the partnering process.

SECTION B: FITTING THE PARTNERING EFFORT TO THE CURRENT PROJECT STAGE

Partnering is a universal tool that can be used in many different businesses, at many different levels of management, and during most stages in the development of a project. In the planning, design, and construction profession every project goes through several well-identified steps on the way to its completion. These are defined in the macro sense by the elements of the line of action described in Chapter 2.

The seven major stages in this design and construction line of action are:

1. *Conceive and communicate.* During the conceptual stage the needs of the owner and user are visualized and captured in some rough form. The needs may be for an increased size facility, larger dollar volume produced, more effective handling systems, or a variety of others. The form may be a pencil sketch or a model, or may remain just an idea in someone's mind. Whatever the need or form, it is here that the project sees its origin. These early ideas often carry through the entire project. A good conceptual grasp is essential if the project is to be completed successfully. An important element of the conceptual stage is the need to communicate the idea to others. No project can ever be brought to life without it being communicated. An excellent project, well conceived and properly communicated, starts out with a high probability of success.

2. *Program and articulate* (see the description and discussion of a facility program in Chapter 6, Section A). The program is a narrative statement of the needs and character of the proposed user operation, the requirements of the user and owner, the nature of the environment to be planned, designed, and built, and the corresponding characteristics of the space that will satisfy these needs and requirements. The program is sometimes called the brief. In writing the program the needs expressed by the concept are put into easily understood, specific data form—so many square feet for storage and offices, so much clear height in the receiving and shipping areas, and the other use, relational, and space definitions. During programming the actual physical demands of the environment are set forth in the project program or bible.

   Development monies are actually committed in the program phase. The amount of funding needed is determined by financial analyses of income and expense all balanced against what the program defines as the needs of an adequate facility. Once the program is under way it is gradually converted into preliminary construction language suitable for approval by the decision makers. Floor plans are drawn in accordance with the functional needs and space allocations. Materials are called out on elevations, sections, and details—all this to ready the program, the funding, and the graphic translations for the final approval and the go-ahead.

3. *Approve.* This is the most critical point in the line of action. By now sufficient work has taken place so that the decision makers can understand the project and say "I like this" or "I don't," "Change this," "revise that," "let's increase that a bit and let's cut down here". And—finally—"OK, I'm satisfied, let's move on." The approval unlocks the design and construction period.

4. *Design.* Now the products of the previous steps are concurrently analyzed and are assembled to produce a set of working drawings and specifications. These translate the concept into a cohesive whole, combining the steel, concrete, space, and other components that make up a completed facility.

5. *Construct.* Next, the actual environment—the facility—is built. Construction is the point where something tangible appears on the physical site of the facility.

6. *Turn over.* When the facility is constructed to a point where it can be occupied for use as intended, it is turned over to the owner and the user. The turnover is accompanied by release of appropriate operating and maintenance manuals to the occupants. A good turnover is an important step since it ensures that a valuable commodity, the completed facility, is given to those who are to occupy and use it.

7. *Operate and maintain.* The facility is now run in and begins to achieve its full purpose. Good maintenance by the owner and the user helps assure that the new facility, which has been nursed through the previous six stages, will be maintained at its best for those who must use it.

Each of these phases in the production of a finished facility can be further subdivided into activities to which we could apply partnering techniques. The format of these partnering meetings would be heavily influenced by the action steps required for the system to go into service. For example, let's consider the partnering effort needed for Londonderry, a moderate-size shopping center of 200,000 square feet. The project is now in the early design phase where the land planner, the architect, the engineering design team, and an at-risk construction manager have been selected by the owner, a local developer. Partnering practices in this instance would be established based on a charter written by the owner, the user, the architect of record, the site engineer, and the construction manager.

Issues addressed would center on how best to get the project design depicted in a usable and accurate set of construction contract documents. The elements stressed in the charter would focus on how to design the project so that it can be built within the time available and an affordable cost structure. Preparing a simple matrix showing these parameters can help establish the partnering conditions to be met now and in the future. The matrix in Figure 8.1 shows the successive actions and their estimated time frames on the project along the horizontal axis. Participants who have responsible partnering roles on the job during its construction period are shown on the vertical axis. An "X" indicates where participants are expected to assume active partnering involvement on the project.

The matrix indicates that the first partnering meeting should be held and a charter written early in the first month of the job. This charter would concern primarily the owner, the design team, the general contractor, and possibly the testing agencies. The main focus in this effort should be how to move the project through the program, approval, and design phases into construction.

	participant	Month Number & Phase of Work					
		0 to 1 program	0 to 2 approve	3 to 9 design	6 to 22 construct base bldg	18 to 23 tenant work	20 to 24 run in & opening
1	01.00 - Owner staff	X	X	X	X	X	X
2	02.00 - Financing souce		X	X	X	X	
3	03.00 - Tenants				X	X	X
4	04.00 - Architect of record	X	X	X	X	X	X
5	05.00 - Landscape architect	X		X	X	X	
6	06.00 - Site and civil engineer	X		X	X	X	
7	07.00 - Structural engineer			X	X		
8	08.00 - Mechanical engineer			X	X	X	
9	09.00 - Electrical engineer			X	X	X	
10	10.00 - General contractor	X	X	X	X	X	X
11	11.00 - Subcontractors				X	X	X
12	12.00 - Testing lab	X		X	X		X
13	13.00 - Property managment						
14	14.00 - Tenant architects/engineers			X	X	X	X
15	15.00 - Tenant work contractors			X	X	X	X

Figure 8.1 Partnering participation matrix for Londonderry Shopping Center, Point Stevens, Pennsylvania. X indicates sufficiently high involvement to participate in partnering effort as stakeholder. Shaded columns indicate line-of-action points where charter meetings might be held.

The matrix shows a considerable change in job staffing occurring in the third to sixth months of the project. This change in emphasis from design to construction indicates that a second charter might well be written in the sixth or seventh month. At this time most parties to the project will be on board and under contract for the base building. The charter written here would be focused primarily on the actual construction of the base building, site work, and turnover to tenants for start of the store interior work and fixturing.

This construction charter meeting would be attended by the owner, the financing source, the design team, the prime contractors, and the subcontractors. Perhaps the early tenants who have committed to taking space in the Londonderry Center would also have good input to the charter as stakeholders or observers. The construction-oriented agenda would be devoted to defining objectives that would assist in guiding the effective application of construction-phase resources—personnel, money, space, time, talent—during the actual construction operation. In months 18 to 23 the main focus will be on designing and constructing tenant work to go into the base building. Here a partnering effort would be appropriate for the owner of the center, the major tenants, their contractors, and the base building contractor.

Another matrix arrangement depicts the stage of the work in a grid showing the time frame of the project along the vertical axis and the various trades that are needed for the project along the horizontal axis. The phase of the work for a particular trade design or construction is shown at the intersection of the appropriate month and the trade work to be done. An example of such a matrix is shown in Figure 8.2. The Cranbrook Machine Systems Company is about to embark on a 17-month program to install a new tooling line of self-contained robots. Functions to be accommodated in the various phases of the expansion work include:

- Structural
- Architectural
- Civil and site
- Fire protection
- Mechanical
- Electrical
- Process work
- Controls
- Instrumentation

The phases of the project as established by the project manager include:

- Project program preparation
- Design
- Bidding

month	struct, arch, civil	fire prot, mech	electrical	process	controls, instrumentation
1	program			program	
2 •	design	program		program	
3	design	design		program	
4 •	bid	design	program	design	
5 •	const	design	design	design	
6	const	bid	design	design	program
7 •	const		bid	design	program
8	const	const		bid	design
9 •	const	const			design
10	const		const		design
11 •	change	const	const	install	design
12 •	finish	const		install	bid
13 •		const	const	install	
14 •		change	const	install	
15 •		finish	const	install	install
16			change	install	install
17			finish	install	install

Figure 8.2 Function and staff participation at monthly progress points, Cranbrook Machine Systems. Dots indicate new charter or charter revisiting points.

- Construction
- Installation of process equipment
- Installation of controls
- Installation of instrumentation
- Building work changes to accommodate final process, control, and instrumentation work

Figure 8.2 displays the relation of function and phase to monthly time points on the project. The matrix shows that if we are considering external or internal partnering, key times to develop a new charter or revisit the existing charter are in the second, fourth, fifth, seventh, ninth, and eleventh through fifteenth months.

SECTION C: SELECTING THE PARTICIPANTS IN THE CHARTER MEETING

All key project management, including the ultimate decision makers (UDMs), should attend the partnering charter meeting. Included in this decision-making group are project team members from the owner's staff, the facility users, the planning and design team, the prime contractor(s), and the major subcontractors and vendors. All will be active participants who will actually write the charter. (The partnering charter meeting is discussed in detail in Chapters 9 to 11.) Professionally related special guests might be invited to observe how the partnering process works. This is acceptable and desirable as long as the members of the specially invited group clearly understand they are observers, not participants.

The invitation letter to the charter meeting should come from the directing management on the project. This is so that all participating in the charter preparation understand that the partnering effort has the support of top management. Those who are candidates for an invitation to the partnering charter meeting might be classified by their technical specialties and by the positions they occupy on the project. A good set of guidelines for deciding on a prospective participant's worth to the process is to ask the following questions:

1. Does the person have the authority to make decisions relative to project work?
2. Is the person able to be active on the project to the extent that his or her position warrants?
3. If the person did not come to the partnering charter meeting, would it leave an unfilled gap in the project decision-making process?
4. Is the person's organization financially at risk on the project?
5. Is the person authorized to make a decision to sign or not to sign the charter?
6. Will the person's contribution to the charter meeting, the issue resolution process, and the project evaluation system improve project performance?

If the answer to all the questions above is yes, it is a good indication that the person should be invited. Of course, there is a limit on how many people should be at a partnering charter session. I have had as many as 50 and as few as 15. This is a helpful sample range from which to judge what the actual number should be. My experience indicates that a good meeting chair can lead 20 to 35 people through a partnering charter session in a single day. This includes discussing the key questions needed to develop project missions, goals, and objectives, preparing a mission statement, preparing a set of charter goals and objectives, and signing the charter. It also allows the signed charter to be printed.

The issue resolution process and the project evaluation system should be prepared and implemented a short time after the charter is written and signed. This gives the stakeholders a soaking-in or gestation period in which to think and formulate best approaches to the issue and evaluation methodology. A list of potential attendees from which to select is given below. Please note that on most projects only some of those listed might have positions on the project where their presence would make a significant contribution. Individuals or organizations other than those named might have to be added where special talents or demands would make their attendance desirable. The list is divided by major organizational function and subdivided by organizational interior functions and individuals.

Owner's Staff

- Chief operating executive
- Executive in charge of project
- Project manager
- Director of facilities
- Director of administrative services
- Resident architect
- Resident engineer
- In-house technical managers
- In-house public relations managers
- Technical specialists consulting to the owner
- Procurement director or project staff
- Contracts manager
- Real-estate staff (where their actions affect project performance)

User's Staff

- Facility manager or director
- Executive of users' staff responsible for facility operations
- Emergency services staff (fire, pubic safety, security, and similar operations)
- Maintenance staff
- Property management staff
- Leasing staff

Programmer's Staff

- Program writer

Urban Planner's Staff

- Principal or executive in charge of project
- Project manager
- Land planners

Architect of Record's Staff

- Executive or principal in charge
- Project manager
- Site representative
- Key technical staff members
- Technical specialists consulting to the architect

Design Architect's[3] Staff

- Lead design architect

Landscape Architect's Staff

- Landscape designer
- Project manager
- Site representative

Interior Design Staff

- Interior designer
- Project manager
- Site representative

Communications Consultant's Staff

- Project manager

Computer Systems Consultant's Staff

- Project manager

Food Service System Staff

- Project manager

Engineer of Record's Staff[4]

- Site and civil engineers
- Structural engineers
- Mechanical engineers

---

[3] On design-sensitive projects a highly qualified design firm may be retained to do the conceptual studies and set the design standards, from which an architect of record prepares working drawings, specifications, and other construction and contract documents.

[4] Includes site and civil engineering, structural engineering, mechanical engineering, electrical engineering, and technical specialists consulting to the engineers.

- Electrical engineers
- Technical specialists consulting to the engineers

Preconstruction Service Staff[5]

- Project manager

Construction Manager's Staff (if using CM project delivery system)

- Executive or principal in charge of organization
- Executive or principal in charge of field operations
- Project director
- Field superintendent
- Project manager

General Contractor's Staff

- Executive or principal in charge of organization
- Executive or principal in charge of field operations
- Project director
- Field superintendent
- Project manager

Subcontractor's Staff[6]

- Executive or principal in charge
- Field superintendent
- Project manager
- Key supplier and vendor representatives

Observers[7]

- Regulatory agencies staff
- Utility company staff
- Financing participants
- Present clients

---

[5] Preconstruction assistance is often provided to the owner, user, or architect/engineer, as a professional service in early estimating, material selection, planning, scheduling, value engineering, life-cycle costing, and contract document packaging.

[6] Includes such major subcontractors as excavation, site utilities, paving, concrete, sheet metal, plumbing, electrical, structural steel, miscellaneous metals, masonry, roofing, fire protection, food service equipment, furniture/fixtures/equipment, communications, painting, carpentry, electronic equipment, glazing.

[7] Observers may occasionally be invited as participants when their efforts significantly affect the supportive, executive, functional, or project efforts of the design and construction team.

- Labor representatives
- Legislative representatives
- Affirmative action staff
- Insurance company staff
- Political entity representatives

Testing Staff

- Hydrologic testing and design
- Soil sampling and testing
- Abatement
- Remediation

9

SETTING THE PARTNERING CHARTER MEETING FORMAT

The format of meetings for stakeholders to interact effectively should be established early, preferably when you are scheduling the partnering meetings. Meeting structure will vary as the needs of management vary. As an example, an early orientation session might concentrate on team building as part of the partnering process. Where internal or external quality issues may significantly affect project performance, some partnering-related training sessions could focus on total quality management, quality assurance, or quality circles. Many of these subjects concern management techniques designed to provide improved results within an organization's macro business and technical boundaries. Most are helpful but not essential in a partnering effort.

A separate meeting agenda related to partnering might be directed toward discovering major points of difference among the stakeholders. Once these differences are identified, the partnering effort could focus on using these few major points to develop detailed procedures for working with and resolving the differences. Techniques to be followed in a partnering charter meeting usually are determined by multiple influences. These range from the nature of the project, the makeup of the stakeholder group, the location of the ultimate decision makers, on through to the abilities of the charter meeting chair or the document type to be produced.

Many project characteristics relate directly and indirectly to the subject of partnering. These characteristics should be studied carefully and selected to structure the partnering format and agenda. This will help assure that the results of the meeting will be what is desired and intended. The list of influences below provides a good range of project features that shape the meetings needed to develop a partnering system:

Factors That Influence the Format of a Partnering Meeting (listed in random order)

1. Results desired from use of the partnering system
2. Stakeholders' experience with partnering
3. Job cost and time growth potential
4. Project claim-prone potential
5. Current stage of the project (as of the partnering charter writing)
6. Location of the project and participants
7. Follow-up requirements of the partnering system
8. Project type and characteristics
9. Project delivery system being used
10. Size of group participating
11. Time available for charter meeting and follow-up sessions
12. Visitors to be accommodated
13. Meeting leader
14. Location of charter meetings and of task force meeting
15. Facilities available for meetings
16. Technical, managerial, and educational backgrounds of participants
17. Size of project

Not all the influences listed above carry equal weight in determining how to approach the subject of partnering. The discussion below reviews some of the main features that exert critical influences on the format of a partnering session.

1. Results Desired from use of the Partnering System   The complex elements of partnering systems make the process a multidisciplinary effort. Preparation of a sound charter-meeting agenda demands that desired subject material be pinpointed so that those attending can come prepared to contribute positively in dialogue. The results of a partnering meeting might be perceived differently by those active in the many different aspects of a project. Partnering is a design and construction standard as well as a high-value operating system. Expectations resulting from the use of partnering include:

- Improved job working relations
- Improved project communications
- Heightened awareness of project mission
- Lowered insurance costs
- Quick, fair resolution of project disputes
- Higher respect for the needs of others on the job

- A reduction in the number and intensity of contested claims
- Increased profitability

Subjects to include during a partnering charter session that help produce results as noted above are certainly of interest to stakeholders in a partnering effort. Some of the more important of these might include (listed in alphabetical order):

- Alternative dispute resolution
- Approval processes
- Being a good off-site neighbor
- Being a good on-site neighbor
- Closing out the project
- Communicating with others
- Construction document quality
- Cost growth
- Good work site
- Job staff morale and attitude
- Legal entanglements
- Organization, authority, and responsibility patterns
- Paperwork
- Partnering evaluation and effectiveness
- Payment processes
- Personality types involved
- Planning and scheduling
- Policies and procedures
- Processing revisions
- Procurement
- Profitability
- Program conditions
- Project administration
- Project constructibility
- Project cost structure
- Quality management
- Safety
- Stakeholder interrelations
- Submittal processing
- Team building
- Time growth

- Total quality management
- User-group interactions

Meeting formats must be carefully matched to the results desired. The end product will, in the main, dictate the subjects to be addressed in the meeting lectures, workshops, seminars, and symposium sessions. As an example, assume that the sponsor of a partnering effort desires to produce a team building effort. In addition, the sponsor wants to develop a total quality management emphasis in his user, design, and construction project team. Time limitations alone dictate a partnering charter-writing format that concentrates on team building, TQM, and partnering. The sponsor will probably plan to cover this territory in three or more separate sessions.

Why is this? Team building and TQM are subjects conveyed by those who have superior knowledge and experience. Attendees at the team-building session desire to be shown and acquire the related knowledge and understanding of the experience of the instructor to use in their day-to-day work. Team building and TQM knowledge is to be shared by experts with the partnering team members. They then build this acquired knowledge into the partnering system.

Partnering is not a subject to be learned in a vacuum; it is a technique to be used. Partnering synthesizes learned knowledge. It grows from an organic body of knowledge: it has its roots in the abilities, memories, and experiences of those participating in the partnering session. The separate roots come together to form a trunk, branches, leaves, twigs, and if everything goes according to form, fruit. The knowledge and experience possessed by the stakeholders creates a resourceful (resource-full) working environment. Knowledge about a subject is coupled with experience and molded into the partnering charter, an evaluation system, and an issue resolution process. This process relies heavily on intuition and experience in the use of new knowledge.

Suggestions:

1. The planning, design, and construction business is a results-oriented activity. Match subject content with results desired.
2. Keep the format loose enough to accommodate subjects not on the conventional agenda.
3. Set procedures and timetables that are reasonable. Don't try to cram more into an agenda than you can cover in the time available.
4. Allow for adequate time to train and educate in the disciplines needed to write the charter, the issue resolution system, and the evaluation procedures properly.
5. Ensure that those responsible for teaching or leading discussions about the subject material are knowledgeable about the subjects and know what conceptual results they are to achieve.

6. Provide all charter-meeting attendees with a set of handouts or a notebook containing the content of material to be reviewed and discussed.
7. Distribute written material to be used in the meeting before the session begins to allow the stakeholders to preview what is expected of them.

2. *Stakeholders' Experience with Partnering* Whether or not the stakeholders have experience with partnering will have an impact on the format and content of the partnering charter meeting. Inexperienced participants often need extra coaching and explanations about what is to be accomplished and how it is to be done. Those who previously have gone through the partnering process are usually well prepared to work with partnering concepts and may even take leadership positions in the charter meeting. Proper format balance between the veteran and the newcomer is important. You do not want to lose the benefits of the veteran stakeholder's experience or the enthusiasm and fresh contribution of the novice stakeholder.

A thorough briefing of newcomers often provides them with the information they need for partnering success. In such cases, where unbalanced partnering experience is to be joined in a common project effort, a separate and early briefing meeting for the newcomer can provide a valuable orientation to partnering. It is important to bring relatively equal knowledge and experience to the charter-writing table. This helps ensure that the actions leading to desired results are addressed early, effectively, and in a balanced manner.

Suggestions:

1. In workshop sessions, mix experienced partner participants with the inexperienced to keep the teams balanced in the degree and nature of their bias. Extensive individual experience within any team may lead to collectively biased problem statements and slanted recommendations.
2. When presenting a subject that may be well known to some in a mix of experienced and inexperienced participants, try to find new approaches to the presentation of the subject that might cast an unconventional or different light on the matter.
3. Explain the purpose of and results expected from the partnering efforts. If done well, the diversity and complexity of design and construction will add flavor to the approach of both inexperienced and experienced.
4. Allow the experienced people some time to show off their familiarity and abilities in the subject matter to the newcomers. If done properly, this adds zest to their efforts.
5. Do not discourage the beginner from trying to use the material creatively. Keep stressing the contribution of innovative thinking that adds value.
6. Stress the need for random thinking and nonjudgmental thinking, particularly to the experienced people. Remember that creativity and innova-

tion will seldom flourish in a prematurely judgmental environment where the potentially good thought is stamped out before it is fully stated.

3. Job Cost and Time Growth Potential   If one of the desired results of implementing a partnering system is to contain costs and to control deviations from scheduled dates, the tools for containment and control must be available to the stakeholders. A hard-money, fixed-time project (e.g., [construction project delivery system A3, B3, C1, D1a(1, 3, 4); see Chapter 2, Section C]) already has built into it the scope of work and the time of completion. In such a conventional project it is not likely that the stakeholders can reduce project costs subsequent to contract awards. If they do, the rewards from an improved and authorized application of good practices will usually belong to the contractors. In such cases an intensive discussion and incorporation of charter objectives dealing with cost growth may be counterproductive. In extreme cases the inclusion of such provisions might be considered potential maladministration.[1]

Time growth objectives in a charter pose similar problems. The name of the claim potential in this case is constructive acceleration.[2] If it is desired that contractors and others in the stakeholders group participate in cost and time control beyond contract boundaries, there must be provisions to define contractually the obligations of both the owner and contractor.

Suggestions:

1. Where there seems to be a high potential for increased project costs as the design and construction work proceeds, those selecting the project delivery system should specifically recognize the impact of cost growth as early as possible. Recognizing the impact of possible cost increases on projects will help to better assign risk to those most capable of assuming it. This action is one of the fundamental prevention techniques used in alternative dispute resolution.

2. The types of projects most susceptible to cost growth are large facilities depending on complex and evolving systems. In them the type and cost of equipment to be used may not be set firmly until close to installation and turnover. The costs for support space can escalate rapidly as the needs of such equipment are defined. Typical of these are large medical buildings, forerunner research complexes, or highly complex manufacturing plants.

---

[1] Maladministration: The interfererence of one contract party with another contract party's rights that prevents the latter from enjoying the benefits of least-cost performance within the contract provisions.

[2] Constructive acceleration: An action by a party to the contract that forces more work to be done with no time extension or the same amount of work to be done in a shorter period of time.

3. A good program is critical to success of a project with high-cost growth potential. Even though the final nature of the facility working elements is not always possible to set with full confidence, the performance definition should be such as to allow the performance and cost flexibility needed to stay within budget demands.

4. The management mechanism for accommodating cost growth should be established by participation of the entire project team. These are the experts who will ultimately be lead stakeholders in developing partnering standards of performance. Early knowledge of the potential for cost growth can help to establish a control process that will minimize destructive conflict during design and construction. It is better to simulate and model the problem and its solution than to wait for the actual experience to establish the manner in which responsibility for cost overruns are to be allocated.

4. Claim-Prone Potential   If early planning and design work has built claim-prone job characteristics into the project, it is unreasonable to expect the partnering system to remedy all of these through the charter. On the other hand, one of the reasons we use partnering is to try to defuse claim-prone influences on a project. If it appears there is the potential for claims against the project, an earlier and smaller meeting prior to the partnering charter meeting might be to discuss these potentials and how to handle them in the charter meeting. At this early meeting the upper field and office management of the key stakeholder groups could express their concerns privately. These concerns might then be carried forward and addressed in the break-out sessions during the charter meeting.

The claim problem is often a very sensitive issue. However, it is better to anticipate and discuss the problem ahead of the partnering charter meeting than to have it brought up unexpectedly and see it develop into an unresolvable issue as the charter is being written.

Suggestions:

1. A responsible decision maker for the principals in a project should evaluate the degree to which a job is claim prone. The material in Chapter 5 provides 23 characteristics of claim-prone jobs that can be used to evaluate the problem job.

2. If the job fits three to five of the characteristics of a claim-prone project, it is a cause for concern. In such a case the principal stakeholders would be well advised to have a prepartnering conference to decide how best to alleviate any real, imagined, or potential problems.

3. Preplan the charter meeting so that the claim matter is discussed if it appears to others as a potential problem. Be certain to alert the chair

about the problem so that he or she can cope properly with the matter as it arises.

4. Remember, the partnering system is designed to alert stakeholders to potential problems and prescribe remedies before a problem occurs. If you see present or future problems, discuss them either before or during the partnering charter meeting.

5. *Current Stage of the Project* As described in Chapter 2, Section A, a project passes through several well-defined transitions on the way to completion (the transitional steps are described in more detail in Chapter 8, Section B). To review, these transitional steps include:

1. Conceive and communicate.
2. Program and articulate.
3. Approve.
4. Design.
5. Construct.
6. Turn over.
7. Operate and maintain.

Any of these steps in the production of a finished facility can be broken into subactivities to which partnering techniques can be applied. For example, let's view a design phase situation where the land planner, the architect, the engineering design team, and a preconstruction services contractor have been selected by the owner to plan and design a new building. The format of a design charter meeting with these stakeholders would be considerably different from that for the construction partnering meeting since no construction contractors are yet on board. At the design level there would be mostly talk about how to get the construction documents completed and relatively little talk about how to build the facility.

To carry the example further, let's look at a partnering system being instituted on the same project where construction contracts have been let and the contractors are ready to start field operations. The construction charter meeting would be attended by the owner, the user, the design team, the construction advisor (now the at-risk construction manager), and the contractors. A construction-oriented agenda would be devoted to how properly to expend the construction phase resources—personnel, money, space, time, talent—on the actual construction operation.

Suggestions:

1. Tailor the agenda, the participants invited, and the subjects addressed at the partnering meetings to the stage of the job.

2. Revisit and possibly revise the charter when new organizations and people who may require changes to the original partnering agreement join the project. An obsolete partnering agreement should not be allowed to guide project noncontract behavior.

3. Be alert to where and when the project crosses critical transition points. These are usually points where the project experiences significant changes in patterns of responsibility, authority, staffing, construction sequencing, or other pivotal events. These critical transition points indicate where revisiting the partnering charter is to be considered. Examples of critical transition points are when construction begins after design is complete, when tenant work begins in a shopping center, or where extensive equipment startup is initiated in a manufacturing facility.

6. *Location of the Project and Participants*   If the project is geographically distant from the operational base of several stakeholders the cost of several separate meetings to develop a partnering system and its related subsystems might be prohibitive unless done in a grouped set of sessions. For a design or construction project in which the stakeholders have come from many different distant locations, the briefings on partnering-related subjects might be held the first two days of the total partnering development time. The partnering charter session might be held the third day of the session, and the development of an issue resolution system and a partnering evaluation system defined on days 4 and 5. Prepare and publish documents regarding items required by the newly written charter could also be completed on days 4 and 5.

The essential ingredient here is that the partnering system not be allowed to price itself out of business by becoming a heavy overhead item in which the penalties outweigh the value added by partnering. Such overhead costs as living expenses and excessive travel from remote locations could diminish the appeal of partnering.

Suggestions:

1. Select a block of time well in advance and designate it for partnering planning. Wise scheduling of these meetings will keep travel time and expenses to a minimum.

2. Set a specific agenda for the partnering and related work that keeps everybody as busy as possible and saves downtime due to travel.

3. If training and education are critical to developing the partnering charter properly, conduct such briefings before the charter writing session.

4. Select competent, experienced people to conduct the training and to chair the developmental meetings, such as for charter writing and follow-up documentation.

5. Take steps to ensure that complete notes are taken and transcribed of all sessions. These should be distributed to the participants while they are all still working on the partnering matters.

6. It is appropriate, if time allows, to close out the partnering planning sessions with a regular project meeting. In this meeting the partnering and supplementary work can be presented to the stakeholders and related to the project work directly.

7. Follow-up Requirements of the Partnering System   Directed and well-managed follow-up after the charter-writing session is of prime importance to the success of a partnering effort. Prime follow-up items are the partnering evaluation process and the issue resolution system. Others that may emerge out of the charter meeting are the preparation and publishing of documents, defined in point 6 above and in the glossary, Appendix A.

Preparing and publishing documents often includes definitions of project authority and responsibility or guidelines for payment policies, revision processing, planning and scheduling, and similar items that must be defined, usually in writing. Typical preparation and publishing guidelines from actual project charters include:

- Prepare and publish a chart of channels for communication, responsibility, and authority.
- Prepare, publish, and commit to a set of punch list and close-out guidelines, and conduct user orientation as part of the close-out process.
- Prepare and publish conflict issue resolution guidelines[3]:
  a. For internal issues
  b. For contract parties
  c. For our customers
  d. That address the problem, not the people
  e. That resolve problems promptly and at the originating level where possible

Where distances among stakeholders are small, the follow-up process is relatively simple. Communication can effectively be maintained and interaction among stakeholders is relatively simple to establish. However, when communication channels lengthen and become more obscure through distance gaps, the follow-up process begins to weaken. If this is a possibility during the life of the project, take procedural steps to ensure follow-up attention and continuity.

Suggestions:

1. Schedule periodic stakeholder meetings at which the charter is reviewed in detail. This is often done in the partnering system evaluation meetings,

---

[3] Paraphrased for clarity and to conceal the identity of the stakeholder group.

which require periodic analysis and evaluation of the performance expected by the project charter.

2. Follow up promptly on "prepare and publish" items required by the charter.
3. Follow up promptly on task force assignments such as the issue resolution policy and the partnering evaluation process. These are important to project health and must be implemented as soon after the charter is written as is possible.
4. Assign dedicated, responsible, interested people to activities requiring critical follow-up. Don't leave follow-up to chance.

8. Project Type   A very broad set of facility types is outlined below. The boundaries between types is often blurred, but for our purposes the categories listed will serve to help define some of the basic differences in partnering charter format. The facilities can be built in the private, public, or volunteer sectors of our society:

- Commercial
- Industrial
- Residential
- Institutional
- Recreational
- Site work
- Heavy construction
- Transportation

The project type affects the charter meeting format by imposing differing demands on the participants and on their types and degrees of interaction with other stakeholders. For instance, if the project is a highly confidential job demanding a secure design, such as a competitive retail outlet, the final location of which is still not public, the charter participants in an early partnering effort might have to be carefully screened. They may also have to provide a nondisclosure affidavit. In this case, there could be serious doubt cast on the effectiveness of such a partnering effort. Partnering focuses considerable attention on the project and a cautious owner may object to having the mission and charter information written and disseminated in a partnering system.

Frequently, a public facility is a candidate for partnering. In public work construction is often overseen by public-sector staff and it is possible that these employees will be prevented by local government policy from signing the charter. An example of this occurred recently when a state agency overseeing the expenditure of funds for a new community college supported the preparation and implementation of a partnering system on the project. They sent three key project people to the meeting, including the funding project manager, the funding project inspector, and the funding project engineer. The three actively participated in a full day of rigorous charter writing and discussion.

As the signature copy of the charter was being printed, the three, very reluctantly, announced that they were not allowed by the attorney general to sign any partnering documents produced that day. They then packed up and left a rather startled project team. The stakeholders were now wondering if the funding agency would actually support the partnering effort. Erosion of management credibility can result from a stakeholder who is not permitted by his employer to participate fully and to exercise judgment as to whether or not to sign the charter.

Facility type influences the charter-writing process in a manner similar to the influences exerted by the scope of work. However, the boundaries between facility types are often sharper and more definitive than the boundaries set by the scope of work. Types of facilities on which partnering is used range from tunneling through a remote mountain for a new highway to constructing a large federal courthouse in the heart of a major metropolis. An important trade on the tunneling job may be only a minor participant, or not involved at all, on the courthouse facility. Many of the courthouse trades would have no role whatever on the tunneling project.

Yet partnering systems and methods are equally valuable to the tunnel and to the courthouse if the differences between the two are correctly recognized. These differences usually occur in several categories:

- Contract conditions
- Project delivery systems
- Estimating systems
- Types of trades contractors
- Locational demands on subcontractors (on a remote location the project team is generally away from home—on an urban project many of the project team live at home)
- Equipment used
- Planning and scheduling methods and terminology
- Salary scales
- Support staff requirements
- Degree of project management autonomy
- Design emphasis (the tunnel is basically an engineering structure with few architectural design features; a metropolitan federal courthouse is likely to require the application of both outstanding architectural and engineering design skills)

The type and scale of problems encountered and resolved on the tunnel will be different than those encountered and resolved on the courthouse.

These must be taken into account as the charter, the evaluation system, and the issue resolution system are prepared and implemented. As an illustration of how the differences in project type are reflected in the charter writing we can compare the charter written for the design and preconstruction phase of a midwestern health campus facility to the charter written for a large truck terminal in New England. First we should look at the mission statements for each of the two projects:

- *Health campus:* "Design an effective and flexible community-based outpatient centered facility that provides for present and future quality health care services."
- *Truck terminal:* "Through teamwork, cooperation, commitment, and communication, we will build a truck terminal with quality workmanship, on schedule, safely, within budget, and with profitability for all partners."

The two missions are dissimilar. Community service is stressed in the charter for the health facility. The truck facility charter mission stresses schedule, safety, budget, and profit. Both missions fit the type of facility to which they apply very well. Costs and budgets for the health facility are addressed by the following provisions of the charter:

- Maintain control of design costs and construction budgets.
- Prepare and publish design development-based total target cost.
- Prepare and publish FFE budget.
- Prepare and publish life-cycle costing guidelines.
- Prepare and publish preconstruction costing guidelines.
- Prepare and publish payment policies.

In the truck terminal charter the cost problems are addressed specifically by a single provision:

- Review, agree upon, and implement an acceptable payment process at all contractor levels, including change orders and close-out processes.

Other truck terminal charter provisions address cost factors indirectly but nevertheless with unmistakable firmness. Some of these provisions are:

- Do it right the first time!
- All parties will strive to reduce excessive and unnecessary resubmittals of shop drawings and other submittals.
- All parties will strive to close out the project in a timely and efficient manner.

The differences in the two charters are caused to a large extent by dissimilarities in the project type. This causes stakeholders to place emphasis on the factors most important to them when setting the moral ground rules by which they wish to operate.

Suggestions:

1. If possible, make certain that the key participants in the meeting are not prohibited from signing the charter by their organizations or by the nature of the facility. This is not imperative, but it helps maintain the spirit of partnering if all present can sign the charter if they wish.

2. Be certain to inform the stakeholders of the handling and disposition of any confidential information brought up during the charter meeting. Do not assume that everyone present knows the information is privileged—tell them.

3. Recognize the importance of the major trades or disciplines normally found in varying types of projects. For instance, a museum being built in the heart of a very large urban and academic community should have major representation from the architectural design office. A large storm sewer being built through the same area would require major representation from the consulting utility and structural engineers designing and building the project.

4. Try to appoint a discussion leader or facilitator familiar with design and construction of the type of facility for which the charter is being written. A well-founded appreciation of the characteristics of the project is of help in encouraging and critiquing the materials prepared in the workshops. It is also of great help in actually writing the final charter provisions.

9. Project Delivery System Being Used   The project delivery system describes the business structure by which the team is going to carry the work through from concept to turnover and operation of the facility. It affects the operations in each of the four major characteristics of a project agreement:

1. Agreement premises
2. Authority limits
3. Payment method
4. Scope of services

These four characteristics, described in detail in Chapter 2, Section D, establish what is called the project delivery mode during any of the phases along the line of action from concept through construction and completion of the job. The *agreement premises* have a strong impact on the degree of formality with which the partnering system will be initiated. In a design or construction contract won through a highly competitive, hard-money bid effort, the participants will usually guard the expenditure of the fee or fixed price with great care. Neither the design team nor the contractors are going to spend money unless they are satisfied that the return on the investment will justify the effort. Often, the use of new and relatively untried management methods will meet with resistance that in the case of partnering may lead to reluctant or lukewarm participation. This reduces the probability of partnering success.

In negotiated work where the participants are paid for what they do on a unit basis with either a fixed or a variable fee, innovative techniques that show promise of improving performance are often welcome. They frequently will encourage and bring out the team spirit needed for successful partnering. The *authority limits* placed on the design and construction firms also affect the amount of participation in project partnering. If the firms during design or construction are limited to a contractor role,[4] it requires the strong effort and support of the owner to initiate successful partnering. The owner's design and construction consultants and contractors can certainly recommend a partnering system be used. However, that owner must commit the resources needed for the effort.

If either of the design or construction advisors have a partial or, very unlikely, a full agency agreement that allows them to commit for the owner, they can spearhead the partnering effort. However, it is critical to remember that for the partnering system to work the owner's top managers must support it and participate in the process.

The *payment method* set by the project delivery system indicates how tightly the predesign or preconstruction documents must require the participants to dedicate funding to a partnering effort. If we consider that a project partnering effort will require one to two days of focused effort on the part of 20 to 40 key people, the need for a financial commitment becomes apparent. An average cost of middle- and top-management personnel, including direct salaries, overhead, profit, and a consideration of the time lost to other profit-making efforts, might amount to as much as $150 per hour. Just the attendance of 20 people for one day could cost participating companies as much as $24,000. This forces partnering to produce a high value-added result. The investment in partnering must be made worthwhile for the stakeholders to buy into the system.

Finally, the scope of work must be considered. If a potential participant's contract amount is for a very small percent of the total cost of the work, these team members will probably choose not to participate. This could happen where the number of contracts is increased considerably to gain a larger than usual participation of smaller firms. High participation is most likely when the number of major contracts is in keeping with the traditional contract work scope.

---

[4] A contractor is one who agrees to do a certain amount of work, either design or construction, for a specific agreed-upon price. The contractor does not represent the principal as an agent.

Suggestions:

1. If a nontraditional project delivery system is being used, the partnering meeting may be somewhat less formal than with traditional systems that have been competitively bid on a hard-money basis (see Chapter 2, Section D, for a description of the nontraditional project delivery system). In the nontraditional system, the client, architect, and contractor may tend to resist setting out additional guidelines than they feel already exist in the construction documents and understood in their negotiated agreement.

2. Payments to companies at all designer and contractor levels are often more clearly defined in traditional project delivery systems than in the nontraditional systems. The stakeholders in a nontraditional project partnering meeting might be eager to clarify the formalizing of their payment processes, particularly if the group includes consultants to the designer of record or subcontractors to a prime contractor. Such clarification requests are legitimate in most cases.

3. Negotiated work agreements often fail to establish well-identified clearly stated administrative procedures for the project. Plan to spend some time in the partnering workshops discussing organization, communication, submittal processing, and other activities that are usually tightly specified on a traditional hard-money project. They may get pushed aside in the charter discussion.

10. Size of Group Participating   Project partnering with more than 40 people in attendance is difficult to manage well. To start with, it forces the use of mass communication techniques and loses some of the individual attention so important to creating a trustful meeting environment within the spirit of partnering. Partnering strives to create an atmosphere of confidence in the synergistic potential of the full team. If the size of the group brings a loss of individual or group identity, the full benefits of the team approach may be weakened.

The size of the group is strongly affected by the stage of the job. Early line-of-action partnering meetings tend to be more lightly attended than are meetings held when all the participants are under contract. High-visibility projects, such as an urban-area downtown high rise with strong aesthetic appeal to the public and the profession may draw a higher level of attention from the potential stakeholders than more mundane projects such as a pipeline, a police station, or an off-price shopping center. The latter tend to be of limited interest to all except those at high risk, and attendance is likely to be smaller at a routine job charter meeting.

Attendance of all key personnel on a project is critical to the success of a partnering effort. I usually gauge if the attendance need is satisfied by asking the question: "Is there anyone not present at this charter preparation meeting whose decisions are critical to implementing the project?" If the answer is no from the potential stakeholders, the chances are high that the meeting will be productive. If there are missing management decision makers, the charter produced may be less than effective.

The problem of potentially missing attendees should be addressed early in the planning for the partnering meeting. The list of those to be invited should be reviewed carefully and the question asked above of the attendees should be asked of the potential attendees: Are any key decision makers missing from the final list of invitees? If the answer is no, the list can stand as is. If the answer is yes, the list must be revised and augmented until it contains all of those whose decisions and input are vital to project excellence.

Suggestions:

1. Determine the key ultimate decision makers early in the project and make certain they are invited and will attend the charter-writing session.
2. Plan carefully how the visitors and observers at the charter meeting are to be allowed to participate in the workshops. Excessive numbers of nonstakeholders at the worktables tends to inhibit stakeholder discussion. It may also reduce actual stakeholder input into the formulation of the charter. Remember that the purpose of the charter meeting is to write a charter.
3. Be certain that stakeholders and guests have a real interest in the project. Those who attend the meeting should be either supportive of partnering or be conservative in remarks that may damage the chances of partnering success. With very large groups, negative influences may go unnoticed and could damage the potential for success. In these cases the workshops should be watched carefully for such unwarranted disruption (see item 12 below).

11. *Time Available for Charter Meeting and Follow-up Sessions* A considerable difference of opinion surrounds the length of time that should be devoted to a partnering meeting. In today's organizational world, most effective men and women are trying to reduce the amount of time spent on nonproduction work. Meetings are frequently perceived as unproductive, particularly if the matters discussed are not directly applicable to the work of the participants. My experience in chairing many partnering charter meetings indicates that a good charter can be written in one full day. Usually, it is wise to budget at least eight hours or more to the effort and get it done while the subject is fresh and the people are at hand. Once the pattern of continuous discussion is interrupted, or one or more of the stakeholders departs, as often happens when a meeting is adjourned until the next day, the spirit of partnering as a team fades. Achieving partnering excellence is fueled by a sense of urgency about getting the system in place.

Sessions dealing with construction and management techniques ancillary to the partnering system should usually be held separately from the charter writing session. Such subjects include team building, quality management, quality control, qualification-based selection, project management, organizational planning, alternative dispute resolution, and other people and management related subjects. Improved knowledge of these topics can help a stakeholder make a substantial contribution to the process of partnering. However, they are best applied after the base document of the partnering system—the charter—is in place.

Suggestions:

1. For the most part keep the partnering charter-writing session separate from training and educational sessions. They are related but not interdependent.
2. Prepare a well-rounded agenda for the charter writing session and stick to it!
3. If you are leading the workshops, prepare as many standard forms as possible to save time during the meetings. Some of the documents that can usually be prepared in advance include:
   - Partnering information notebook
   - Note-taking template in which to record results of the partnering workshop discussions
   - A list of those attending, showing their names, positions, company name, address, and phone number (the list can then be corrected quickly by the attendees and reissued at the meeting)

12. Visitors to be Accommodated   Partnering efforts tend to be of interest to those who are newly learning the technique or who are attempting to influence others to use the system. Often, we find that these people want to attend the partnering charter meetings or other working sessions as observers. Visitor presence is normally not a deterrent to writing a good charter. In fact, if the visitors and the stakeholders agree, the observers might even be invited to sit in on the various breakout sessions to discuss specific job problems and their resolution.

Of great importance in inviting guests is to ensure that they understand their role at the meeting and that they maintain an observer position only. Usually, the observer or visitor is not privileged to sign the charter or to participate in the task force meetings set to carry out specific activities called for by the charter provisions. Such activities might include preparing and publishing an issue resolution system, developing and implementing a charter evaluation system, or developing a set of soft policies to expand on potential troubles centering on payment practices, paperwork processing, and other support features not fully covered by the contract.

At one recent charter meeting, the financing source for the project was invited by the owner to attend and participate. This was a bit unusual. However, it served the valuable purpose of showing the funding managers what they could expect in the way of project management, team effort, and attention to financial responsibility. Overall, financial participation in the discussions at this meeting added visible credibility to the project for the subcontractors, provided a different outlook to the design team on a key element of project work, and made information available that is normally not provided to active members of the design and construction team.

If observers or visitors are invited and expected to attend, try to involve them in the discussions. Make them feel that what they see and hear is important to all attending the charter meetings. The better understanding others have about how you approach partnering and how you use it to improve performance on your projects will benefit all the stakeholders in the long run.

Suggestions:

1. Encourage visitors to sit at working tables during the problem identification sessions. The visitor who is not a stakeholder may often offer a unique outsider point of view of a situation. Permit the team at whose table he or she is sitting to incorporate the visitor's perceptions if the team so wishes.
2. Provide the visitors with the same handout materials as those given the stakeholders.
3. Encourage the visitors to look over the charter reference set of contract documents and to ask questions of importance to them.

13. Who Will Chair or Facilitate the Meeting   The selection of who is to chair the partnering meeting is vital to the success of the charter writing effort. Most successful partnering efforts are chaired by an outside party retained specifically for that assignment. Chairmen or chairwomen who are drawn from within one of the stakeholder organizations usually come burdened by too many biases, instructions, and preconceived notions to be fully effective in their leadership of the meeting.

An important element when considering a member of any of the stakeholder organizations on the project as a possible chair or facilitator is the potential for creating personality clashes. These may be based on past experiences with or perceptions about the chair. These perceptions may detract from creating the partnering atmosphere so critical in the charter-writing process. An outsider tends to bring an objective sense of balance to the meetings. He or she focuses the attention of the stakeholders on the issues and can direct or divert destructive conflict without being a part of the conflict. In addition, the outside party may bring valuable views of the job that produce fresh ideas about how it might be best managed.

I strongly recommend that the chair of the meeting be someone who is well respected in the design and construction community and who is very knowledgeable about the construction industry. He or she should not have any influencing biases about the project at hand and be able to lead and motivate the group to a writing success within the time frame allowed. A good chair and an enthusiastic body of stakeholders is one of the best assurances that a partnering charter-writing effort will be successful.

Suggestions:

1. Contact your local office of the American Arbitration Association (AAA) or any of the architectural, engineering, or contracting associations in your vicinity for the names of discussion leaders from their lists of standing neutrals.
2. The fees for discussion leaders are sometimes paid in part or full by the insurance companies of the stakeholders. This is particularly true with underwriters of professional liability insurance. Check with your architect, engineers, or contractors about this.
3. Try to vary the partnering leader assignments from project to project. It is possible that exposing the stakeholders on your projects to other facilitators can introduce beneficial changes and newer or different techniques to alternative dispute resolution and group dynamic methods.

14. *Location of Charter Meetings and of Task Force Meeting*   Usually, the briefing meetings, the charter-writing sessions, and the task force meetings should be held close to the job site location. This allows the participants easily to visit the site before or after the meeting if there are field observations to be made that are pertinent to the partnering work of the group. The meeting should be held away from the the actual site itself to minimize phone interruptions and casual drop-in visitors.

Partnering meetings are usually stressful and it is helpful to provide a physical environment that has amenities such as food and drink, pay telephones, and comfortable surroundings. Also, the stakeholders may wish to converse privately with each other during breaks or after lunch or dinner if these spaces are provided. This kind of an environment is found best in a hotel, motel, conference center, or other meeting facility designed specifically to accommodate presentation and workshop meetings.

In partnering charter writing it is necessary to have nearby duplicating equipment for copying partnering material as it is developed in the workshops. Such meeting amenities are readily available in those facilities tailored for group meetings. Other equipment, such as audiovisual equipment, is also more conveniently available and of high quality in the professional meeting locations.

The meeting format is heavily influenced by the location of the meeting. When all facilities are close at hand and a service staff is available to serve the needs of the stakeholders and the meeting leaders, the partnering effort gets off to a running start, with all the up-front and physical operating details handled by meeting experts. Another consideration is the travel needs of the stakeholders and visitors attending the partnering meetings. A location in a motel or hotel near the site provides the time-pressed traveler quick access to other resources that he or she needs to call upon during their trip to the partnering sessions. This is especially critical when the stakeholders come from widely dispersed geographic areas.

Suggestions:

1. To reemphasize, make the meeting location convenient for all stakeholders and visitors attending the sessions. The attendees are friendliest and most productive when they are fresh and working in a convenient location.
2. Plotting the map locations from which the attendees must travel can often give a good view of the best location for a meeting location. Try to equalize the distances or time each participant must traverse to and from the meeting.
3. The meeting should be held (when possible), near a home or branch office of one of the sponsoring organizations. When office help, equipment, or supplies are needed in a hurry, they can be obtained without delaying the meeting.

15. Facilities for Meetings   The facilities available for partnering meetings may range from a small space in a job trailer to a conference room in a fine resort hotel. The place where the charter writing is to be held and the subsequent task force meetings are to take place should be selected carefully. Charter writing should be done in a neutral location, large enough for the group to work comfortably for an entire active day. The room arrangement for full attendee participation should be classroom style. The classroom arrangement has all tables facing forward toward the leader and placed to ensure adequate individual note taking and table work room for each participant. Full attendee meetings are held to brief the entire group and to discuss results of the workshop breakout sessions.

The stakeholders will also be working from time to time in separate workshop teams made up of people with similar business interests and disciplines. For the workshop sessions it is best to use round tables that comfortably seat three to five people. Generally, the number of people per workshop should be kept at either three or four. If space allows, the workshop tables are best located to one side of the room, with the full attendee meetings held on the other side or end of the room. Materials, equipment, and supplies that will be needed for the charter meeting should be assembled ahead of time and a premeeting check made to ensure that the required items are in place or readily available. (see Chapter 10, Section C, for detailed infomation about the materials, equipment, and supplies needed for the meeting).

Suggestions:

1. Discuss the meeting arrangements ahead of time and in detail with the people in charge of rooms, equipment, meals, and staff at the location to be used.
2. When the meeting has been completed, consider tipping the manager or the staff that serviced your conference. You may be back to the facility in the future and the relatively small amount of a tip will help ensure even better service in the future.
3. Try to use facilities that have contributed to the business success of one or more of the stakeholders on the project. This allows a return of favors and helps assure good service at the facility if the business relation was friendly.

16. Technical, Managerial, and Educational Backgrounds of Participants   The level of knowledge and experience of the stakeholders will heavily influence the format and agenda of the meeting. If the project is a very simple job that requires knowledge and experience of only a rudimentary level, those attending should be kept within groups that have similar knowledge and experience. This is particularly so in the breakout and workshop exercises. Here the need is to obtain from those of similar functional interests and backgrounds a good reading of what they as a group feel are appropriate problems and solutions. For instance, a team identifying problems and their solutions in the installation of complex controls and instrumentation should be composed of people having considerable technical training and experience in these systems. Putting a young, inexperienced electrician apprentice in with this group would be counterproductive and perhaps seriously affect the quality of the ideas and recommendations. This would be so particularly if the inexperienced person had strong leadership abilities and a tendency to influence even where his or her experience did not justify such action.

The process of partnering is a synthesis of the knowledge and experience of a number of people who are responsible for implementation of the project plans and specifications. Inadequate experience, education, and training, strongly expressed as a correct opinion and course of action, may not be adequately challenged in a partnering session. Therefore, the partnering leaders and decision makers must make certain an intelligent and workable blend of experience, education, ability, and training characterize the attendees to the partnering session. Further, these must be blended correctly in the breakout workshops so that a dominant but in-error opinion does not overly influence the work of the workshop group.

Suggestions:

1. Balance the experience and managerial levels of the stakeholders invited to attend the charter meeting.
2. Try to include a few bright and promising young people from the various organizations in the charter meeting as either participants if they are connected with the job, or as observers if they are not. This helps ensure continuity of partnering concepts among possible future stakeholders occupying key positions.
3. Try to avoid inviting people who tend to form cliques that might weaken or disrupt the established management and communication channels being set by responsible job management.
4. Be certain that all key ultimate decision makers are invited and attend the partnering meetings. Any decision needed to run the project properly should be available from the group attending the charter-writing session.
5. Encourage higher management personnel to attend the meeting in person. Discourage them from delegating information gathering and note taking to others when critical decision making will be needed in project related matters.

17. *Scope of Partnering Work to be Done*   The scope of work for a partnering effort can range from a relatively modest partnering program during the design period to the institution of a full-scale, wide-ranging partnering effort when construction is under way with all subcontractors on board. In most cases the scope of partnering work is set by the position of the project on the line of action (see Chapter 2, Section A, for a description of the line of action). The project remains relatively constant in size, type, and cost while the scope of the partnering work will vary according to where the project is on the line of action (see Chapter 8, Section B, and Chapter 10, Section A, for examples of how partnering efforts must be tailored to the project position on the line of action).

Scope of work—as defined by the number and range of activities to be considered as the outside boundary of the partnering effort—also influences the format of the partnering preparation sessions. When the project scope is such that the user's work is itself highly important and somewhat independent of the base building contract, such as tenant-installed work in an office building or tenant work in a shopping center, a separate partnering system might be appropriate for each. Those invited to the base building work session would be a different group with a different set of problems than those invited to the tenant work partnering meetings. Actually, the number of people at a tenant partnering session could be larger than for the base building meeting.

Also, spearheading a tenant work partnering meeting might shift from a single-point responsibility as might be furnished by the general contractor in a hard-money base-building construction effort to a single-point responsibility centered on the owner or the manager of the facility. Another center of action for tenant work could be one of the larger prime tenants. If there is a tenant coordinator[5] active on the project, the partnering effort could well be assigned to this man or woman.

Another management focus on partnering in a shopping center project could be the merchants' or tenants' association. This is often composed of active tenants who have a set of business goals similar to those of the developer or owner of the center but considerably different in scope. The tenants need to get their individual space on line as quickly and economically as possible so that it can begin generating sales. The developer needs to get the total center on line as quickly and economically as possible so that it can begin generating rental income. The tenant has a micro need; the developer has a macro need. This produces two different scales of work at separate critical times on the project.

Other situations where partnering systems may be influenced by the scope of work could include:

- Installation of a complex processing system such as a paint line in a new manufacturing plant. Here the installation of the paint line may be sufficiently independent so that it can proceed with its specialized contractor needs considered as a separate construction project.
- Construction of several buildings in a complex, such as a new grade school and high school campus, being managed by one at-risk construction manager and being built by several prime and subcontractor organizations working under the construction manager. In this scenario macro partnering might be appropriate for the total effort, while each separate facility would have its own partnering system developed by the stakeholders on that particular construction project.
- Construction of a new office facility that will be turned over as the work is completed to allow the occupants of an existing facility to move into the new building. Their move will, in turn, allow the vacated space to be remodeled for a move of another set of occupants into the remodeled space. Such occupancy moves are called domino moves. They often are carried out by a separate set of prime or subcontractors than those involved in the main building. This is particularly so if the construction documents for the two separate-but-related programs are not issued at the same times.

---

[5] The title usually given to the development's owner representative responsible for integrating and directing the lease execution, construction process, tenant move-in, and operational startup of tenant spaces in the base building.

- A stormwater retention basin might cost as much to build as a moderate-size college steel-framed multistory laboratory building. However, the trades involved will probably be fewer and of a different magnitude for the retention basin. For instance, the basin will usually have a large ground coverage and sizable depth. This will require a larger amount of earthwork and grading than for the smaller footprint of the multistory laboratory. Retention basins require large amounts of cast-in-place concrete, whereas the laboratory building structural steel frame will require very little cast-in-place concrete. Laboratory interior finishes are completely different from any of the retention basin work. In fact, it is entirely possible that the basin will require none of the trades used for interior finish work in a building.

Suggestions:

1. Early in the project life prepare a charter participation matrix such as those shown in Chapter 8, Section B. The matrix will provide a picture of the density of stakeholder participation at any given phase of the project. From this the potential for value added by partnering can be determined.

2. Select people carefully who are to be invited to the partnering session. For example, an earthwork subcontractor would probably have little interest in attending a partnering effort held during the design period. If a member of the project team is invited but is not involved deeply in the work for which the partnering charter is being written, you may have difficulty getting the person back to the charter meeting in which he or she will play a critical role.

3. Try to obtain a good mix of knowledge, experience, and education when inviting several people from the same organization. Usually, one to five people attend from any one organization, depending on the degree of risk and involvement their organization has in the project and the charter. Above all, be certain that the prime ultimate decision makers, such as the owner, the user, the architect, the engineers, and the prime contractors, are represented in enough depth to make sound business, technical, and operational decisions that may affect how the charter is written. Usually, attendance at any charter meeting should be kept under 35 people.

18. Size of Project Project size often will determine how many stakeholders are needed to write a meaningful project charter. On a $500,000 prefabricated, preengineered metal truck terminal, the number of stakeholders and the respective trades represented who would be invited to the construction phase partnering meeting might include:

- Two from the owner: president and operations manager
- Two from the structural steel, siding, roof panel, close-in contractors: project manager and field superintendent
- Two from the site excavation, utilities, and paving contractors: estimator and field superintendent
- Two from the general contractor: president and project manager
- One from the mechanical contractor: project manager
- One from the electrical contractor: project manager
- One from the architect/engineer of record: principal in charge

Total stakeholders attending = 11 people

If the manufacturing facility were to cost $5,000,000 in total construction and to be built with a conventional structural steel frame, the number and trades to be invited at this same phase of the project could very well include:

- Three from the owner: president, facilities manager, and operations manager
- One from the structural steel and metal deck fabricator: sales representative
- One from the joist fabricator: sales representative
- Three from the structural steel, joist, and metal deck erector: president, project manager, and field superintendent
- Two from the site excavation contractor: estimator and field superintendent
- One from the site utilities contractor: project manager
- One from the paving contractor: project manager
- Three from the general contractor: president, project manager, and field superintendent
- One from the painting contractor: project manager
- Two from the heating, ventilating, and air-conditioning contractor: project manager and field superintendent
- One from the plumbing contractor: project manager
- One from the fire protection contractor: superintendent
- One from the electrical contractor: project manager
- Four from the architect/engineer of record: principal in charge, project manager, mechanical engineer, and electrical engineer

Total stakeholders attending = 25 people

The number and disciplines of those attending a partnering session will vary greatly depending on the true needs for input to the charter process. As mentioned previously, I usually request the person in charge of selecting participants to ask the crucial question of the lead project management team: "Is there anyone not invited that can make a final binding decision in any critical project-related matter?" If the answer is yes, I inquire further to ensure that all key decision makers are invited and are going to be there.

Partnering success demands that top decision makers be active participants in the process of building the partnering system. The size of the project usually determines how close these decision makers are to the actual day-to-day work on the project.

Suggestions:

1. The size of the project will often determine how far the ultimate decision makers are managerially located from day-to-day management of planning, design, and field operations. On very large projects those with the most clout might be farthest away from the arena in which the stakeholders must operate. You must make certain that those you invite to a large-cost project charter meeting will exercise their management strength by delegation to lower echelons of management in partnering matters. It is better to have third-tier managers who will accept the responsibility for managing lower-tier operational work attend the charter meetings than to have second-tier managers who will ultimately have to delegate operational functions affecting the partnering system.

2. On large-dollar-value projects it might be necessary to have more partnering charter meetings than on smaller jobs. A well-prepared project participant matrix will help determine the optimum points at which to write and revisit the partnering charter. A good rule of thumb is to write a charter just as the operational work to be partnered is about to begin. In smaller work there are usually fewer major phase points than on very large jobs. This difference in cost might strongly affect the entire partnering structure.

10

BUILDING A PARTNERING SYSTEM

SECTION A: THE OLANTA DATA SYSTEMS EXPANSION PROGRAM—A CASE STUDY IN PREPARING PROJECT CHARTERS

A successful partnering charter meeting requires good planning. The Olanta Data Systems expansion program described below is a composite picture of many projects upon which partnering has been used. It embodies most of the ingredients encountered in live projects, and the conditions, characteristics, and people described in it give a very real picture of how partnering charters are actually prepared. In Chapters 10 to 13 I shall use it to help describe how a project team for a successful business enterprise brings a partnering charter and its related systems into being. The descriptions should help stakeholders and other participants move through the entire process with confidence and with a good understanding of all elements of project partnering. The company selected for the case study is a moderate-size communications-service private business organization in Travis, North Dakota. It bears the name of the family that founded it, Olanta Data Systems, Ltd.

Company Background Olanta Data Systems, Ltd., usually known by its initials, ODS, is a very successful, privately owned business devoted to providing economic-related data to management throughout the world. Olanta collects, classifies, analyzes, and disseminates hard, highly statistical, and very

176

accurate information in 80 countries. Its clients include private businesses, political units, educational institutions, and organizations of every type.

Olanta's success is rooted in the ability of its operational staff to identify current and potentially successful information-using organizations early and then to enter into well-crafted service agreements with them. Most of Olanta's income stems simply from the fees it is paid to provide authentic, accurate, timely information, and analyses specifically tailored to bring rapid success to these organizations using its services. The information that ODS collects, processes, and uses is concerned primarily with how information, wealth, value, currency, or other equivalents are used to allow organizations to interact with the risk marketplaces in which its clients do business.

The company maintains information collection and market operations from field offices located in 25 cities worldwide. Their home office is in Travis, North Dakota. Travis has a population of 120,500 people. Of these, 1500 work for the Olanta corporation. Data are collected continuously by active representatives in the 25 offices. Information is coded, classified, and forwarded to Olanta–Travis daily to use in updating existing files and supplying new material for Olanta's clients. About 50% of the data received from the field offices are transmitted electronically. All data analysis is done at the home office and dispatched to the point of use electronically or by mail, courier, or special messenger. Sixty percent of the dispatch volume is electronic.

The mission of the company as defined by its founders, Charles and Lee Olanta, is: "To derive useful micro to macro economic information from data collected worldwide, and to provide this information to our clients in highly accurate, easily used, and high value-added form." The company is family founded, owned, and operated. Family members have been actively involved in the direction of the firm for 55 years.

Facility Background   ODS found that it has reached the limit of its effectiveness in its existing facility. The increasing need for rapid communications, quick response times, and the resultant new equipment complexity and requirements helped the Olantas make the decision to expand. The business volume of ODS had increased to a point where their executive staff have decided to make a major commitment to continued growth in the community.

The company plans to construct a new building and to occupy it in stages, gradually vacating the existing building division by division. As the larger blocks of moves are made to the new building, the existing building will be remodeled. The proposed new facility is to be located adjacent the existing building. The Olanta site consists of 50 acres of gently rolling terrain on the outskirts of Travis. The existing facilities have been well maintained by Olanta and are in excellent shape. Most maintenance on the building has been by a full time in-house maintenance and operations staff.

Relations between Olanta and the community are excellent and the general business climate in the town is fair to good. There is some mild and often humorous mystery and confusion among the Travis residents and businesspeople about what ODS actually does. However, the firm is a good customer of local businesses and circulates a good share of its income in the community. Recently there has been an increase in the volume of design and construction work in the Travis area, and this has brought several new architects, engineers, and contractors to town. Business relations among people in active design and construction are generally good. However, increased numbers of construction practitioners has increased the potential for conflict on upcoming planning, design, and construction programs in the region.

The management members of Olanta's project team have discovered in their work to date on the new office building that they can work most effectively by identifying and resolving destructive conflict early in the process of implementation. Olanta's facility staff, user group, and its architect–planner, Loring and Metzer, assembled the project program, the early pro forma cost analyses, and the financial plan several months ago. These were presented to the Olanta board of directors, who approved the backup material and authorized preparation of contract documents for the new building.

In its most recent meeting on January 3, the board approved construction of the new office building, including award of a prime construction contract to Tiltsen and Greene, general contractors. Tiltsen and Greene were also selected as construction consultants for remodeling of the existing building. The board gave the project a high funding priority, which means that it is of critical importance to Olanta.

The new building is expected to take about 25 months to complete and occupy. Moves from the existing building are expected to start in about the twenty-first month of the new building construction and to continue for about another 14 months. This gives a total construction period of about 35 months. Field work on the new building is expected to begin in March and to continue over three winters. The project is the largest commercial design and construction program ever proposed for Travis, and local architects, engineers, and contractors are excited about the project.

The Olanta executive staff—at the urging of Travis Loring, the architect of record—has also given serious thought to the use of project partnering. Olanta has been using strategic partnering agreements for many years and is convinced that project partnering can be applied successfully to the new facility's design and construction. They have asked Henry Leinenklugle, ODS's facilities manager, to investigate project partnering and provide them with recommendations and a specific work plan by which a program of project partnering can be implemented. Mr. Leinenklugle has suggested that partnering be considered as the project moves into active construction. The working drawings are complete and although he feels that the partnering effort could have been used at the start of programming and the initiation of schematic studies, he believes that there are still four other phases of the project at which a charter would be of great help:

1. Now, as phase 1 construction of the new building is about to begin
2. At the beginning of the phase 1 occupancy of the new building
3. At the start of phase 2 remodeling of the existing building
4. At the start of the phase 2 occupancy of the new building

The four phase points with the participants involved in each are shown in Figure 10.1. Shaded vertical columns 7, 8, 10 and 11 are the points in the expansion and remodeling identified by Mr. Leinenklugle. These occur where either the nature of the work changes considerably or where new key players are brought into the project. The matrix in Figure 10.1 has been submitted to the Olanta management, and all agree it represents a desirable and contributive project partnering target to achieve. Mr. Leinenklugle has been asked by Lee Olanta to plan the first of the partnering efforts shown in column 7 of Figure 10.1. The actual work to be done in arranging for the effort by Mr. Leinenklugle is described in Figure 10.2. A detailed description of this plan of action is given later in this chapter.

Proposed facilities consist of two major elements, a new building and a complete remodeling of the existing building. Details of the improvement program are outlined below.

Physical Characteristics of the Project

- *Phase 1 work:* new office and data processing center building
  - To be built on a part of the site currently being used for employee parking
  - 200,000 square feet on three floors and a lower level
  - Drilled-in caissons, and grade beam foundations
  - Steel-framed structure with concrete floor
  - Patterned masonry exterior skin and panelized curtain wall
  - Full amenities for employees and visitors
- *Phase 2 work:* existing building, to be remodeled after third and second floors of new building are occupied
  - 160,000 square feet on two floors and a lower level
  - H pile driven foundation with grade beams
  - Structural steel frame with concrete floors
  - Plain face brick exterior skin with punched windows; good appearance
  - Minimal amenities for employees and visitors
- *Phase 2 work:* remodeled building scope of work
  - Each floor to be gutted and remodeled completely
  - Full amenities to be provided for employees and visitors; amenities to be compatible with new addition
  - Exterior skin fully renovated, pointed, and cleaned

stakeholder \ function	col 1 conceive	col 2 program (comp)	col 3 finance	col 4 schematic (comp)	col 5 project approval	col 6 des dev & wk dwg p1	col 7 const phase 1	col 8 occupy phase 1	col 9 des dev & wk dwg p2	col10 const phase 2	col 11 occupy phase 2
00.01 - Olanta facility staff	X	X	X	X	X	X	X	X	X	X	X
00.02 - Olanta user staff	X	X	X	X	X	X	X	X	X	X	X
00.03 - Loring and Metzer - Architects & Engineers of record	X	X		X	X	X	X	X	X	X	X
00.04 - Toonk & Smith - Structural Engineers		X		X		X	X		X	X	
00.05 - Frank Wilson & Sons - Mechanical & Electrical Engineers		X		X		X	X	X	X	X	X
00.06 - Varlent Engineering - Civil Engineers		X		X		X	X	X	X	X	X
00.07 - Strendel - Geotechnical Engieers & Testing		X		X		X	X	X	X	X	
00.08 - Mechelct - Mechanical & Elect. Balancing & Commissioning - Phase 1				X		X		X	X		X
00.09 - Tiitsen & Greene - Construction Consultants & Advisors - Phases 1 & 2	X	X	X	X	X	X			X		
00.10 - Tiitsen & Greene - General Contractors - Phases 1 & 2							X	X		X	X
00.11 - Brown Mechanical - Mechanical Contractors - Phase 1							X	X			
00.12 - Powers Electric - Electrical Contractor - Phase 1		X			X		X	X			
00.13 - Efficiency Design - Fixtures & Furnishings Contractor - Phase 1							X	X			
00.14 - Subcontractors for Phase 2 construction work										X	X

X - indicates sufficiently high involvement to participate in partnering effort as stakeholder. Shaded columns indicate phases of project where partnering charter meetings should be considered.

Figure 10.1 Olanta Data Systems expansion, Travis, North Dakota: project phase. X indicates sufficiently high involvement to participate in partnering effort as stakeholder. Shaded columns indicate phases of project where partnering charter meetings should be considered.

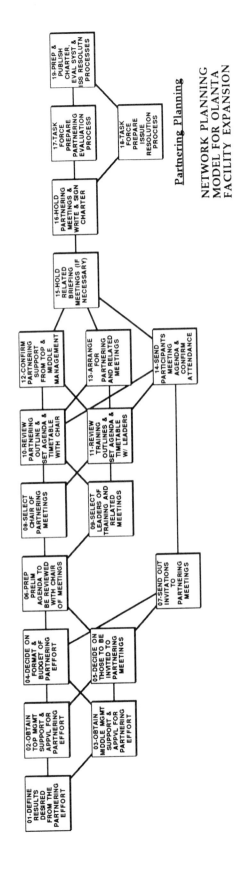

Figure 10.2 Network model for partnering planning work, Olanta Data Systems, Ltd., Travis, North Dakota.

- *Phases 1 and 2:* site work for new office and for remodeled building
  - Construct new surface parking lots, 1050 cars.
  - Rebuild existing surface parking lot, 500 cars.
  - Construct new retention pond
  - Construct new employee recreation area

Those Involved in Project (all have been requested to attend the partnering meetings)

- *ODS staff:* owner and user
  - Charles Olanta, president and chief operating officer, ODS
  - Karl Largo, vice president of operations
  - Henry Leinenklugle, facilities manager
  - Peter Blasco, security manager
  - Sylvia Goldsmity, office manager
  - John Tarkington, data processing manager
  - Joyce Hallmark, public relations manager
- *Loring & Metzer:* architects of record
  - Travis Loring, president and chief operating officer, architect
  - Ling Metzer, architectural designer, architect
  - Lisa Jones, project manager, architect
  - Al Comstock, field inspector, architect
  - Patrica Eames, interiors, interior design
- *Toonk and Smith, Inc.:* consulting structural engineers
  - Marc Smith, principal and project manager
- *Frank Wilson and Sons, Inc.:* consulting mechanical and electrical engineers
  - Frank Wilson, Jr., project manager, mechanical
  - Crystal Strode, project manager, electrical
- *Varlent Engineering, Inc.:* consulting civil engineers
  - Eugene Knowlton, civil engineer
- *Mechelct:* mechanical and electrical balancing and commissioning
  - Fred Taylor, project engineer
  - Claret Ellis, project engineer
- *Strendel:* geotechnical engineering and testing
  - Tobin Strendel, vice president
  - Mark Jones, field and project engineer
- *Tiltsen and Greene:* general contractor
  - Pat Greene, president
  - Charles Klamber, vice president of operations

- Marla Jinns, project manager
- Charles Flyer, field superintendent
- *Brown Mechanical:* mechanical contractor
  - Lincoln Brown, president
  - Raymond Strayhorn, estimator and project manager
  - Ted Cherbourg, field superintendent
- *Powers Electric:* electrical contractor
  - Gerald Powers, president
  - Peggy Roethler, project manager
  - Alfred Joiner, field superintendent
- *Efficiency Design, Inc.:* fixtures, furniture, and equipment contractor
  - Larry Ibis, president
  - Mable Tattler, project manager for design and installation
  - Dean Albertson, field superintendent
- *Dining Specialists, Inc.:* food service equipment
  - Harry Wolfson, vice president and project manager
- *Datacomp, Inc.:* computer systems contractor
  - Mary Charles, computer hardware project manager
  - Mavin Olelsky, computer software project manager
- *Chair of meeting* (chairman, facilitator, and meeting leader)
  - Robert Shea, engineering consultant

Contract Parties and Types

- *Olanta Data Processing Systems, Inc.:* owner and user
- *Loring & Metzer:* architects of record
  - Contract with ODS
  - Partially qualified, selected and negotiated from prequalified list prepared by ODS facilities manager
  - Authority limits: as limited agent
  - Payment method: payroll costs × 2.75, plus expenses with cap on total cost
  - Single responsibility for building and site design work, all in-house with outside consultants; computer systems design work, furniture and fixtures design, and food service equipment contracted separately by ODS
- *Toonk and Smith, Inc.:* consulting structural engineers
- *Frank Wilson and Sons, Inc.:* consulting mechanical and electrical engineers

- *Varlent Engineering, Inc.:* consulting civil engineers
  - Contracts ,with Loring and Metzer, architects and engineers of record
  - Partially qualified, selected and negotiated from prequalified list prepared by Travis Loring, and Henry Leinenklugle, facilities manager for Olanta
  - Authority limits: as contractor
  - Payment method: payroll costs × 2.75, plus expenses with cap on total cost
  - Single responsibility for building and site design work of their disciplines
- *Mechelct:* mechanical and electrical balancing and commissioning
- *Strendel:* geotechnical engineering and testing
  - Contract with ODS
  - Partially qualified, selected and negotiated from prequalified list prepared by Travis Loring, Pat Greene, and Henry Leinenklugle, facilities manager for Olanta
  - Authority limits: contractor
  - Payment method: time and material with fixed fee and guaranteed maximum price (GMP), no share in savings over GMP
  - Single responsibility: provide all labor and materials for work under their contract
- *Tiltsen and Greene:* general contractor
  - Contract with ODS
  - Partially qualified, selected and negotiated from prequalified list prepared by Mr. Olanta and facilities manager
  - Authority limits: contractor
  - Payment method: time and material with fixed fee and guaranteed maximum price (GMP), share in savings over GMP—80 to ODS : 20 to T&G
  - Single responsibility: provide resources or manage all subs to provide resources for all building and site construction work
- *Brown Mechanical:* mechanical contractor
- *Powers Electric:* electrical contractor
  - Contract with Tiltsen and Green
  - Partially qualified, selected and negotiated from prequalified list prepared by Pat Green and Mr. Leinenklugle, facilities manager
  - Authority limits: contractor
  - Payment method: time and material with fixed fee and guaranteed maximum price (GMP), share in savings over GMP—80 : 20
  - Single responsibility: manage all subcontractors and suppliers to them to provide and install labor and materials for all building and site work

- *Datacomp, Inc.:* computer systems contractor
  - Contract with ODS
  - Partially qualified, selected and negotiated from prequalified list prepared by data processing manager
  - Authority limits: as contractor
  - Payment method: fixed cost
  - Single responsibility: provide all management, design, materials, and equipment, and install all materials and equipment
- *Efficiency Design, Inc.:* fixtures and furniture contractor
  - Contract with ODS
  - Partially qualified, selected and negotiated from prequalified list prepared by Mr. Olanta
  - Authority limits: as contractor
  - Payment method: time and material with fixed fee and guaranteed maximum price, no share in savings
  - Single responsibility: provide all management, design, fabrication, and installation of fixtures and furniture
- *Dining Specialists, Inc.:* food service equipment
  - Contract with ODS
  - Partially qualified, selected and negotiated from prequalified list prepared by Mr. Olanta and Henry Leinenklugle
  - Authority limits: as contractor
  - Payment method: time and material with fixed fee and guaranteed maximum price, no share in savings
  - Single responsibility: provide all management, design, fabrication, and installation of food service equipment
- *Robert Shea:* consultant and chair for partnering work
  - Contract with ODS.
  - Partially qualified, selected by Mr. Leinenklugle, Mr. Loring, and Mr. Green
  - Authority limits: as contractor
  - Payment method: time and material for work as defined by agreement
  - Single responsibility: provide all consulting partnering services as required by Mr. Leinenklugle

Status of Project as of February 20, 19xx

- Contract documents for new building complete
- General construction contracts for new building awarded
- Construction subcontracts for new building awarded
- Testing contracts for new building awarded

- Remodeling for existing building in design development
- Construction consultant contract (professional services) for existing building awarded to general contractor for new building, to be converted to guaranteed maximum construction contract as design proceeds
- Specialty subcontractors for new building in favored position for existing building remodeling, if they perform well on new building; this is well known by the subs

SECTION B: PREPARING FOR THE PARTNERING CHARTER MEETING

Once the commitment to partnering is established and the other influences on partnering success have been weighed and considered, it is necessary to initiate active planning for the partnering effort. The steps to be taken in properly preparing for a partnering system are shown in the nonquantified network model, Figure 10.2. The activities shown are arranged in sequence with each activity being restrained by, or restraining, one or more other activities.

A brief description of each action in Figure 10.2 is given below.

*Activity 1: **Define the results desired from the partnering effort*** The partnering effort should be carefully planned so as to be as predictable as possible. Before embarking on the preparation of a partnering system those responsible for the work should conduct adequate study and research to determine what benefits and products of value will be added to the program by partnering. Some of the benefits of partnering are shown in Chapter 7, Section A. If we are considering using partnering on a new service industry main office addition and remodeling, we would prepare a statement of results we expect as part of the background material for determining the value that might be added by partnering.

Following is the statement of results expected whose preparation is described in activity 1 of Figure 10.2. It was prepared by the Olanta facilities staff to introduce the material being assembled for executive staff information.

The use of a partnering system on the building expansion and remodeling program for Olanta Data Systems, Ltd. is designed to accomplish the following:

1. Allow stakeholders to meet the other stakeholders on the project and become better acquainted with the role each must play and the work to be done by them.
2. Provide an issue resolution system that stimulates early settlement of disputes at the originating level and in a manner that avoids residual destructive working relations.

3. Provide a forum for all decision-making levels of project management to meet, discuss behavior on the job, and reach mutual agreement on job-related noncontract matters.
4. Improve the methods and quality of communications among people and organizations on the job.
5. Determine early in the project how various responsible individuals feel they should interact with other stakeholders to achieve their individual and organizational missions by completing the expansion program successfully."

This clear, succinct statement of desired results can be a powerful argument to the Olanta board of directors for establishing and implementing a partnering system. The results statement should be carried through into the partnering meeting so that participants have a solid owner-based concept upon which to build their discussions.

Activity 2: Obtain top management support and approval for the partnering effort Partnering can succeed only when it has the support of top decision makers in the sponsoring[1] organizations. The support can be gained only by the sponsors providing the decision makers with sufficient proof that the project will benefit from the use of partnering. This should start by preparing and submitting the statement of results expected as described in activity 1. Often, approval of top management will be informal and contingent on the quality of the preparations for partnering and the support the approval gains from potential stakeholders and their management.

Activity 3: Obtain middle-management support for the partnering effort (*Note:* Activities 2 and 3 may follow each other, be concurrent, or may even, in some instances, be reversed in the sequence: it all depends on organizational and project characteristics.)

Support from middle management is critical to partnering success primarily because it is middle management that must implement the system. The sponsor should make certain that adequate information is provided to the middle management staff concurrently with gaining top management support. If those affected wish to attend, it is certainly appropriate to hold one or two internal orientation sessions on partnering.

Additional support may be gained for the use of partnering if some early discussion and information is conveyed about related subjects. Topics might include basic information about team building, quality management, project management, or other useful middle-management subjects. Once top and middle management support is given in activities 2 and 3, the work of formally planning the partnering effort can proceed.

---

[1] A sponsor is a person or group who supports or champions an activity or assumes a responsibility for an action plan.

Activity 4: Decide on the format and budget for the partnering effort
Influences on the format and the methods of establishing how best to structure
the partnering effort are reviewed in detail in Chapter 9. Careful attention
should be given to matching the meeting format to the factors described in
Chapter 9. Successfully matching actual conditions to the influence of these
factors is critical to maintaining the ongoing interest of stakeholders, and
effective implementation of partnering throughout the project life.

It is also good at this point to consider if the cost of the partnering effort
is to be shared with others on the project. If it is, the respective financial and
other responsibilities for those sharing should be reviewed and set with those
concerned. Costs of the partnering program vary considerably depending on
factors discussed earlier in the chapter. The expense components of partnering
usually include:

- *Leader fees.* These vary with the skill and the abilities of each person.
  Professional leader fees will run from $80 to $130 per hour. The charter-
  writing session should be completed in a one-day session of eight to 10
  contact hours. Normally the leader or facilitator will require from four
  to eight hours of preparation time to prepare handout material and plan
  the charter meeting.

    Training sessions, if held, should be tailored to the need and paid
  for by those who actually are interested in partnering, management, or
  technically related instruction. Special training sessions are usually very
  expensive. I suggest that the sponsors of the partnering effort concentrate
  on the project partnering portion of the effort unless they have a competent and experienced in-house trainer and coach.

- *Leader out-of-pocket expenses.* For an out-of-town facilitator, expenses
  will usually include air fare, car rental, mileage allowance, meals, tips,
  lodgings, and similar expenses. These vary considerably from place to
  place and will have to be estimated for each case.

- *Stakeholder social expenses.* Before the partnering charter meeting starts
  many sponsors like to provide the opportunity for a early social event.
  Attendees are often nervous, might not know many of the others on the
  project, and may like to get familiar with their surroundings before the
  meeting. A good morning ice-breaker is a roll and coffee stand-up snack.
  During the day I suggest that two full coffee breaks be provided, one in
  the morning and one in the afternoon.

    After the the heavy work of the day is done and the charter is signed,
  it is often a good idea to have a complimentary social hour for the
  stakeholders. The day's work is stressful, occasionally there are conflicts
  and arguments as issues are resolved, and there is the constant pressure
  to produce a superior charter. The welcome change of pace afforded by
  some relaxed socializing can do wonders for the morale of those participating.

- *Operating and room expenses.* These include the costs for rental of the meeting room, rental of the audiovisual equipment needed, coffee breaks, and expenses such as copying or other special services for which the conference facility charges a fee.
- *Videotaping and sound recording expenses.* Occasionally, the owner or other stakeholders will want a video or voice record of the proceedings. The equipment used can vary from simple internally owned and operated recording machines to very sophisticated, professionally operated recorders and cameras. Costs vary according to the product quality desired, the quantity of hours to be expended, and competence level of the recorder.

  On very significant projects, sponsors might wish to create a graphic history of the entire process of partnering for public consumption. Or, when a sponsor is conducting project partnering for the first time the stakeholders might wish to videotape the charter meeting to provide other members of their organization with some instructional and orientation material for use in the future.
- *Related workshops.* Many sponsors and stakeholders involved in a project partnering effort desire additional information to help them use the partnering system in the most effective manner. These supplementary training and information sessions can be held either before or after the charter has been written in the sessions being discussed here.

  Some of the other subjects which are related to project partnering but which are not usually covered in the one-day partnering charter meeting include:
- Team building
- Strategic partnering
- Quality management
- Project management
- Effective field administration
- Personal communication skills
- Time management
- Effective leadership
- Alternative dispute resolution

  These are only a few of the partnering-related topics that possibly can help improve individual, team, project, and organizational skills, and consequently improve the probability of project success. Extra training is expensive and the value gained must be carefully balanced by the expense of training as well as the staff cost for time lost to current productive work.
- *Partnering performance evaluations.* Stakeholders or their designated representatives should meet regularly during the course of the project work to evaluate the status and quality of the project work. The meetings

should be scheduled well ahead of time, preferably at an off-site location. One organization, deeply and successfully involved in partnering implementation, hosts breakfast at a local hotel where the stakeholders meet each six weeks to review the grades they have given the project in the period just past. The sponsor pays the cost of the meeting room and the breakfasts. It has proven to be a good investment that the sponsor credits for a return on the investment of several times the cost of the effort.

Activity 5: Decide on those to be invited to the partnering meeting   As part of this activity the sponsor should prepare a mailing list of the potential attendees, their positions, organizations, addresses, phone numbers, fax numbers. Also included should be any data about each stakeholder's knowledge, experience, abilities, education, special training , and credentials. This special information can often be of help in utilizing the stakeholders' talents in making the partnering program more effective.

Activity 6: Prepare a preliminary agenda for the partnering meeting   The partnering agenda should include a draft of the suggested meeting plan for the charter-writing meeting. The agenda items should include outline drafts for any related orientation meetings to be held in the interest of improving the partnering effort. The agenda outlines should be explicit enough to discuss with whomever you, as the sponsor, select to lead each of the meetings. The number of related meetings to be held to complement the charter-writing session should be kept to a minimum. Spacing related meetings, which are often instructional in nature, allows stakeholders' main effort to be fully focused on the objective of the charter meeting—*to write and sign a sound, workable charter.* Providing time between meeting also allows the attendees a mental absorption time for the instructional material to be considered and understood.

A discussion draft of the Olanta project agenda sent to Mr. Olanta and the facility task force for approval is shown below.

Date: January 15, 19xx

Re: Discussion draft of suggested agenda for project partnering meeting for
        Olanta facilities expansion program.

To: Mr. Charles Olanta, president and chief operating officer, and building task
        force members

From: Henry Leinenklugle, facilities director, Olanta Data Systems, Inc.

As we discussed in our recent design and construction review meeting, below is a draft of the proposed partnering meeting agenda for the Olanta home office expansion and remodeling. Those items yet to be fully decided on are shown in parentheses. Please return the agenda with your comments by January 25, 19xx. Thank you for your cooperation and assistance in this important effort.

Agenda Draft

Location of meeting: Snowbird Inn, 3233 Carlsbad Road, Travis, North Dakota

Date of meeting: February 24, 19xx

Those attending: To be determined; will consist of representatives from each of the major participants in the program

Meeting chairman: Robert Shea

Purpose of meeting: To develop and adopt a partnering charter for the guidance of Olanta's project team.

Results expected from meeting: The use of a project partnering system on the building expansion and remodeling program for Olanta Data Systems, Ltd. is designed to:

1. Allow stakeholders to meet the other stakeholders on the project and become better acquainted with the role that each must play and the work to be done by them
2. Provide an issue resolution system that stimulates early settlement of disputes at the originating level and in a manner that avoids residual destructive working relations
3. Provide a forum for all decision-making levels of project management to meet, discuss behavior on the job, and reach mutual agreement on job-related noncontract matters
4. Improve the methods and quality of communications among people and organizations on the job
5. Help determine early in the project how various responsible individuals feel they should interact with other stakeholders to achieve their individual and organizational missions by successful completion of the expansion program

Timetable (preliminary, for review and comment)
7:00 to 8:00 A.M.   Rolls and coffee, River Bend Room, first-floor level
8:00 to 10:00 A.M.

- Introductions by Charles Olanta, Pat Greene, and Travis Loring
- Brief review of project characteristics by Henry Leinenklugle and Travis Loring
- Orientation and review of day's activities by meeting chairman
- *Breakout workshop 1:* "What actions do others take that create problems for us?"

10:00 to 10:15 A.M.   Coffee break
10:15 A.M. to 12:00 noon

- *Breakout workshop 2:* "What actions do we take that cause problems for others?"
- *Breakout workshop 3:* "What is the most important achievement for my organization to gain by completing this project successfully?" This is an individual mission statement by each stakeholder.
- *Breakout workshop 4:* "In light of the problems stated in workshops 1, and 2, and the missions defined by this team, what recommendations can we make that could help improve relations and performance on the Olanta Data Systems project?"

12:00 noon to 01:00 P.M.   Lunch, Dakota Room, second-floor level
12:30 to 1:30 P.M.   Mission-writing task force (overlaps workshop 4).

- Concurrent with the latter part of lunch and workshop 4, a selected task force of three stakeholders will meet apart from the main stakeholder group to prepare a first draft mission statement for the project.

1:00 to 3:00 P.M.   (overlaps with mission-writing task force work)

- *Breakout workshop 4:* "Recommendations," continued to 1:30 P.M.
- *Stakeholder workshop 5:* Mission-writing task force presents first draft of project mission to full stakeholder group for discussion.
- *Stakeholder workshop 6:* Full stakeholder group set specific project objectives from the results of workshops 1, 2, 3, 4, and 5. These are to be designed to help ensure that excellent relations and performance are maintained on the Olanta project.

3:00 to 3:15 P.M.   Refreshment break
3:15 to 4:00 P.M.

- Stakeholders continue workshop 6, merging results of previous workshops into a first-draft charter and refine mission statement.

4:00 to 4:55 P.M.

- *Full stakeholder workshop 7:* Combine all previous workshop discussions into a charter.
- Print final draft of charter.

5:00 P.M.   Sign charter and receive award memento.
5:15 P.M.   Adjourn to social hour celebration sponsored by Olanta, and Tiltsen and Greene.

The tentative outline agenda is sent to Mr. Olanta and the building task force members for comments. After their review the draft is corrected and enclosed with the invitation letter described in activity 7 and sent to the potential stakeholders.

*Activity 7: **Send out invitations to the partnering meetings*** The invitation letter to potential stakeholders is a critical document in setting the stage for the partnering charter meeting. The letter should be from the highest executive in the major decision-making organization connected with the project. In most cases this is the chief operating executive of the conceiver, the owner, or the user. Occasionally, the invitation letter might come from the chief operating executive of the principal prime contractor. However, in the Olanta project, since many of the recipients are within the client's organization, a letter from its top executive is a powerful stimulus to middle management acceptance and commitment to partnering within the client organization.

If the agenda has been set by the time the invitation is sent to the potential stakeholders, it can be included with the letter. However, the invitation should be extended to the potential stakeholders far enough ahead of the meetings to allow them to plan their time conveniently. Usually, at the start of the project most of the project team is busy, issuing drawings, letting subcontracts, ordering materials, getting the job started in the field and other critical job-related actions. Providing team members with adequate lead time so they can plan properly for the partnering sessions is an appreciated courtesy. The early notification may also help gain support for the entire partnering effort.

A copy of the invitation from Mr. Olanta to those invited to the charter meeting on the Olanta facilities program is reproduced below. Note that a current version of the meeting agenda is to be enclosed.

Sample Invitation Letter to Partnering Charter Meeting

Charles Olanta
President, and Chief Operating Officer
Olanta Data Systems, Ltd.

January 31, 19xx

Ms. Mary Charles, Project Manager
Datacomp, Inc.
239 Federal Street
Rapid City, South Dakota 07072

Re: Implementation of partnering system for Olanta expansion and remodeling

Dear Ms. Charles:

The Olanta Data Systems, Ltd., staff and I request your presence on February 24, 19xx, at a meeting in the Snowbird Inn to prepare a partnering charter to guide construction of our new headquarters building and the remodeling of our existing facility.

Enclosed with this letter is a packet of partnering explanatory materials. I would appreciate it if you would read this material prior to the meeting. Mr. Pat Greene of Tiltsen and Greene, the general contractor on our project, tells me that you have already attended a briefing session on partnering conducted by his project management staff.

I fully support partnering—as do Olanta's senior executives and its facilities staff, Mr. Travis Loring, president of Loring and Metzer, the architects/engineers of record, and Mr. Greene.

Below are outlined the meeting objectives and agenda for the January 9, 19xx session. We appreciate your interest, participation, and efforts to help improve the probability of success for you and for us on this significant project.

                                                            Sincerely yours,

co:see   Charles Olanta
                                                            President and Chief
                                                            Operating Officer

---

Activity 8: Select a chair for the partnering meetings   The chair for a partnering charter meeting is sometimes called a facilitator. This title tends to blur the true responsibility held by the chairman or chairwoman in a partnering effort. It is a task that requires objective, skilled, and intensely focussed attention on the end product, a workable partnering system. We have already suggested that a facilitator is best retained from outside the stakeholders' organizations so as to bring unbiased leadership and thinking to the project team. To do this, the facilitator must:

- Have an in-depth knowledge of the design and construction business and profession
- Have a reputation for fairness and a balanced use of judgment
- Know how to exert leadership at times when consensus and facilitation are about to fail
- Have the endurance of a draft horse to keep alive the faith in the mission when the stakeholders are on the verge of quitting

People with these attributes are not easy to locate.

Occasionally, we are tempted to use internal resources from one of the organizations. The caveat again is this—it is extremely unlikely that anyone with a sufficiently unbiased outlook will be found in any of the stakeholders' organizations. An exception might be found where the sponsoring organization is large enough to have functional divisions far removed from the project. Here the party selected must still be fairminded and possessed of the technical, professional, and soft skills needed to lead a part nering effort.

Several independent organizations have compiled lists of those they consider capable of acting as partnering leaders and neutrals. Some of these organizations are:

- The Associated General Contractors of America
- The DPIC Companies, a professional liability carrier
- The American Arbitration Association
- The Michigan Society of Professional Engineers
- Construction Industry Dispute Avoidance and Resolution Task Force

Selection of an appropriate partnering meeting chair could spell the difference between success and failure of the system. The selection must be wisely made. One of the most reliable systems is to discuss your needs with those in the industry who you trust and respect. Generally, if you have maintained your industry and professional network as it should be, someone of your acquaintance will know, or know of someone who knows, a chair and leader who fits your special requirements.

Activity 9: Select leaders for training and orientation in partnering-related subjects Factors similar to those to be considered in selecting a partnering leader also apply in picking the people who are to lead the related training programs. However, these sessions are basically to impart information, past experience, and resource material to those attending. Technical excellence and teaching skills are somewhat more important than the qualities of design and construction knowledge, industry reputation, leadership, and durability so important for a partnering charter chairwoman or chairman.

Selection of such specialized assistance is best accomplished by reviewing the immense number of teaching, training, and coaching offerings from private consulting and training firms, colleges, and associations throughout the world. A good first step is to discuss your needs with local professional and technical associations active in design and construction in your locality. Most vocational schools, colleges, and universities also have sizable resources available to assist in finding training help.

A repeat word of warning—keep the partnering effort focussed on the prime objective of writing and implementing a system that allows the stakeholders to execute a specific project well and with minimal outside interference or legal complications. *The training and education effort is ongoing—the project partnering effort is explicit and discrete.*

Activities 10 and 11: Review partnering and training outlines and set agenda and timetable with chair Now that the chairmen, chairwomen, leaders, teachers, and trainers have officially been brought into the partnering effort, the leaders should collectively review what has been done to date, make revisions that the sponsors and they think are appropriate, and put the material into final form for the invitation to potential stakeholders. Documents prepared during this task are critical to capturing the interest of those who are to build the partnering system. In activity 7 the highest UDM executive sent out the invitation and a summary outline. Now the entire group must put final details of the initial sessions together and ensure that the partnering packet is ready for the stakeholders.

One product of activities 10 and 11 is a notebook containing all the pertinent material to be used in the charter meeting. Almost any binding method is suitable for a partnering meeting, although I prefer the loose-leaf binder. The notebook encourages participants to keep the many pages of loose material generated during the meeting in a central reference file.

Activity 12: Confirm partnering support from top and middle management Once the partnering outlines and agenda have been set to the satisfaction of the charter meeting leaders, the material should be provided to the sponsoring top management for their review. At this time those responsible for the partnering planning work should reaffirm that the sponsor support is still there and available as needed.

Activity 13: Arrange for partnering and related meetings It is now time to make arrangements, or to firm up any early arrangements already made, to accommodate the partnering charter meetings. This would include all related conference space, meals, incidentals, and other operational needs for holding the meetings. The requirements for this activity are described in detail in Section C. The importance of the meeting facilities suiting the conference cannot be overstresssed. Acoustics, electrical outlets, heating and cooling controls, table service of water, glasses, pencils, and paper are only a few of the needs encountered in a working conference of this type.

Charter meetings are often very stressful. To some stakeholders who have never been through similar sessions to forge management plans and tools, the prospect of participating is frightening. A friendly environment in which the partnering process can be encouraged and supported contributes greatly to easing their concerns and assisting others in a successful charter-writing effort. If supplementary training or conferring on project-related matters is being held concurrently, make certain that the partnering stakeholder group is able to work intact for the full charter-writing period. The stakeholders must, to the greatest degree possible, participate for the full duration of the charter preparation.

Activity 14: Send a detailed meeting agenda and confirm attendance at meeting Concurrent with activities 12 and 13, the agenda and available informational material on partnering should be distributed by posted mail, E-mail or fax to all those on the invitational list. Once this material has reached the participants—probably within one week of the transmission—all participants should be contacted personally to confirm their attendance. Confirmation of attendance serves several purposes: it allows an accurate head count for the coffee and meal service; it allows an accurate head count for obtaining the correct number of meeting mementos (more about this later); and by its personal nature it stresses the importance of the participant's contribution by evidencing concern for his or her presence. Most important, it serves as a reminder to all individuals that they are expected to attend—and that expectation comes from the sponsoring group, usually the top ultimate decision makers in the group.

Activity 15: Hold related briefing meetings (if necessary)   As mentioned previously, there may be a desire by the sponsor and the potential stakeholders to hold precharter orientation and informational conferences. This decision will be heavily influenced by the level of knowledge held by potential stakeholders about the project, about partnering, and about the desires of the sponsors. If the related meetings being considered in this activity are essential to the success of the partnering effort, they should be held prior to the partnering charter meeting, as is shown in the network model in Figure 10.2. These might include:

- A familiarization meeting for the management of the major stakeholders to acquaint them with the basic concepts of partnering
- A social evening to introduce the potential stakeholders to each other just prior to the partnering charter-writing session
- A partnering business meeting to decide on the responsibilities and cost sharing for holding the charter-writing session and implementing the charter system

Meetings for the purpose of educating or training in subjects not critical to writing the partnering charter should be kept separate and distinct from the charter-writing session.

Activity 16: Hold a partnering meeting, and write and sign the partnering charter   This meeting is the actual conference and workshop where potential problems are identified, recommendations for resolving these problems proposed, the project mission written, and the partnering charter prepared and signed. The end product of activity 16 is a signed moral agreement on noncontract matters: the partnering charter (detailed instructions on how to conduct the partnering charter meeting are provided in Chapter 11). This initial partnering meeting should establish the machinery for appointing a partnering evaluation task force, and an issue resolution task force. The task forces should meet and formulate their evaluation system and their issue resolution process as soon as possible after the charter is written and signed, preferably within two weeks.

Activity 17: Prepare and publish a partnering evaluation system   This activity is usually accomplished by meeting and preparing a detailed set of instructions for periodic measurement of current partnering effectiveness. A sample partnering evaluation system is described in Chapter 12. The task force should be appointed by one or more of the key stakeholders and be charged with the assignment for implementing the entire evaluation system.

Activity 18: Develop a partnering issue resolution process   Here the task force develops a partnering issue resolution process. The task force's responsibility extends to selecting a route of alternative dispute resolution than can be followed effectively to identify and resolve or dispose of contested disputes between stakeholders. The task force should be made responsible for implementing the process among the stakeholder. Additional details on activities of this task force are described in Chapter 13.

Activity 19: Prepare and publish a charter evaluation system and issue resolution process   After completing development of the evaluation system and the issue resolution process, the task force should issue a detailed written set of procedures to the other stakeholders.

SECTION C: GETTING READY FOR THE WORKSHOPS

The purpose of the charter meeting is to identify the project operating model and to write and adopt morally binding behavioral guidelines for the owners, designers, constructors, and users during the period the stakeholders will be involved in the project. The charter is the base document of the partnering system. Its proper production is of the utmost importance.

How best to select those who should be invited to the charter meeting is outlined in Chapter 8, Section C. To reiterate a critical point made there— it is essential that all key project ultimate decision makers attend the meeting. All decision makers must be active participants; they will actually write the charter. Professionally related special guests who are interested in observing how the partnering process works may also be invited. This is appropriate, as long as the guests clearly understand that they are observers, not participants.

Several different kinds of meetings will be be held during the charter-writing day. These include meetings attended by all attendees, meetings of related functional attendees, and meetings of special assignment attendee groups. All participants will meet for subject briefings, to review workshop comments, and to write the charter. The table arrangement for full attendee meetings should be classroom style with participants facing the front of the room. Participants will be working from time to time in separate workshop teams made up of people with similar functional and business interests. For the functional workshop sessions it is best to use round tables that comfortably seat three to five people. The total number of tables should be limited to a maximum of 10.

More than 10 tables increases participant communication difficulties beyond the comfort level and makes effective work among table groups difficult. If space allows, workshop tables should be located at one side or one end of the room. Full attendee meetings can then be held on the other side or the other end of the room. By the morning of the meeting the physical arrangement of the meeting rooms for the partnering session should be completed. I suggest the entire setup be checked the night before so that any missing elements can be acquired in time for their use the next day. If the room is not available the evening before the session, the room check should be made very early in the morning on the day of the meeting.

A comprehensive room inventory checklist should be prepared in advance and used to ensure that the room is ready to go by the start of the session. A well-prepared checklist, used intelligently early at the site, can help avoid embarrassing and sometimes damaging disruptions to the charter meeting. The sponsor of the meetings and the stakeholders responsible for the project partnering effort should prepare their respective checklists as arrangements for the meeting are made. The chair of the meeting may also have his or her own set of lists and reminders. A comprehensive preparation list is important to both.

The lists can be mental if the meeting sponsors, managers, and leaders are experienced and never forget anything; or they can be written and carried to the meeting location for a final check if a fail-safe procedure is desired. Each meeting location and format will have its own characteristics, but for the most part the Olanta checklist can serve as a template for both the sponsors and the chair.

Premeeting Checklist for Olanta Data Systems, Ltd., Partnering Charter Workshop

Equipment to Be Provided at Each Workshop Team Table

A. Easels and flipcharts (by Olanta)
  - Floor mounted preferable
  - Table mounted acceptable
  - At each workshop table:
    - Number required: one
    - Sheet size 27 in. by 34 in.: standard size
  - In front of room at leader's table:
    - Total number required: one
    - Size 27 in. by 34 in.: standard size

B. Felt markers for team flipchart work (by Olanta)
   • Number required: one set per table
   • Type: highlighters or transparent ink felt markers
   • Colors suggested:
     • Black: for main flipchart notes
     • Red: for visual emphasis
     • Blue: for visual emphasis
     • Green: for visual emphasis
C. Transparency blanks at each workshop table (not needed if individual table flipcharts are provided) (by Robert Shea)
   • Number required: five per table
   • Size $8\frac{1}{2}$ in. by 11 in.
D. Erasable transparency markers for each table (needed only if transparency blanks are to be used) (by Robert Shea)
   • Number required: one set per table
   • Colors recommended:
     • Black: for main flipchart notes
     • Red: for visual emphasis
     • Blue: for visual emphasis
     • Green: for visual emphasis
E. Table identification standards and placards (by Olanta) (these allow numbering or lettering the worktables for quick identification and communication among the groups)

Equipment to Be Provided for Leader's Table at Front of Room

A. Overhead transparency projector (by Olanta)
   • Total number required: one
   • Should have high-intensity projector bulb
   • Should be equipped with spare bulb in the projector
B. Projection screen for use with overhead transparency projector (by Olanta)
   • Total number required: one
   • Preferably 6 feet by 6 feet or larger
C. Rectangular work tables for use of discussion leader (by Olanta)
   • Total number required: three
   • Preferable size about 36 in. by 60 in.
   • Covered with tablecloths
D. Chalkboard or marker board for temporary notes: optional (by Olanta)
   • Total number required: one

- If using chalkboard, need soft chalk and one eraser
- If using marker board, need erasable markers and eraser
- Colors suggested:
  - Black: for main notes
  - Red: for visual emphasis
  - Blue: for visual emphasis
  - Green: for visual emphasis

Supplies and Equipment to Be Available to Participants for Participant's Use as Needed

- $8\frac{1}{2}$ in. by 11 in. note paper, three-hole punched (by Olanta)
- No. 2 pencils with erasers (by Olanta)
- Pencil sharpener (by Olanta)
- Three-hole oversize punch (by Olanta)
- Stapler and staple remover (by Olanta)
- Individual name tags (by Olanta)
- Memento to be given each stakeholder as they sign the charter (by Olanta)

Reference Materials to Be Available

A. Meeting notebook for each stakeholder, stapled, bound, or in individual loose leafs; should provide about four or five extra for future use (by Robert Shea)

B. Set of construction drawings and specifications (by Olanta)

C. Site plan or other overview plan of total project. (by Olanta); to be displayed on wall or on easel for reference

D. Project model and renderings or other visual depictions of project as available (by Olanta)

Facilities to Be Available at Site of Meeting

A. Copying equipment close at hand—must be capable of quickly producing moderate-quality copies (by Olanta)   This is a critical operational item! Depending on the style of the chair and his or her method of recording results of the sessions, there will be an ongoing need to provide hard copies of resulting material to each participant as the sessions proceed.

B. Wall space for display of flipchart material (by Olanta)

C. Adequate electrical outlets for fixed and movable equipment (by Olanta)

D. Extension cords for remote electrical equipment (by Olanta and by Robert Shea)   In many hotels, motels, and conference centers, one or more of the enclosing walls are demountable or movable partitions. Usually, no electrical outlets are located in these walls. If the meeting is being held in a room where both sidewalls are demountable, electrical outlets will usually be found only in the front or rear walls or in the floor.

Equipment for Meeting Leader to Use for Note Taking and Publishing the Charter Meeting Newspapers

Techniques and equipment used by session leaders and facilitators vary from person to person. The equipment listed in the checklist below assumes that the meeting will be run on a continuous communications basis. This means that the notes and workshop discussions will be transcribed by computer into a word processor program, printed, duplicated, and distributed to the stakeholders as the meeting proceeds.

A. One or two laptop computers capable of receiving inputted note data from the individual workshop sessions (by Robert Shea)
B. Software to be used for note taking (by Robert Shea)
C. Portable printer for direct printing of note material inputted to computer during the workshop session (by Robert Shea)
D. An LCD projection panel board to use for converting computer screen images into large images by showing the image through the transparency projector (by Robert Shea)
D. For large partnering sessions, an assistant to help record proceedings for printing and duplicating; the general qualifications of this person are outlined below in the workshop descriptions (by Robert Shea)
E. Laser pointer for identifying screen items at a distance (by Robert Shea)

Reference Material to Be Available for Meeting

A. A list of all participants, showing their names, organizations, positions, phone numbers, fax numbers, and addresses (by Olanta)
B. Note template for structuring the partnering meeting records (by Robert Shea)

The template is a preprepared form that allows critical information about the meeting, the attendees, and the workshops to be inserted as the charter meeting proceeds. It is usually prepared by the meeting chair ahead of the session and filled in by the chair or the assistant as work proceeds.

A copy of the Olanta meeting template is provided below. Note that the participant roster is not shown in the sample. These names should be copied to the note template just prior to the meeting to ensure that an up-to-date list of the people is placed in the template.

Note Template for Charter Writing Meeting

A. *Date of charter meeting:* February 24, 19xx
B. *Time of meeting:* 8:00 A.M. to 5:15 P.M.
C. *Location:* Snowbird Inn, 3233 Carlsbad Road, Travis, North Dakota
D. *Those attending charter meeting:* names to be listed just before meeting
  1. By organization
  2. Individually: names to be listed alphabetically just before meeting
E. *Table work résumé*
  1. Break out groups, their comments and their recommendations
- *Table 1:* owner and user, group A
  - Group members (ODS staff: owner and user): names to be listed just before meeting
  - Problems others cause us (ODS staff: owner and user)
  - Problems we cause others (ODS staff: owner and user)
  - Recommendations for improvement (ODS staff: owner and user)
- *Table 2:* owner and user, group B
  - Group members (ODS staff: owner and user): names to be listed just before meeting
  - Problems others cause us (ODS staff: owner and user)
  - Problems we cause others ODS staff: owner and user)
  - Recommendations for improvement (ODS staff: owner and user)
- *Table 3:* architects of record and computer systems, group A
  - Group members (architects of record and computer systems): names to be listed just before meeting
  - Problems others cause us (architects of record and computer systems)
  - Problems we cause others (architects of record and computer systems)
  - Recommendations for improvement (architects of record and computer systems)
- *Table 4:* architects of record and computer systems, group B
  - Group members (architects of record and computer systems): names to be listed just before meeting.
  - Problems others cause us (architects of record and computer systems)
  - Problems we cause others (architects of record and computer systems)
  - Recommendations for improvement (architects of record and computer systems)
- *Table 5:* consulting engineers, group A
  - Group members (consulting engineers): names to be listed just before meeting
  - Problems others cause us (consulting engineers)
  - Problems we cause others (consulting engineers)
  - Problems we cause others (consulting engineers)
  - Recommendations for improvement (consulting engineers)

- *Table 6:* consulting engineers, group B
  - Group members (consulting engineers): names to be listed just before meeting
  - Problems others cause us (consulting engineers)
  - Problems we cause others (consulting engineers)
  - Recommendations for improvement (consulting engineers)
- *Table 7:* general contractor
  - Group members (general contractor): names to be listed just before meeting
  - Problems others cause us (general contractor)
  - Problems we cause others (general contractor)
  - Recommendations for improvement (general contractor)
- *Table 8:* mechanical and electrical contractors, group A
  - Group members (mechanical and electrical contractors): names to be listed just before meeting
  - Problems others cause us (mechanical and electrical contractors)
  - Problems we cause others (mechanical and electrical contractors)
  - Recommendations for improvement (mechanical and electrical contractors)
- *Table 9:* mechanical and electrical contractors, group B
  - Group members (mechanical and electrical contractors): names to be listed just before meeting
  - Problems others cause us (mechanical and electrical contractors)
  - Problems we cause others (mechanical and electrical contractors)
  - Recommendations for improvement (mechanical and electrical contractors)
- *Table 10:* fixtures, furniture, and food service equipment contractors
  - Group members (fixtures, furniture, and food service equipment contractors): names to be listed just before meeting
  - Problems others cause us (fixtures, furniture, and food service equipment contractors)
  - Problems we cause others (fixtures, furniture, and food service equipment contractors)
  - Recommendations for improvement (fixtures, furniture, and food service equipment contractors)

F. *Individual mission statements prepared by participants*
  - Mission statement 1 prior to workshop 1
  - Mission statement 2 after workshop 2
G. *Project charter mission drafts*
  - Draft 1
  - Draft 2
H. *General meeting notes*
  - Time started

- Notes on introduction
- Time completed

I. *Charter for Olanta Expansion Program*
  - Mission
  - Objectives

The template is of great use when the meeting notes are being taken on a word processor and transcribed, printed, duplicated, and distributed as the meeting proceeds. As described earlier, this style of note taking is referred to as technography.

SECTION D: MEETING HINTS AND TECHNIQUES

Most discussion leaders, facilitators, instructors, and other professional experts and consultants have developed presentation techniques and styles adapted to their special kind of work. Many of these techniques are generic and will work well in leading a group through a partnering charter preparation. Others must be adapted specifically for a partnering charter-writing session. This book is not on how to chair a meeting effectively. Often, however, those well suited professionally and technically to help a group prepare a good charter are not skilled in meeting techniques. The guidelines below are intended to provide a few tips on some of the specialized methods and processes to help you work more effectively in charter preparation meetings.

1. *Sessions should be well planned and structured around an agenda. The agenda should be provided to the participants before the meeting.* Construction people are very sensitive to how they spend their nondesign and nonconstruction hours. They want to know ahead of time how what they are doing will benefit themselves and their organization. The meeting agenda should be clear on these points.

2. *The agenda should be carefully followed unless there is some unexpected and overriding outside impact on the program schedule.* Participants in a partnering session generally do not like surprises. They want a structured agenda, and they want the discussion leaders to follow the outline proposed.

3. *Participants gain most from the small breakout sessions by having their efforts and comments in the workshops quickly transcribed for individual reference, study, and reference.* To do this in a timely manner, I recommend that the leader arrange to have the handwritten flipchart or transparency workshop comments transcribed into a conventional word processor computer template as the workshops proceed (see Section C of this chapter for a sample template). This material can then be printed on one of the many small portable printers on the market today, with copies made for each attendee.

4. *Discussions should be designed to lead the team members and teams gradually into a mental pattern that attributes both positive and critical comments to the entire stakeholder group, not just special discipline groups.* Partnering encourages stakeholders to take individual responsibility for making a team effort successful. This can be done only if the end objectives are team *and* individual oriented.

5. *Place time limits on each discussion period.* Time allocations set in the agenda must be followed to keep the workshops effective. Often, more time could be used in a workshop, depending on the project, the subject, the people, and the local and project conditions. However, time allocations given in the sample agenda or derived from good experience should be adequate to generate enough material from which to write a charter. So, set time limits and tactfully but firmly end the discussions when the time limit is reached.

6. *Give the participants enough data in the workbook and notes to force them to think hard about what they are to do on the job.* Try to convey early information to stakeholders that allows them to think creatively about what is required of them in the meetings. Then discipline the work in the discussion periods so that the efforts of the stakeholders are focused intensely on the end product of the breakout meeting.

7. *Keep all idea-generating workshops as nonjudgmental as possible.* Remember, it is always difficult and often impossible to be creative if we are being judgmental at the same time. Allow judging and micro analysis to occur only *after* a proper subconscious thought period has followed the statement and the recording of an idea.

8. *Encourage full participation of all members at each team workshop table.* The partnering discussion process is one built from the ground up— good solutions usually spring from an individual idea nursed successfully into generic solutions for similar problems. Team workshops should give individual stakeholders their best opportunity for expressing ideas without premature challenges and opposition.

9. *Be willing to suggest changes to statements of the total group if project interests are better served by the change.* However, always give the total group an opportunity to review any changes made by the chair or leader. As a recorder and leader, do not bias your note keeping unless you explain clearly to the stakeholders what you are doing and why.

10. *Develop and use recording and feedback techniques that allow a quick turnaround of the information developed in the partnering workshops.* In a typical one-day project partnering session the participants will produce literally hundreds of pieces of information. Much of this data is most valuable if it is recorded in some easily referenced format to allow retrieval and selective arrangement of the information.

General Comments on Techniques Available for Taking Notes in Charter Meetings In a charter-writing session the most common media used for recording comments and decisions are flipcharts, transparencies, conventional notepaper, and computer translations. A brief description of each is given below.

Method 1. Prepare flipcharts displaying problem statements as they are discussed. The chart is the written record of the table team's work. Flipcharts should be displayed continually throughout the day as the charter evolves from the statements from the discussion floor to the final writing of the charter.

Advantages
- Requires very little additional record keeping by the facilitator.
- Does not usually require copying and distributing material to stakeholders during the meeting.
- Flipcharts are available at all times for the stakeholders to refer to and read.
- In small groups (12 to 20), the single focus of the displayed flipcharts tends to bring the group together quickly in a unity of purpose.
- Equipment and materials are inexpensive and easily available.

Disadvantages
- Slows the discussion process because of the note-taking time required and the size and amount of paper to be handled with the flipcharts.
- Requires a meeting room with good sight lines to display and read the flipcharts properly.
- Stakeholders must remember location of material from a dimensionally wide spread of displayed data.
- Viewing distance of normal flipchart text is limited.

Method 2. Write the problem statements on transparency blanks and keep the transparencies as the written record of the table team's work. In this method transparencies are kept in sequence within the workshop subject and displayed on the overhead screen as they are discussed or as the information contained is needed for reference purposes.

Advantages
- Requires very little additional record keeping by the facilitator.
- May not require copying and distributing material to stakeholders during the meeting if the display method on the overhead projector is adequate.
- Visible at long distances to larger audiences.
- Transparencies are easily handled if they have been kept in the proper order.

- Transparencies are easy to refer to as needed and can be reviewed by individuals without overhead projection.
- Transparencies are easily copied to plain paper as needed.
- Equipment and materials are inexpensive and easily available.

Disadvantages

- Transparencies are difficult to keep in order and to find as needed.
- Notes must be projected one sheet at a time or copied to be used easily in group discussions.
- Editing and making changes on transparencies may be difficult and time consuming.
- Multiple projectors are needed if more than one transparency at a time is to be displayed.

Method 3. Write the problem statements on note paper and keep these as the only written record of the table team's work. In this method the notes must be kept in sequence within the workshop subjects and presented orally or read individually as they are discussed and the information contained is needed for reference purposes.

Advantages

- Requires very little additional record keeping by the facilitator.
- Material is easily kept in the proper order for reference during the charter meeting.
- Notes are easily referred to as needed and can be reviewed by individuals.
- Material is easily copied to plain paper as needed.
- Equipment and materials are inexpensive and easily available.

Disadvantages

- Notes must be copied and distributed to be used easily in group discussions.
- Referencing to hand written notes may be difficult if individual table secretaries do not identify the material clearly.
- Produces very large amounts of paper to be shuffled in a very short time.
- Notes may not be legible and require rewriting.

Method 4. Prepare and display the table notes on flipcharts, transparencies, or on notepaper as described in methods 1, 2, and 3. As the notes are available, the facilitator or an assistant copies them to the computer template. When a unit workshop discussion is complete and the notes from it have been typed to the disk file, the material is printed and copies made for the use of all stakeholders in their ongoing work. This document is known as a workshop newspaper. The newspaper should be printed only after a workshop is completed so as to allow for possible changes to the notes during the workshop.

Advantages

- Permits almost instantaneous feedback to all stakeholders of material being written by the team.
- Encourages valuable cross fertilization of ideas from team to team and workshop to workshop because of the relatively rapid communication of ideas being expressed in each of the table groups.
- Provides a permanent written record of the base statement of problems that will be used to generate recommendations in workshop 4.

Disadvantages

- Requires a competent facilitator or assistant who is able to type accurately into a computer word processing program under heavy time pressures.
- Requires a reasonably extensive variety of electronic equipment, including a laptop computer, a portable printer, and a copier capable of quickly producing moderately good copies in moderate volumes.
- May require extra personnel to assist the meeting facilitator.
- May detract from the facilitator's time to pay careful attention to the content and statements of each table team.

In workshops where 10 to 25 people are working in teams of three to five participants, flipcharts are excellent working tools. For a group as large as the Olanta meeting where there are nearly 40 participants the limited-visibility range of flipcharts may be a deterrent to their use in the larger group session. In groups of this size it is usually essential that the facilitator supplement flipchart work with the use of transparencies and an overhead transparency projector.

Lettering for a flipchart presentation should be done in black ink applied by a broad-tip felt pen. It must be easily read and should be of such size that the viewer can see and interpret the text easily from as far away as 30 or 40 feet. When using large numbers of flipcharts wall display space becomes a serious consideration. For instance, as each set of charter objectives is written, usually starting on a fresh flipchart sheet of paper, the previous material should be displayed for reference use by the stakeholders. The amount of space needed for such a display is seldom available in a conventional meeting room.

Transferring ideas, opinions, and processes to the minds of the stakeholders must be done by competent people. Timely preparation, copying, and assembling partnering workshop material is critical to this transfer in a good partnering meeting effort. Fortunately, on the Olanta project, Mr. Leinenklugle, the facilities manager, has provided Mr. Shea, the charter meeting leader, with one of the department's bright young college interns, Lisa Carter. Ms. Carter has the interest, skills, and temperament needed to type and record data electronically during a stressful meeting. She also has the mental and physical stamina to work consistently well for the long hours often needed to conduct partnering meetings properly.

Mr. Shea uses a technique of recording events and decisions called technography. Technography employs easily available computer hardware and software by which the recorder takes electronically stored notes as the information and data are being generated. Some people who are proficient in this method apply it to partnering during the workshops, the mission statement development, and the final writing of the charter objectives. A useful tool in the practice of technography is the liquid crystal display (LCD) projection panel. This is a device through which the computer screen is converted into an LCD image on a flat transparent display panel, setting on the lamp bed of an overhead transparency projector. The image on the LCD is projected and magnified, just as a transparency is projected and magnified, onto a larger screen for easy viewing by relatively large groups of people.

I have found that the LCD projection panel can produce a readable image on an 8-foot by 8-foot screen that can be seen clearly by as many as 40 or 50 people. Room lighting and projector brightness, of course, affect the visibility range. For partnering meetings of up to 40 people the LCD system works very well. Several good audiovisual specialty firms manufacture LCD panels that can be operated with practically any make or type of computer. The LCD projection panel techniques allows continuous preparation of charter documents as the workshops and discussions proceed. In partnering meetings the leader must record large amounts of information rapidly, accurately, and in a displayable format.

The amount of information developed in a partnering session is huge. For instance, in a problem statement workshop each table might produce as many as 30 or 40 problem statements. In the Olanta project partnering meeting we could expect to generate as many as 400 problem statements from the 10 worktables. Guiding 30 to 40 stakeholders through discussions leading to the resolution of these 300 or 400 problems requires that the facilitator have audiovisual equipment to display, revise, manipulate, and recombine data into several forms for effective use by the stakeholders. As with all other actions taken in preparing for the charter meeting, careful planning of how to manage the day's work so as to produce accurate, useful records is critical to partnering success. Time spent in action planning is seldom wasted as long as the plan is used to guide the action.

11

CONDUCTING THE CHARTER MEETING

SECTION A: LAUNCHING THE PARTNERING CHARTER MEETING

A solid base for any meeting is set best and most easily by preparing and following a meaningful agenda. The Olanta Data Systems case study in Chapter 10 provides a good foundation for developing an effective meeting plan of action. Mr. Olanta, president of the company, had sent out the invitation letter along with a draft agenda for the charter meeting (see Chapter 10, Section B, for the invitation letter and other meeting information). The draft agenda was refined progressively as planning for the partnering charter meeting continued. The final agenda that was included in the Olanta charter notebook is reproduced below.[1]

Olanta Data Systems, Ltd., Partnering Meeting Agenda

Date of meeting: February 24, 19xx

Location of meeting: Snowbird Inn, Travis, North Dakota

---

[1] This agenda is derived from several meeting plans that have been used in a wide variety of actual partnering charter conferences.

Session A. 7:00 to 8:00 A.M.   Continental breakfast and morning social, Whitehall
  Room, first-floor level
Session B. 8:00 to 10:00 A.M.:   Welcome, orientation and workshop #1

- Welcome by Charles Olanta, president and chief operating officer, Olanta
  Data Systems, Ltd.
- Introduction of participants by Travis Loring, principal, Loring & Metzer and
  Pat Greene, president, Tiltsen and Greene
- Review of project characteristics by Henry Leinenklugle and Travis Loring
- Orientation on partnering and briefing on workshops by Robert Shea, meeting
  chair and facilitator
- *Workshop 1:* "What actions do others take that create problems for us?"
  Break out for discussion and presentation of findings.

Session C. 10:00 to 10:15 A.M.:   Coffee break
Session D. 10:15 A.M. to 12:00 noon: Workshops #1, 2, 3, and 4.

- *Workshop 1:* continued.
- *Workshop 2:* "What actions do we take that cause problems for others?"
  Break out for discussion and presentation of findings.
- *Workshop 3:* "What is the most important result for my organization to
  achieve by completing this project successfully?" Each participant prepares an
  individual mission statement. These are to be collected, typed, and published
  by 12:30 P.M. to assist the project mission task force write the first draft of the
  project mission.
- *Workshop 4:* "In light of the problems stated in workshops 1, and 2 and the
  individual missions defined by us, what recommendations can we make that
  would help improve relations and performance on the Olanta Data Systems
  project?"

Session E. 12:00 noon to 1:00 P.M.:   Lunch, United States Room, second-floor level
Session F. 12:30 to 1:30 P.M.:   Break out mission-writing task force (overlaps with
  lunch and with workshop 4)

- Concurrent with the latter part of lunch and workshop 4, a specially selected
  task force of three to five stakeholders will meet apart from the main
  stakeholder group. Their assignment is to prepare a 25-word (or less) first-draft
  mission statement for the project. Individual mission statements from
  workshop 3 will be used by this group as they begin their work.

Session G. 1:00 to 3:00 P.M.

- *Workshop 4:* "Recommendations," continues to 1:30 P.M.
- *Full stakeholder workshop 5:* Full stakeholder group discusses and revises the
  first-draft mission statement.
- *Full stakeholder workshop 6:* Full stakeholder group discusses what specific
  project objectives can now be set from the results of workshops 1, 2, 3, 4, and

5 that will help ensure excellent relations and performance on the Olanta Data Systems project.

Session H. 3:00 to 3:15 P.M.:   Refreshment break

Session I. 3:15 to 3:45 P.M.   This discussion to be merged into workshops #6 and #7.

- Review principles of alternative dispute resolution (Robert Shea).
- Review principles of partnering performance monitoring and evaluation (Robert Shea).

Session J. 3:45 to 4:55 P.M.

- *Full stakeholder workshop 7:* Combine all previous workshop discussions into a charter ready for stakeholder signatures.
- Print final draft of charter.

Session K. 5:00 P.M.:   Sign charter and receive award memento.

Session L. 5:15 P.M.:   Adjourn to social hour celebration sponsored by Olanta, Tiltsen and Greene, and Loring and Metzer.

There are many other agenda outlines by which a charter meeting can be guided. The one above has repeatedly proven successful and will form a good base for most project partnering efforts. The meeting timetable will vary depending on the size and nature of the information pool needed by the participants and on the style of the discussion leader. The Olanta agenda provides a pattern that allows stakeholders to complete writing a good partnering charter in one full day. It permits the work to move along at a good pace while giving participants adequate time to apply their experience and intelligence. Most important, it fits the work into a compressed time frame that is vital to maintaining partnering enthusiasm.

Thirty-seven people, including the discussion leader, have been invited to attend the ODS project charter writing (see the Olanta case study in Chapter 10, Section A, for a full list of attendees). This is a larger number than usual. However, when many concerned people are deeply involved in a project and display a strong interest in the process, I normally recommend having them all attend rather than cutting back.

Each agenda component serves a specific purpose in the production of the final charter. The format of the Olanta outline has again and again been found to address the true objectives of partnering effectively. Partnering in a friendly workshop environment helps identify stakeholder stances, actions, and positions that have high potential for causing conflict. It then coaxes the participant into solving problems with a teamwork approach.

Partnering is a deceptively simple system. It is best illustrated by studying and analyzing the agenda items above. Each step makes increased demands on the stakeholder but produces correspondingly more valuable information. The system and process promotes understandings that ultimately help melt the classic causes of destructive conflict in our construction profession. To see how this works, let us examine each agenda item step by step and identify the dynamics and details of the process.

- 7:00 to 8:00 A.M.   Rolls and coffee morning social, Whitehall Room.

Here is where the first movement toward making partnering work for the stakeholders occurs. It is strange, but key decision makers and important operational personnel on a job do not often get together to talk informally, socially, and professionally. The relaxed atmosphere of an early morning social hour can generate initial feelings of trust and interest that carry on throughout the charter day. Many people like to begin their work day by smiling.

A simple example may illustrate the value of a relaxed and enjoyable social time. At a dinner celebration of a successful charter application, a young man approached me and said he had participated in the charter writing and had found it a unique experience. He explained that when he attended the charter meeting, he was a spanking-new superintendent for one of the major subcontractors on the project. He continued, saying that the most impressive part of the entire charter process was that for almost a full day he was in constant social and business contact with key high-level people who he normally saw only infrequently and with whom he almost never had the opportunity to discuss the job.

When he was asked during the charter meeting about what owners, architects, and general contractors did that caused him problems, he said: "I was petrified. I didn't know whether we were serious or not: whether or not to say anything. I took the discussion leader at his word, and expressed some ideas about the job that were later incorporated into the charter." The charter day was an eye opener for this young manager and it gave him a new look at his vocation and profession, one that probably resulted in giving the job a far better product from his labors that if he had not been involved as a stakeholder.

This type of dialogue is repeated time and time again about the value of the social coming together that is promoted by the nonthreatening environment that surrounds a well-run charter meeting. A social tone is often set in the early morning coffee hour. The coffee hour ambience is important!

- 8:00 to 10:00 A.M.   Agenda item B.

Introduction of the subject and the participants by sponsor principals and the facilitator. A proper introduction of the workshops and the participants in the charter meeting is critical to ongoing success during the working sessions.

Knowing the background and interests of each person participating helps stakeholders individually, and as a group, better address the objective of writing the charter—to develop and adopt a partnering charter for the guidance of Olanta's project team.

A principal in the lead sponsor's organization (very often the client) should be assigned to start the formal meeting. This person's view of the total program is often the first time that many of the stakeholders have been briefed by so high and authentic a source. In the Olanta Data Systems project, the client's president, Charles Olanta, has been asked by the project team principals to set the initial tone of the gathering. The total time allocated to introductory remarks is about 15 minutes. Five minutes go to Mr. Olanta, the owner, 5 minutes is assigned to Pat Greene, general contractor, and 5 minutes is assigned to Travis Loring, principal architect. Subjects to be presented by each of the principals might include:

- Mr. Charles Olanta, owner
  - The mission of Olanta Data Systems, Ltd.
  - Characteristics of Olanta, Ltd.
  - History of Olanta, Ltd.
  - What he hopes the new facility will do
  - Names of the staff of Olanta attending as stakeholders
- Mr. Pat Greene, general contractor
  - The mission of Tiltsen and Greene
  - Characteristics of Tiltsen and Greene
  - Description of the construction phase project delivery system being used, insofar as it sets work relations on the job
  - Names of the staff of Tiltsen and Greene attending as stakeholders
- Mr. Travis Loring, architect of record
  - Characteristics of the project
  - Characteristics of Loring and Metzer
  - Description of the design phase project delivery system being used, insofar as it sets work relations on the job
  - Names of the staff of Loring and Metzer attending as stakeholders

One of the main objectives of these early statements is to convey information about the organizations on the job while creating an atmosphere conducive to effective team play and good project spirit. The introductions should encourage building a partnering framework, as opposed to imposing the framework on the people in the meeting. Once introductory remarks are completed, the chair or facilitator of the meeting should ask each stakeholder to introduce himself or herself. Self-introductions should be very brief, no more than a fraction of a minute for each person. Key information from each stakeholder should include the person's:

- Name
- Title
- Company
- Role in company
- Role on project

Review of project characteristics by Henry Leinenklugle, facilities director, Olanta, and Travis Loring, president, Loring and Metzer, architects of record. The technical design experts should be called upon in session B to provide a 10- to 15-minute walk-through of the project, describing its major characteristics. Sometimes having a rendering and a site plan is of great help in such a discussion. The purpose of the presentation is to familiarize those who have not been involved on the project for long to gain an overview of the total project. This discussion, in conjunction with a set of the construction documents in the meeting room, should increase comprehension of the work at hand and allow cross-discipline discussions to occur with easily referenced, authentic documentation at hand.

SECTION B: PARTNERING ORIENTATION AND OVERVIEW OF RELATED SUBJECTS

The start of the actual partnering effort is an important milestone in the agenda. The facilitator should have his or her remarks well outlined and in hand so that the workshops following the orientation can run with a minimum of interruptions to the flow of stakeholder ideas. Orientation comments should be limited to about 30 or 40 minutes. Many subject outlines from which the facilitator could draw would be suitable for this session. Each discussion leader will probably have an individual style and methodology of proceeding with the orientation. Within any format however, the end product of the day's work should be a signed charter.

The orientation session should be tailored to bring the partnering launch portion to a close about $1\frac{1}{2}$ hours after the start of session B. Time limits will be set to a large extent by the experience and background of the stakeholders. If all present have past experience with partnering and alternative dispute resolution, the introductions and project characteristic presentations may be all that are needed for the session B work. Let us, however, assume that most of the Olanta participants are new to partnering and that they will be provided a full partnering orientation. I suggest the discussion leader and the sponsoring group for the Olanta project build their orientation session around the following outline:[2]

- *Topic A   Introduction*   8:40 to 8:50 A.M.
  1. Explanation of the workbook
  2. Key definitions (Appendix A)
  3. Ingredients of a successful partnering effort (Chapter 7)
  4. What can be expected from a partnered project (Chapter 7)
  5. Situations in which partnering might not work well (Chapter 8)
- *Topic B   The mission statement*   8:50 to 8:55 A.M. (Chapter 11)
  1. Role and importance of the mission statements
  2. The basic question from which the mission is formulated
  3. Types of mission statements
     a. The individual for the person in relation to the project
     b. The individual for the project in relation to the project
     c. The task force for the project in relation to the project
     d. The stakeholders for the project in relation to the project
- *Topic C   Workshops: how we produce the charter*   8:55 to 9:05 A.M. (Chapters 9, 10, and 11, and Appendix C)
  1. Format of the workshops
  2. Team groupings for the workshops
  3. How to formulate problem statements
  4. How to formulate the project mission
  5. How to formulate improvement recommendations
  6. How to write the project charter
  7. How the charter provides a partnering measurement standard
- *Topic D   Alternative dispute resolution*   9:05 to 9:10 A.M. (Chapters 6, 7, and 13)
  1. ADR types
  2. The use of ADR as an adjunct to partnering
  3. Setting the issue resolution system with ADR
- *Topic E   Specific procedures to follow in workshops*   9:10 to 9:20 A.M. (Chapter 11)
  1. Appoint a chair for the sessions.
  2. Appoint a secretary for the sessions.
  3. Appoint a representative to present findings.

---

[2] References to chapters and appendices shown in parentheses define where detailed descriptions of the outline subject may be found.

A knowledge of each of these topics is essential to effective understanding and use of partnering, whether you are a stakeholder, a discussion leader, or just an observer. Although the time allocated to each subject seems short, it is important to keep the objective of writing a charter in clear view at all times, particularly in the opening of the partnering session. Even an hour and a half may be too much time to allocate to the introduction, subject review, and orientation. However, if the chair conducts the meeting well, the material discussed in the workshops will supplement the early presentation and provide stakeholders with a well-rounded knowledge of partnering. Chances are that it will also produce a good charter. That is our main job in a partnering meeting.

Each of the five main topics to be covered in the partnering introduction and an overview of how each relates to the workshops are summarized below.

Topic A: Introduction

Item A1: Workbook. The first subject to be covered after introduction of the charter meeting leader or chair is to review quickly the contents of the workbook to be used in the workshops. This book will usually contain material to help the stakeholders achieve the specific objective of the day—to write and sign a charter. The workbook should be prepared by the sponsors and the chair in advance of the meeting and distributed at the beginning of the meeting. I suggest that the materials be put in a loose-leaf binder. This format allows considerable flexibility as other information is added to the stakeholder's file of information.

The loose-leaf binder, if of the proper form and filled with useful information, will probably be saved and used for reference more effectively than if the information was distributed in loose, unbound packets. The book should contain:

- A title sheet showing key data about the meeting
- An index of the contents
- A detailed informational agenda for the guidance of the stakeholders
- Samples of mission statements
- Samples of problem statements
- Samples of recommendations for improving project performance
- Samples of charters to use as models
- Sample issue resolution guidelines
- Sample partnering evaluation guide
- Blank, lined paper

These documents will be used throughout the day as a set of baseline references for workshop participants as the various sessions proceed.

Item A2: Key Definitions. During the session the participants will be using many words that do not appear with great frequency in a designer's or contractor's lexicon. Quick, accurate, effective communications are essential to the success of any partnering session. A comprehensive, available glossary of terms will allow all stakeholders to better speak a common language. The chair might wish to abstract and publish a workshop glossary from Appendix A to use as a reference document for the stakeholders.

As emphasized previously, the charter-writing meeting is a complex simulation exercise of real situations cast entirely in a world of words. Therefore, all stakeholders must have a shared understanding of what the words mean. A definition of frequently used terms assists in achieving a shared understanding of each other's needs, problems, and frustrations.

Item A3: Ingredients of a Successful Partnering Effort. The charter-meeting chair should, in the introduction, touch on the six essential elements needed for a successful partnering effort:

1. A project to which partnering can be applied
2. An intelligent and competent project team
3. Knowledgeable technical and management leaders
4. A management that can meld the desires, needs, and leaders of the project organization into a successful effort
5. A willingness to take the risk that people are truly interested in doing a good job and managing conflict
6. A set of tools by which the partnering system can be built

These critical parts of the partnering effort should be conveyed in clear, concise language to the entire stakeholder group. The need for good ingredients from which to bake the partnering cake is paramount to partnering success.

Item A4: What Can You Expect from a Partnered Effort? Expectations of any system are the standards by which the outcomes are evaluated and their worth defined. The orientation presentation of the stakeholder group should set some benchmarks to aim at in their efforts. The results expected from the Olanta partnering effort have been defined by the Olanta facilities staff and are listed in Chapter 10, Section B. In essence, the results are the goals and objectives of the stakeholders as they begin to assemble and implement their partnering system. They should appear in the stakeholder's notebook.

Item A5: Situations in Which Partnering Might Not Work Well. One hundred percent success is difficult, perhaps impossible, in most management efforts. In fact, one of the most critical elements of management excellence is *not to try to manage the impossible.* An intelligent approach to partnering is to look at those situations that have proven by experience to resist and sometimes damage efforts to partner a project. Many of these are listed in Chapter 8, Section A.

By the time the charter meeting has started the project will have been screened for damaging influences and have received a go-ahead from the ultimate decision makers. However, project managers and stakeholders must be ever alert to any erosion of the factors that contribute to partnering success. Stakeholders should be made a part of this early alert effort and be aware of what impacts tend to destroy or weaken a partnered project effort. Good briefing up front by a capable facilitator can help strengthen the partnering system and bolster the stakeholders resolve to keep it healthy.

Topic B: Mission Statement   The concept of the mission should be initiated in the opening remarks of the charter writing session and carried through the entire day's work. The mission is the thread that knits together the efforts of the stakeholders. The basic statement of mission is, very simply, an identification of the most important result to be achieved by the effort—in this case, partnering—being successful. During the orientation the leader should summarize how the project mission evolves from the individual statements of subjective purpose and is built into a collective stakeholder concept of project success by the stakeholders. A mission statement is needed early in the preparation of the partnering charter. Therefore it should be made a part of the initial orientation by the chair.

Topic C: Workshop Overview   The workshops should begin immediately following the orientation given by the chair. Workshops consist of the entire stakeholder group divided into table groups of three to five people. The division of the people to be grouped is best established by major responsibilities within the group. For instance, all stakeholders representing Olanta (the owner) should be seated at the same table. If a similar-interest group is too big for one table, it should be divided further so that the table group sizes are somewhat similar. Stakeholders at the tables should have similar interests and be able to speak a common language. The facilitator should have made a preliminary breakdown of those sitting at each table. In the briefing session he or she should provide details of the makeup of the tables and the duties and responsibilities of each group of stakeholders. The table work to be done will be clear as we continue to walk through the workshop portion of the charter meeting. For meeting hints appropriate for this portion of the opening remarks, see Chapter 10, Section D, and for the Olanta meeting table makeup, see the partnering charter meeting template in Chapter 10, Section C.

Workshop subjects will cover three major areas of work: problems others cause us, problems we cause others, and the recommendations those in the breakout groups feel will help reduce the frequency, intensity, and damage potential of each problem area. Further instructions to the stakeholders at the breakout tables should be provided by the facilitator as the separate workshops proceed.

Topic D: Summary of Alternative Dispute Resolution Once instructions for conducting the table work have been completed under the Olanta agenda, the discussion leader should very briefly address the subject of alternative dispute resolution and its relation to the workshops and the ideas they produce. Generally, no major discussion of ADR is appropriate at this time in the meeting except to remind stakeholders that issue resolution and ADR are critical follow-up items for them after they have written the charter.

Topic F: Specific Procedures to Follow in Workshop 1 The essential ingredients for a good start to the workshop operation are for the table stakeholder to appoint a discussion leader, a secretary to take notes, and a spokesman or spokeswoman to present findings of the group. At this point the total group is ready to break into the table-size workshops and to begin their active work of formulating answers to question 1: What do others do that cause problems for us?

To recapitulate, the subjects covered and the time frame of the initial full stakeholder session for the Olanta project up to the beginning of the workshops are as follows:

25 minutes	Introduction of company, systems, and individuals
15 minutes	Review of project characteristics
40 minutes	Orientation and briefing on partnering and related subjects
80 minutes	Elapsed time from start of meeting at 8:00 A.M.

Under this schedule, the group session briefing would be completed at about 9:20 A.M. Allowing participants about 10 minutes to stretch, find their team tables, and become acquainted with each other while getting their assignments, brings the meeting time table to 9:30 A.M. This is where the actual workshop sessions begin. This is the point where the stakeholders begin to build their information pool from which to write the charter.

SECTION C: THE CHARTER WORKSHOPS—BUILDING THE INFORMATION POOL

The workshop questions asked of the stakeholders in workshop sessions 1, 2, and 4 in the charter meeting are simple, straightforward, and reasonable. However, they are questions that are seldom, if ever, addressed in any rational, organized, or intelligent manner in the day-to-day activities of the construction business. Instead, construction practitioners have for years assumed that the less said about the problems we mutually cause each other, the less likely we are to get anyone who is important mad at us.

Similarly, when we try to presolve problems to avoid trouble and destructive conflict, we are forced to define those problems. And by all means, those managers working in design and construction avoid publicly admitting that they ever cause anybody else connected with the project any problems that might damage the job in any way. The fear of litigation, of permanent public relations damage, and of a damaged ego are often the barriers that prevent serious and objective queries to be made.

The table breakout sessions in a charter-writing session should be designed to encourage free interchange of discussion between people in the same or closely related work efforts. For example, it is best to put the architectural production team at one table. If one table will not accommodate the group break it up into a table A and a table B. A table arrangement that has been proven to work well is shown in Chapter 10, Section C, the note template for charter writing. However, every job is different and often the success of a partnering meeting hinges on how well the discussion leader formats the table assignments.

Try especially to keep trades or disciplines traditionally antagonistic toward each other at separate workstations. The attendees at the separate tables will generate more realistic and useful problem and mission statements than if coupled with others with whom they argue and disagree over design and construction ideas and behavior. As pointed out frequently, a time for reconciliation comes gradually in the partnering charter. Don't force it too early until each attendee begins to see where he or she is actually or potentially damaging the job by their behavior.

The partnering workshop sessions offers a method by which potential job difficulties can be identified within a friendly environment. They then, under the ground rules of good partnering, can be presented without criticism or negative judgment by others. In effective partnering the problem statements of one group are merged with the problem statements of another group, and gradually, through the process of writing the charter, the problem statements lose their individual and organizational origins. They become just another problem the project team, the stakeholders, must solve.

The teams are now at their breakout tables and have a clean sheet of paper up on the flipchart or a clean overhead transparency and marker pen on the table. The recording secretary is ready to take notes, and the team-appointed chair is in charge. The charter meeting leader, Robert Shea, announces that the secretary is to put the team workshop table designation in the upper right corner of the flipchart or the first transparency, today's date under the table identification, and the workshop number WS 1 under the date. All problem items are to be numbered consecutively. If neither flipcharts nor transparencies are available for presentation use, the table team secretary can take notes on regular paper and the teams can discuss the workshop products from copies or memory. This is a difficult technique to use and I do not recommend it as a technique to be considered.

With the materials at hand and the people eager to begin, the workshop is officially launched and under way!

Workshop 1: "What Actions Do Others Take That Create Problems For Us?" (9:30 A.M. to 10:50 A.M.) In this discussion period the participants are to list problems that they feel they might encounter on a project similar to the Olanta job. Table leaders should stress to their groups that problem listings should be drawn from the stakeholder's past experience. The projects from which the statements are drawn can be of any nature, for any owner, in any location, and of any size. The important element here is that problems that may be encountered on Olanta should fall in the list somewhere.

The meeting or table leader can often move the discussion along by asking the group to define what constitutes a problem. Usually setting definitions of terms forces serious thinking and reasoning about the assignment. A list of problem categories can be provided to the stakeholders as a reference to get them started on the actual problem statements. Following is an alphabetical list of 45 common problem types that stakeholders normally encounter in partnering workshops. Definitions of each category are given in Appendix D, Section D.

1. Approval processes (apv)
2. Back-charges (bch)
3. Being a good off-site neighbor (ofn)
4. Being a good on-site neighbor (onn)
5. Closing out the project (clo)
6. Communicating with others (cwo)
7. Constructibility (cbl)
8. Construction document quality (cdq)
9. Contract interpretation (coi)
10. Cost growth (cgr)
11. Decision making (dma)
12. Documents and documentation (doc)
13. Equipment and material problems (emp)
14. Financial problems (fin)
15. Inspecting and testing (ite)
16. Issue, conflict, and problem resolution (ire)
17. Job management (jma)
18. Labor conditions (lab)
19. Legal matters (leg)
20. Maintaining regular project evaluations (mpe)
21. Organization, authority and responsibility (oar)

22. Paperwork and administrative work (paw)
23. Payment processing (ppr)
24. Personnel quality and problems (pqp)
25. Planning and scheduling (pas)
26. Policies and procedures (pop)
27. Procurement of materials and equipment (prc)
28. Program conditions (prg)
29. Project cost structure (pco)
30. Quality management (qma)
31. Regulatory agency matters (reg)
32. Revision processing (rev)
33. Safety (saf)
34. Staff morale and attitudes (sma)
35. Staffing and personnel (stf)
36. Submittal processing. (spr)
37. Substitutions and alternates (sal)
38. Time growth (tgr)
39. Timely action (tac)
40. Training (tng)
41. User-group interaction (ugi)
42. Value engineering (ven)
43. Warranty conditions (war)
44. Weather conditions (wea)
45. Work-site conditions (wsc)

As the conversations pick up speed and become more intense, the table secretary must make certain to record the problems as the individuals and groups at the table discuss and agree to them being included in the list. The flipchart writing should be legible, within the context of the individual statements of the problem, and phrased in such a manner that the problems are reasonably self-explanatory for ease of discussion at the completion of the workshop. Workshop 1 should proceed across the coffee break, with stakeholders having their break as they continue discussing the content of the problem statements. For a start of 9:30 A.M., I recommend that workshop 1 be allowed to continue with a working coffee break until about 10:20 A.M. As the problem statement sheets are completed they should be posted on the walls adjoining the workstation in the order in which they are prepared.

A few examples of typical problem statements encountered in partnering sessions are given below.

- Insufficient time allowed to make required decisions
- Poor communication (inappropriate amount of written/oral communication)
- Missed dates and not keeping schedules
- Late or incomplete submissions
- Directed to proceed without change order
- Clear chain of command not established
- Closed minds (preconceived solutions)
- Putting self-interest above overall project goal
- Improper contract documents
- Late submission of shop drawings and quotations
- Poor interfacing of trades
- Lack of timely notification of required shutdowns
- Short-notice scheduling of meetings, equipment demonstrations, and testing
- Untimely delivery of owner equipment
- Unreasonable payment requests for payment
- Poor planning schedule and sequence
- Failing to follow plans and specifications
- Feeling that the owner has an open checkbook; no regard for budget
- Unreasonable demands by regulatory agencies
- Failure to meet project deadlines
- Maintaining clean job site in residential area
- Slow decision making
- Changing scope of design
- Environmental problems
- Slow review of proposed operations manuals
- Capital improvements budgeting is not well understood by all departments
- Too many chiefs
- Delayed payments to contractor
- Late changes
- Slow change orders and general paperwork
- Not demanding expected performance

As the problem work sheets are being prepared, I suggest that the discussion leader or facilitator begin filling in the template prepared before the meeting. A suggested format for the template is shown in Chapter 10, Section C.

Our Olanta meeting has now proceeded through topics A, B, C, D, and E, and is well into workshop 1. Space should be left under the main headings in the template for recording the results of the individual workshop discussions for each of the tables. There are many methods of tabulating results. Some of the methods are listed later in this chapter.

When workshop 1 table work is completed (about 40 minutes after its start) each team arranges a display of its end product in the most favorable visual order on the wall near their table station. Then team members begin a round of presentations of their ideas and statements to the other stakeholders. This is best done by allowing the viewing teams to stand or sit in whatever locations are convenient from which to view the presentation material of each of the other teams. Encouraging the viewing teams to move around allows them a little muscle stretching and brings them into brief contact with some of the other stakeholders.

The presentation of each team's work is made by the table leader or other selected representative from the group. It is made directly from the work product of the session. If the statements have been made on transparency blanks, these are displayed on the overhead projector. If problems have been recorded on notepaper, they are presented orally. Viewing and listening stakeholders should pay careful attention to each problem statement from each of the other teams. They are allowed to take notes, ask questions, or seek clarification. However, they are not allowed at this time to criticize the problem statements or to argue about the origins or validity of the statements.

The purpose of the workshop is to provide a clean, straightforward view of potential problems for the purpose of finding methods to manage them so they do not ultimately result in damaging the job. Each team should be allowed from 4 to 7 minutes to show and explain their set of statements. This is a tight presentation schedule but does have the advantage of helping minimize criticism and argumentative discussion. Opportunity for such conflicts in views are provided later in the group discussions following table team work.

As the teams are making their presentations, the facilitator and the assistant, if available, should be completing their electronic recording of the flipcharts or transparency notes. These notes are then printed, copied, and issued to the stakeholders as the workshop 1 newspaper. At this point workshop 1 is complete and the teams ready themselves for workshop 2.

Workshop 2: "What Actions Do We Take that Cause Problems for Others?"
(10:50 A.M. to 11:30 A.M.) Workshop 2 is very much like workshop 1 in format and discussion sequence. However, there are some major differences in tempo, intensity, and output. First, the teams will usually be surprised that the chair or facilitator actually wants them to criticize themselves. Once they realize that this request is for real, the team members usually go to work and gradually produce a timidly evolving list of problem statements totaling about half of the number they produced in workshop 1. Depending on the characteristics of the team members, the comments will range from duplicating or paraphrasing the problems that others cause us given in workshop 1, extending all the way to a completely different set of problems unique to the disci-

pline of the table group (see the list of problem statements from workshop 1 listed above).

When the workshop 2 problem statements are completed, the same presentation and discussion format is followed with the other stakeholders, as in workshop 1. Each table presents its list with a brief comment on the ideas behind the problem statement. The others listen and are allowed to question or ask for clarification. However, as with workshop 1, no criticism or editorializing is permitted. Meanwhile, the facilitator and the assistant should be recording the problem statements and getting ready to print and duplicate the workshop 2 newspaper. If the material for workshop 1 newspaper is not yet completed, it should be held for a few minutes and the results of workshops 1 and 2 combined into one publication.

Workshop 3: "What Is the Most Important Result for Me and My Organization to Achieve by Completing this Project Successfully?" (11:30 A.M. to 12:00 noon)   Now that the first two intensive workshops are completed, the direction of the charter conference is changed slightly. It is time to move out to the macro view of the job and determine, from the material discussed and presented to this point, what the mission of the project should be. We often look on mission statements as being philosophical and somewhat esoteric in nature. For some reason, they never seem to relate to our day-to-day efforts to be good and to achieve those things that are important to our organization's and our own success. All too often there is a tendency to write a mission statement for reading by those outside the organization. Sad to say, many of the organization's internal staff whose mission statement is on display have never read it. It simply does not seem to relate to their daily duties and responsibilities.

In this workshop the efforts of the group are directed toward synthesizing the problem statements presented to date and seeking methods of improvement. The vehicle is the individual mission statement, where the statement of mission points directly to the person's organization and this specific project. The statement moves the person and what she or he can do into sharp focus. An analysis of individual workshop mission statements from many partnering meetings indicates that the quality of the individual concerns for favorable outcomes of project work may actually forecast how successful the total project actually may be. Mission statements are important parts of every charter.

A few examples of individual project mission statements taken directly from a wide variety of project teams are paraphrased below:[3]

- To complete all phases of the project on time and within budget while providing the maximum quality of product for our client. Quality means

---

[3] Some editing has been done to conceal the identity of the parties involved and to clarify the meaning of portions of the missions. Very few corrections have been made and most statements are as they came from a random list of individual mission statements.

the ability of the buildings to satisfy the program while providing maximum satisfaction.
- Construction of a facility that satisfies the client fully and is satisfying to the construction team.
- To have an owner who is satisfied with the project. To have established good ongoing relationships with the team players.
- To provide a high-quality electrical system to the owner on time and within budget.
- To complete the project within budget, on time, and to the quality standards desired by the owners. To develop a prequalified team that can be used on future projects.
- We recognize the common goal to finish this project with the highest quality, on time, and within budget, and agree to work together safely, as a team with trust and cooperation.
- To work in harmony with all team players to provide a project that everyone can be proud of.
- To complete a quality facility with appropriate benefits to all participants without legal disputes and in a manner that is enjoyable to all.
- We, as a team, seek to create a high-quality product, maintaining both sensitivity and integrity of design, a facility constructed in a cooperative spirit, on schedule, resulting in an equitable and pleasant experience for all.
- To deliver a high-quality product safely, in sequence, on time, and under budget.
- To work together to produce a high-quality project in a win–win partnership.
- To help the joint venture management provide effective leadership and make informed decisions that result in a completed project that exceeds their expectations.
- To complete the project in a manner that results in a win–win for the owner, designers, and contractors in terms of schedule, cost, profit, pride of workmanship, and relationship building.
- To provide accurate, timely, and sound structural design and implementation guidance.
- To provide a high-quality project on time and within budget for the community college and other participating educational groups.
- We have one objective, and that is to complete the project in a manner that will make us all proud to be a part of it.
- That the client will be pleased with our efforts and give excellent recommendations to other potential clients.
- Timely delivery of a competently and soundly constructed facility that meets the operational and mission needs of the operators and clients for which the facility was designed and constructed.

- To complete our portion of the project in a manner that will best serve the owner's needs and be something to look back on and be proud of: to do so in an efficient and profitable manner.
- To deliver to our client a high-quality project, on time, within budget, and to develop a long-term relationship beyond the project completion.
- To conceive and design building systems that will stand the test of time— that is, live up to the expectations of the building occupants over the building life.
- A satisfied client and establishment of a good long-term working relationship with the owner and design team.

Of particular importance for the facilitator in leading the Olanta stakeholders through the mission-writing workshop is to stress brevity and clarity. Mission statements should generally be 25 words or less. They should focus on performance. They should tell what is to be done by the person's company to achieve the desired target and then describe what that end product is going to be. Individual mission statements should be written on separate sheets of notepaper by stakeholders working by themselves. Each stakeholder should print a three-character code[4] in the upper right corner of his or her paper. The stakeholder code should be used by each stakeholder to identify all work that he or she may do during the partnering session.

Mission papers should be folded and turned in to the meeting leader. As the workshop 1 and 2 newspapers are being printed and copied, the facilitator and his or her assistant should begin transcribing the individual mission statements for use by the project mission-writing task force. The mission statements are written into the workshop template section F.

Project Mission Workshop (Purpose: To Write a Preliminary Presentation Project Mission Statement)   After the individual mission statements have been written and turned in, and all presentations and comments have been completed for workshop 2, the meeting leader and facilitator should ask for three to five volunteers to serve on a task force to write the project mission. If none offer to serve, the leader should appoint the stakeholders to the task force.

This group is to reassemble in the meeting room about 30 minutes before the start of the afternoon session. If lunch is at noon and the stakeholders and visitors are given 1 hour for eating, walking, and fresh air, the project mission task force members should be called into session at 12:30 P.M. They can meet in a secluded corner of the main meeting area if the area is relatively undisturbed by the ongoing meeting of the remaining stakeholders starting

---

[4] The code can be any combination of letters and number from one to nine desired. For example, a stakeholder code could be 5HQ. This code should be used on any individual effort during the session to allow material to be anonymous but easily identified by the owner.

up again at 1:00 P.M. If the main meeting room is not suitable, the task force can often find a quiet place in the lobby of the conference location or perhaps an empty meeting room near where the charter meeting is being held.

The task force is to derive a project mission statement from the combined individual mission statements prepared in workshop 3. The job of the project mission task force is not easy. Usually, the effort is a thankless one and occasionally the task force finds itself embroiled in disputes over the true mission as perceived by each of its members. One task force with whom I worked actually came to only an uneasy compromise on their mission statement with which one of the participants vigorously disagreed. However, since the mission statement is to be submitted to the full group of stakeholders, there is always a check-and-balance system at work to ensure that the mission speaks to concerns of the most prevalent points of view, so such disagreements can often strengthen the statement and its intent.

The working method I recommend for the task force requires that the list of individual mission statements developed in workshop 3 be provided to them for their use during their short working conference. Within our Olanta agenda these will probably have to be typed and printed over the early part of the lunch period. The task force should be provided, in addition to the individual mission statements, a flipchart, a few blank overhead transparencies, and a variety of colored transparency marking pens. The first step to be taken in preparing draft 1 of the project mission is for the task force members to identify as many operative words in the individual mission statements as possible. Operative words are the nouns, verbs, and sometime adjectives that vividly describe what it is the writer of the mission was trying to say.

As an example, one of the statements above states that the mission of the individual is:

- To complete the project within budget, on time, and to the quality standards desired by the owners. To develop a prequalified team that can be used on future projects.

Underlining the operative words in these two sentences provides us a further emphasis of the points the writer wanted to make:

- To <u>complete</u> the <u>project</u> within <u>budget</u>, on <u>time</u>, and to the <u>quality</u> standards <u>desired</u> by the <u>owners</u>. To <u>develop</u> a <u>prequalified team</u> that can be <u>used</u> on <u>future projects.</u>

As the operative words are identified in the various mission statements and the frequency of their use noted, the goals and objectives of the collective stakeholders should begin to emerge. Since the stakeholders are the people at risk on the project, and since we are attempting to develop a working nonlegal moral agreement among these people, we should take their statements seriously. The operative words have been found to be their serious words.

Once the task force has identified the important words to include, they must then prepare a first draft of a mission to bring to the meeting floor. The draft 1 statement can be printed on an overhead transparency or written on a blank flipchart, whichever is most convenient for working with the full group toward a final draft of the mission. The task force merges their project mission efforts with the main group work at about 1:15 P.M. Meanwhile, the stakeholders have assembled after lunch, at 1:00 P.M., to continue their partnering efforts.

Workshop 4: "In Light of the Problems Stated in Workshops 1 and 2, and the Mission Defined by Me in Workshop 3, What Recommendations Can We Make that Would Help Improve Relations and Performance on the Olanta Data Project?" (1:00 P.M. to 1:30 P.M.)   Upon reassembling at 1:00 P.M., the Olanta stakeholders should have, or be given, three documents to assist in their efforts:

1. Workshop 1 newspaper: problems others cause us
2. Workshop 2 newspaper: problems we cause others
3. Individual mission statements as provided to the project mission task force

When the stakeholders reassemble, they should take their seats at the breakout tables in their original groups. They have in their hands the problem statements and missions developed in the morning workshops and are now ready to generate recommendations that will help the project team solve these problems and accomplish their defined missions. The workshop session is a one-half-hour discussion in which the teams prepare as many recommendations as time permits. The objective is to prepare a starter list of goals that will be used later in the afternoon during the intensive charter objective definition work. This later session is one in which the full stakeholder assembly participates.

Work at the individual breakout tables should proceed as vigorously as possible and basically address solutions to the problem statements prepared earlier by that team. Recommendations will probably come slowly since often single recommendations will apply to many problem statements—just as multiple recommendations may occasionally be applied to single problem statements. Assembling this material in full by addressing all problems defined is very difficult with the time limitations on the partnering charter session. However, some work on recommendations is essential if the charter is to be written properly.

Examples of recommendations that have been made in workshop 4 during several project partnering meetings provide an insight into the suggested content of these recommendations:[5]

---

[5] Minor editing has been done on some of the items to improve readability.

- Be prepared for progress meetings.
- Provide accurate and complete close out documents.
- Releasing line-item retainer will speed up close-out.
- Be considerate of all parties' roles in the project.
- Resolve disputes fast.
- Prioritize submittals.
- Develop and maintain an organization chart for the project.
- Follow the established chain of command.
- Provide information to develop, maintain, and adjust a project work schedule.
- Know the job.
- Keep disruptions to existing work at a minimum.
- Maintain a safe site.
- Maintain a secure site.
- Stress and encourage pride and workmanship.
- Recognize and reward good work.
- Respect other trades.
- Be a good neighbor.
- Pay promptly.
- Invoice proper amounts.
- Allow for adequate staff input prior to making major project decisions.
- Provide detailed budget and scope information to appropriate internal staff prior to final design.
- Continue to encourage internal staff input and involvement through project meetings, consultant review meetings, and task forces.
- Identify project spokespersons and their backup.
- Work in the spirit of the partnering charter.
- Work hard on communication (timely, forthright, thorough).
- Remember that our customers may be you, your family, or even your future customers!
- Develop an issue resolution system.

The important accomplishment of workshop 4 is to get an idea quickly of how the loose-cannon problems identified earlier can be brought under control and managed so as to benefit the job. These will later be merged with other ideas formulated by the group in a full stakeholder meeting to make up the charter that is signed at the end of the day's work. As with previous workshops, the notes prepared during the session should be recorded on flipcharts, transparencies, or notepaper, and transferred to a medium such as a computer for printing and duplicating. The recommendation publication becomes workshop

4 newspaper. It should be ready for use by the group during the full session of the stakeholders in which the project objectives are set.

Workshop 5: Full Stakeholder Group Discusses and Revises First Draft Mission Statement (1:30 P.M. to 1:45 P.M.)

> Writing this draft mission statement was the hardest work I have ever done.
> —*Vice president of stakeholder company*
> *and member of a project mission task force*

By 1:30 P.M. the table work sessions are completed and the task force has prepared its draft 1 recommendation for the project mission statement. Recall that this statement was derived from the individual mission statements, paying particular attention to the operative words. This should bring the task force's ideas into reasonably good alignment with the intent and desires of the stakeholders.

Many different mission statements have been produced during charter workshops over the past few years. Despite what may appear as an item that would tend to become repetitive from partnering meeting to partnering meeting, there is a surprising variety of ideas, desires, goals, and objectives expressed in the missions. Some examples of first-draft missions by task forces working from the individual mission statements will illustrate how much diversity can exist in project goals, objectives, people, and managements. These were taken from an industrial plant expansion, a community health center, a city stormwater retention basin, a major airport expansion, a family resort facility, and a small community school expansion, each project in different locations and with different staffing.

- The project team's commitment is to construct a high-quality facility, on time and within budget, maximizing safety, communication, and cooperation so that all participants can be proud and profitable in their accomplishments.
- To design a community outpatient-centered facility that provides present and future health care services.
- Beyond the contract requirements, the project partners will achieve a high-quality project, mutual success, and avoid litigation by a commitment to:
  - A safe workplace
  - Effective communications
  - Trust
  - Timely action
- Emphasizing teamwork: we intend to construct this project on time, within budget, and with the least disruption while providing the most value possible to all customers.

- To deliver "our" project safely, on time, within budget, while maintaining quality, a spirit of cooperation, enjoyment, profitability, and a mutual respect for other's interests.
- This partnering team commits to deliver a safe, high-quality project on time, one that is within budget and profitable for all, through continual communication and cooperation

The statement prepared by the Olanta mission task force and presented to the stakeholders on an overhead transparency read:

We, the project team for ODS, commit to plan, design, construct, and renovate the ODS headquarters into a community-based world-class data center safely, on time, and within budget while striving for profitability for all stakeholders.

I recommend that the first round of discussions by the entire stakeholder group be chaired by the project mission task force leader, in this case Marla Jinns, project manager for Tiltsen and Greene, general contractors on the project. Miss Jinns first asks for general comments on the content of the statement. "How well does it seem to express the general feelings of the group according to the individual mission statements?"

All stakeholders now have their own copies of the individual mission statements for reference. Some of the stakeholders say they wish safety to be stressed more, some want a better tie into community concern, others want to emphasize the economic importance of Olanta to the local economy. The design team representatives feel that the significance of the design should be mentioned in the charter and its proper execution be a part of the charter. This discussion should be kept short and lead to a suggestion of specific changes to be made.

Marla should underscore the fact that stakeholders will have other opportunities to revise the mission as the afternoon workshops continue. She should stress that the project mission statement remains a fairly open system, able to be changed until the charter is actually written.

Next, the full group of stakeholders suggest specific changes that should actually be made to the mission to bring it better in line with the desires of the project team. The result of this session is the second draft of the mission statement, which reads:

We, the project team for ODS, commit to design and construct the ODS headquarters profitably as a world-class, architecturally significant, community-based business center safely, on time, and within budget.

The draft #2 mission seems to satisfy most stakeholders temporarily, and Marla wraps up workshop 5 with a moderate degree of satisfaction. She rewrites the mission statement on a fresh flipchart and displays it so that all stakeholders can refer to it in workshop 6.

SECTION D: WRITING THE CHARTER

Workshop 6: Full Stakeholder Group Discusses What Specific Project Objectives Can Now Be Set from the Results of Workshops 1, 2, 3, 4, and 5 (1:45 P.M. to 4:30 P.M.)   Workshop 6 in the Olanta charter-writing session will start about 1:45 to 2:00 P.M. with Robert Shea resuming his facilitator role and with all stakeholders, including the mission task force, attending. Material in hand or available at this point includes:

1. Workshop 1 newspaper: table work on problems others cause us
2. Workshop 2 newspaper: table work on problems we cause others
3. Individual mission statements as provided to the project mission task force
4. Workshop 4 newspaper: table work on recommendations to improve performance
5. Draft 1 of the project mission statement
6. Draft 2 of the project mission statement

The first item of business during workshop 6 is to select and present one objective that should be achieved if the problems discussed in workshops 1 and 2 are to be minimized and resolved promptly within the staff and at the originating level. One problem mentioned by Olanta's staff and about which there was much discussion during workshop 1 was stated as:

1. Areas containing electronic equipment being kept clean and free of noise and vibration during construction

The specifications address this problem partially by describing locations and composition of dustproof partitions to be built around existing areas. However, the owner and the architect/engineer staff admit that they can only specify so much. At some point there is a boundary where moral commitments to a set of clearly implied actions must be called into play. This was the main reason for the alert to the problem by Olanta.

Robert Shea asks the stakeholders (now meeting as a total group) if and how this problem might be addressed in the charter. He first asks a most important question and one that must be repeated over and over during the charter-writing session: "Is this matter covered satisfactorily in the contract documents?" If there is a unanimous "yes!," the item *does not* belong in the charter. One of the basic definitions of a charter stresses that it is an agreement in principle that does not supersede, alter, or supplant the planning, design, and construction contracts in place, or to be written and executed.

The problem of temporary protection for Olanta's present operation requires some additional questions. These include:

- How is temporary protection covered in the full contract documents?
- What additional requirements must be met to accomplish a level of protection satisfactory to the Olanta facilities' staff?
- What protection will be needed that is not presently covered by contract?
- What additional costs might be incurred by whom to provide what the owner expects as compared to what the contractors expect to provide?

Temporary protection is obviously a complex matter and it would be neither desirable nor possible to cover all ramifications of the potential problem and fully answer the questions in this session. A brief discussion reveals that there are still serious and potentially expensive gaps in the temporary protection system. One of the stakeholders offers the suggestion that a charter phrase be written as follows:

All stakeholders on the Olanta Data Systems project commit to:
1. Protect Olanta's existing and new facilities so as to maintain operational continuity within the scope of the contract. A need for and a scope of additional protection will be mutually agreed upon as the need arises.

Objective 1 is entered in the charter draft and the item is checked off the list of problem areas being considered. It may seem at first that this charter provision states the obvious. However, several things have been achieved as the discussion about protection proceeded. First, the potential problem was brought clearly to the attention of the stakeholders. They were made aware of the possibility that the scope and cost of protection work might increase. Second, the stakeholders have recommitted their intent to abide by the contract they have with Olanta and with Tiltsen and Greene. Third, Olanta and the architect have signified their intent to recognize legitimate increases in scope and cost. Fourth, the contractors desire to accept what is their legitimate risk in the matter of temporary protection. However, they seek, and are entitled to, reassurance from their client that extra costs will not be rammed down their throats.

In essence, what at one time in the history of construction would have been a given with no questions asked has now been confirmed in a contemporary construction project where a partnering effort must often create such an attitude. A most important factor in partnering success is that at the end of the session, all stakeholders and signatories to the charter have made a moral pact to behave in a manner as stated in the objectives. In the construction world, this commitment is a very powerful incentive to live up to your word.

An actual event may serve to illustrate how powerful the moral commitment is in most construction professionals. In a partnering session on a project where considerable earth had to be moved, the stakeholders were discussing a suggested charter objective. "No litigation!" was the suggested charter statement. No comments came from the stakeholders, and the group was just about to move to the next provision when a man stood, identified himself as the dirt contractor, and said "If that provision goes in, I don't sign the charter."

Not a sound was heard in the room. The facilitator asked how he could keep the intent of the statement and still write in something the contractor could sign. After a moment's thought the excavator said, "It should read that all stakeholders will strive to avoid litigation on the project." The facilitator asked if the excavator would sign the charter if the change was made. The dirt contractor said "yes," as did the other stakeholders, and the charter signing later went without a hitch.

Two things happened here. The excavator, most properly, did not want to give up his right to litigate if circumstances warranted. However, he recognized that litigation should and must be a last resort. The first happening of importance is that if the excavator signed the charter with no litigation morally agreed to, he would be bound to that provision. He could not do this in good faith and within his obligations to his company. Therefore, if the clause was in the charter, he could not sign the charter—and he very much wanted to sign.

Second, the excavator indicated a willingness to negotiate an alternative phrase. He was saying to the group: "I don't want litigation, but I cannot morally agree to close the door on the possibility." The alternative statement will probably reduce the possibility of litigation as much as would the initial statement. The excavator will be willing to abide morally by a agreement that gives him the option of litigation but obligates him and the other stakeholders to avoid it if possible. Such an attitude usually signals the intent to be a good partner.

In the partnering charter meeting how are the problem statements and the proceedings of writing the charter actually handled? In Chapter 10, Section D, I described an overhead projector system of note taking called technography. In this system a liquid crystal display (LCD) projection panel is used to write directly to an overhead display screen. As the charter provisions are projected on the screen they are discussed, set down, rewritten, and rewritten again by the entire group of stakeholders until the words and the arrangement are right. Then the charter is printed and signed.

In my opinion there is no better current method than the LCD technique to write a well-structured charter that can be developed mutually by a competent leader and a cooperative group of stakeholders. As the charter preparation proceeds, the articulation, writing, and approval of the objectives picks up speed as problem statements are rolled into objectives developed from other problem statements. This tempo requires rapid, flexible, accurate writing that reflects the true intent of the partnering participants . The LCD helps do this.

To show how this process is followed, I have duplicated below a typical sequence of events based on experiences in many charter meetings. Draft 2 of the Olanta mission statement is displayed prominently on a wall of the meeting room and reads:

> We, the project team for ODS, commit to design and construct the ODS headquarters profitably as a world-class, architecturally significant, community-based business center safely, on time, and within budget.

A portion of the problem statements from workshops 1 and 2 that are being considered now include the following. Robert Shea has arranged them alphabetically to blur their source and allow easy tracking of those that have been discussed and incorporated into the charter.

1. Areas containing electronic equipment being kept clean and free of noise and vibration during construction
2. Changes imposed by the client
3. Cleanliness of work-site eyesores
4. Cost overruns
5. Delay in submittals
6. Delays in paperwork
7. Disregard for progress schedule
8. Disregard for safety regulations
9. Equipment turnover training and documentation must be thorough
10. Failure to recognize the urgency of the problem
11. Fear of risk taking as it threatens security
12. Inappropriate clothing on workers
13. Incomplete or delayed billing documents
14. Incomplete drawings
15. Incomplete information on drawings
16. Incomplete response
17. Incomplete submittals
18. Individual cleanup
19. Lack of appreciation for a contractor's problem
20. Lack of response to questions by A and E
21. Lack of timely decisions
22. Long punch list
23. Lose interest in mundane but necessary requirements of contract
24. Making untimely changes to contractual requirements
25. Not keeping on schedule with project
26. Not listening attentively due to distractions
27. Not tracking paperwork from beginning to end
29. Personalize issues
30. Poor design
31. Poor workmanship
32. Poorly defined requirements
33. Poorly defined chain of command
34. Priorities do not mirror contract needs

35. Slow processing of change orders
36. Staffing limitations create personnel shortages
37. Subcontractors/suppliers not ready to start job
38. Traffic flow problem (contractors, lodge customers, lodge vendors, employees)
39. Unsatisfactory management of issues
40. Untimely responses

Earlier in this section we saw how problem statement 1 was translated into a tentative charter objective. The result of the discussion was objective 1 in the charter.

All stakeholders on the Olanta Data Systems project commit to:
1. Protect Olanta's existing and new facilities so as to maintain operational continuity within the scope of the contract. A need for and a scope of additional protection will be mutually agreed upon as the need arises.

The next step is to convert problem statement 2 to a charter objective. The statement "changes imposed by the client" is considered by the stakeholders as being difficult to support since Olanta is certainly privileged to make changes as it desires. However, the stakeholders decide to address the problem by stating that they will commit to:

2. Keep project revisions to a minimum—and when changes are essential, to process revisions promptly in accordance with a detailed procedure to be prepared and published by Loring and Metzer.

Notice that proposed charter provision two contains the phrase "prepare and publish." "Prepare and publish" is a phrase often inserted into the partnering charter to direct the stakeholders or a designated person to write and implement a policy, procedure, or guideline for accomplishing a performance that may be required by contract but whose detailed nature is not specified. In this case, Loring and Metzer, the architect and engineer of record, has been directed to prepare the revision procedures detailing the actual implementation expected by the broader specification requirements. It is worthwhile emphasizing that where prepare-and-publish statements become a part of the final charter provisions, some stakeholder must be assigned the responsibility for implementing the prepare-and-publish request.

Problem statement 3 says "cleanliness of work-site eyesores." The statement is somewhat fuzzy, as will be many of the problem statements resulting from the workshops. Mr. Shea asks the team members identifying the problem to be more specific. The chair of the table writing the problem responds that they were part of the site-work team, and that all too often they must work around junk and debris from interior work contractors. They would like to see conscientious and continuous cleanup by all contractors. He adds: "This would help them all to produce a better job for the owner and the user."

The stakeholders suggest that problem 3 be cast into the following form and made a tentative charter provision 3:

3. Being a good on-site and off-site neighbor, by promptly disposing of their own trash off-site, helping to keep the site clean and safe, and cooperating with all others to maintain the workability, appearance, and usability of the site.

Notice that the wording of the initial three provisions of the charter is not especially elegant or complete. The aim in this pass through the problem statements by Robert Shea is to capture the spontaneous ideas and intent expressed by the stakeholders. Cleaning up the statements will come as the afternoon work moves on and the charter statements are altered to accommodate later problem statements.

The next problem situation, 4, posed by the stakeholders merely said "cost overruns." Mr. Shea points out that there are many different kinds of cost overruns for different kinds of participants. The owner's table chair spoke up and said they had written this problem statement as applying to potentially high pricing on bulletins and the possibility of them pushing the owner's budget beyond allowable limits. A noisy discussion immediately centered on the point that specialty or subcontractors for the most part were working under competitively-bid hard-money agreements with the general contractor, Tiltsen and Greene. Therefore, they had little budgetary control over the owner. They suggested that the charter address the reduction of postcontract revisions to a minimum and to stress that the documents issued should be as complete as possible. Mr. Shea suggested the following statement:

4. Reduce construction cost growth by producing complete construction contract documents that clearly express the intent of the owner. When changes must be made, the stakeholders further commit to preparing and submitting accurate, fair pricing of the revisions, and to prompt, competent, fair evaluations and releases of the work to the field.

The Olanta stakeholders are now well on their way to building a first draft of the charter that will ultimately be used for their project. Problem area 5 is "delay in submittals." Mr. Shea and the stakeholders addressed this problem as follows:

5. Prepare, provide, and process submittals in a timely and fair manner while striving to reduce excessive and unnecessary resubmittals of shop drawings and other documents to which approvals must be provided.

Problem statement 6, "delays in paperwork," is translated by the stakeholders into a charter item that reads:

6. Reduce paperwork to an absolute minimum and to process all paperwork promptly while recognizing the importance of pertinent documentation.

The next few items in the problem statements flow smoothly into the translation process and result in the charter provisions for draft 1 of the charter being set down in quick fashion. The phrase "All stakeholders on the Olanta Data Systems project commit to . . ." is assumed to be prefixed to each charter provision.

Problem statement 7, "disregard for progress schedule," is translated into a charter item that reads:

7. Actively participate in the preparation of project network plans and schedules by Tiltsen and Greene, with due diligence, to construct the project in accordance with the current published plan and schedule of work. Tiltsen and Greene will promptly update and publish long- and short-term plans and schedules as job conditions warrant.

Problem statement 8 is "disregard for safety regulations." I suggest that charter objectives dealing with safety be merged with related objectives rather than being isolated as separate safety items. For instance, the matter of safety is already addressed in the Olanta mission. The mission reads:

We, the project team for ODS, commit to design and construct the ODS headquarters profitably as a world-class, architecturally significant, community-based business center *safely,* on time, and within budget.

Further elaboration on safety, which is largely a technical and regulatory matter might lead to conflicts in charter language with the language of the regulations, ordinances, and laws already governing safety. In addition, the acid-test question to be applied to all charter objectives should be, as we saw in charter provision 1, "Is this matter covered satisfactorily in the contract documents?" The contract documents and the rules and regulations regarding safety make further elaboration vulnerable to multiple interpretations.

Problem statement 9 is "equipment turnover training and documentation must be thorough." Problem 9 results in a rather wordy charter provision 8 which includes the phrase "preparing and publishing." Before the charter meeting is over, Mr. Shea must make a final review with the stakeholders and assign the task to a specific person. The need to do this is often signaled by an asterisk placed in front of the objective number.

*9. Closing out the project in a prompt and timely fashion by:
   A. Preparing, publishing, and following acceptable close-out guidelines including those for equipment turnover training and documentation.
   B. Establishing clearly defined punch-out procedures and standards early in the project.

Problem statement 10 involves "failure to recognize the urgency of the problem." The stakeholders propose that deriving a charter statement from this very general problem statement be deferred until encountering a specific instance where the failure to recognize urgency can be neutralized by a definable action.

> It seems to me that this charter, when we finish it, is actually going to be sort of a report card of how well we are doing our job
> —General contractor field superintendent
> and stakeholder on large retention basin

At this point the stakeholders on the Olanta program have disposed of 10 problem statements resulting from workshops 1 and 2 and as many or more of the recommendations made in workshop 4. The remainder of the afternoon is spent in an intense and tightly structured discussion of the randomly gathered ideas and statements of the stakeholders. The final charter is a statement of standards set by the stakeholders by which the project can be evaluated at any point in its progress. A somewhat automatic provision of project partnering charters is an instruction to the stakeholders to establish and implement a partnering evaluation system and an alternative dispute resolution system.

By about 4:30 P.M. the rough draft 1 of the charter should be completed and Robert Shea will first check to ensure that most or all of the important ideas from workshops 1, 2, and 4 have been covered in the charter. Each stakeholder should check over his or her published list of items from these meetings to see if the items are all checked off. Next, Mr. Shea puts the current draft of the mission statement up on the wall or on the screen and gives all stakeholders another opportunity to revise the statement. This option is available up to the time the charter is ready to print.

The last charter shakedown of the meeting is a pass through the charter draft 1 to allow stakeholders to change, delete, streamline, or add any items they wish to improve that might strengthen the document. This process may become somewhat boisterous as the participants begin to understand that this document they have been carving out of hard rock is actually going to be printed for their signature. If there is considerable disagreement about a charter provision near the end of the meeting, I recommend that the chair take a vote on each disputed point. This may seem high-handed. However, after a full day of discussion, editorializing, and thinking about what partnering is and seeing how the concept is reflected so accurately in the stakeholder's charter draft, it is unlikely that any major dampening of the basic philosophies of the group will be destroyed by a majority vote for or against any proposed charter provision. The time is now 4:50 P.M. and Mr. Shea asks the group to scan the final provisions of the charter. Remember: this is the project team's noncontract report card—check the punctuation, grammar, and spelling, and give him the final OK to print the signature copy.

The raw partnering material for the Olanta project has gone through many inspections, revisions, disputes, and arguments. By this time in the process the origins of most of the criticisms that surfaced during workshops 1 and 2 have been lost and the problems are now the stakeholders' problems. The recommendations and resulting objectives have been merged and are now the stakeholders' objectives. The mission no longer is a statement by a task force—it is the stakeholders' statement of the single most important result of the Olanta project being successful.

The document now on the screen is ready for a final review by the participants. It has been reduced to a set of terse, descriptive words that give the charter a moral and ethical meaning that says: "This is how we will behave in noncontract matters!" Usually, there will be 3 or 4 minutes of miscellaneous comments and some last-minute sharpening of the document—and then usually one or two stakeholders will shout "Print It!" Mr. Shea makes a final spell-check, saves, and then presses the print button. Out of the printer emerges our project partnering charter, ready for the stakeholders' signatures. It reads:

Project Charter for the Olanta Data Systems, Ltd., Expansion Program, Travis, North Dakota, February 24, 19xx

Mission

We, the project team for ODS, commit to design and construct the ODS headquarters profitably as a world-class, architecturally significant, community-based business center safely, on time, and within budget.

Objectives

All stakeholders on the Olanta Data Systems project commit to:
1. Protect Olanta's existing and new facilities in order to maintain operational continuity within the scope of the contract.
2. Keep project revisions to a minimum. When changes are essential, to issue bulletins promptly, to prepare and submit accurate, fair pricing, and to release the changed work in a timely manner.
3. Be a good on-site and off-site neighbor and to cooperate with all other stakeholders to maintain a clean, safe, secure, and well-organized work site.
4. Actively assist to control and, where possible, reduce project time and cost growth.
5. *Prepare, provide, and process submittals in a timely and fair manner while striving to reduce excessive and unnecessary resubmittals of shop drawings and other documents for which approvals must be provided.
6. Keep paperwork to a minimum while recognizing the importance of essential documentation.
7. Actively participate in preparing, monitoring, updating, and using project network plans and schedules and to strive to construct the project in accordance with current plans and schedules of work.

8. *Prepare, publish, and implement a close-out procedure that will provide other contractors and the owner with a facility delivered on time, in accordance with the desired contract phasing sequence, and so that retention is released in a fair and timely manner.
9. *Prepare, publish, and implement an alternative dispute-resolution system that will assist to avoid litigation and will help resolve problems in an effective, timely, and fair manner at the originating or lowest possible operating level.
10. Communicate effectively and on an as-needed basis at all levels by being available, recognizing the need for quality information, and attending required meetings.
11. *Prepare, publish, and implement a project partnering evaluation system.
12. Pay all properly submitted project invoices promptly.
13. Employ intelligent and timely use of risk-allocation and cost/benefit concepts on the project.
14. Respect and remain open to the needs and ideas of other project stakeholders and staff.
15. Orient, train, and educate all project staff in the principles of effective partnering as defined by this partnering charter.
16. Take pride in our work and encourage others to take pride in theirs.
17. Accept responsibility for our actions.
18. Listen well and try to understand others' concerns.
19. *Prepare and publish an organizational chain of command showing communications flow, responsibilities, and authority.
20. Maintain continuity of key personnel throughout the life of the project.
21. Make timely decisions in all project-related matters.
22. Do it right the first time.
23. Have fun.

---

* Indicates item to be prepared by stakeholders.

SECTION E: SIGNING THE CHARTER AND FOLLOWING UP

The charter signature copy is placed on a table and Mr. Olanta is invited to be the first signatory. After he signs, the others follow suit until all the participants have become signatories and full stakeholders in the Olanta project. Most stakeholders will sign the charter. To my knowledge, among the several hundred stakeholders I have worked with in writing charters I have observed only a handful who have not signed the charter. The main reason for these participants not signing has been that their organization's policies did not permit them to sign any agreement without review and approval from their top management.

Personally signing the charter immediately after the meeting is significant. Occasionally, I have found that a stakeholder participant may have to leave a few minutes early for important reasons and cannot stay for the signing. If a stakeholder has stayed for most of the important charter content decision but must leave before the charter is final approved and printed, it is usually permissible to have another stakeholder, preferably from the absent party's organization sign for the absentee. It is best to permit only those who have participated in the workshops and the charter preparation to be a signatory. Late signing by nonparticipants weakens the underlying purpose of the charter and is not recommended. I have found it is best not to try to identify those who do not sign, or to call attention to the act of not signing. The entire concept and principle of project partnering is built around trust and belief in the qualities and abilities of yourself and others on the project team. These qualities are most often emphasized in the charter and expressed by those who have signed it. The moral agreement must extend to our actions toward all members of the project team. Unanimous assent is not a prerequisite of successful partnering.

After Mr. Olanta signs he is presented with a handsome blue ceramic cup inscribed in gold leaf with the simple statement, "I am a stakeholder in the ODS expansion program—February 24."

This is an important moment and it should be recognized as such. Try always to celebrate the successful completion of the charter-writing session by the award of a memento to each stakeholder as he or she signs the charter at the close of the meeting. The blue cup may seem like a simple, ordinary gift, but the possession of a similar cup by the 36 people attending that meeting and participating in hammering out the Olanta charter has bonded the group in many ways.

In the several partnering charter meetings I have had the privilege to chair, gifts have included:

- A very good, very sharp stainless steel pocket knife engraved with the name of the project and its location
- A high-quality 6-foot pocket tape measure
- A personal and professional pocket weekly planner
- A weatherproofed clipboard for taking and storing field notes
- A captain's cap with a personalized logo showing the project and the location
- A crystal jar full of peppermint candies with a pewter label showing the project owner's company name
- A baseball cap with a special badge identifying the project and the date

In some cases the sponsors have not only awarded a gift but have hosted a social hour or two after the charter has been signed. Notice that these mementos are all unique gifts—items that are not available except to the stakeholders at this one-of-a-kind conference and workshop. Remember, as planners, designers, and constructors we are very much like children. We respond positively, emotionally, and with enthusiasm to motivational rewards that signify celebration of success.

These are only a few of the mementos that have marked the celebration of charter-writing success. Some sessions have ended without the memento or a celebration. These have tended to be less successful partnering efforts than those which are marked by a formal recognition of achieving this first milestone along the project partnering path.

One other housekeeping item of importance after the charter meeting is complete is the duplication and distribution of the charter. If at all possible, the final signed charter should be given the stakeholders before they leave the conference area. If a social hour follows, many of those attending will be signatories and the charter copy can be distributed in part at that time.

The immense amount of information gathered during a charter-writing session is of value to the stakeholders in that it allows a world-of-words tracking of how the various workshop sessions lead to the formulation of the charter objectives. For instance, the stakeholders might have identified 50 or more problem situations that could develop affecting the mechanical and electrical engineering design team during the project design, construction and turnover. In the recommendation session, during which the question asked was "In light of the problems stated in workshops 1, and 2 and the missions defined by this team, what recommendations can we make that could help improve relations and performance on the Olanta Data Systems project?", perhaps as few as five or 10 terse overview recommendations would result. These might be boiled down by the discussion to as few as three or four charter objectives dealing with the problem. The verbal trails leading to the charter statements often are valuable reminders of how the stakeholders actually selected their charter objectives for the signature draft.

The other main charter element, the mission, is also important to track. In the full set of notes the individual mission statements generated in an early workshop are reproduced as the stakeholders wrote them. These are then used by the project mission task force to prepare the first draft mission statement. It is surprising to see how the attitudes of the project team and the stakeholders are reflected in their individual answers to the question, "What is the single most important result to be achieved for me and my company by this project being successfully completed?" By analyzing these anonymous statements, the experienced and astute observer of construction professionals can actually determine the degree of success the stakeholders will have in their partnering efforts. In fact, the individual mission statements are powerful indicators of the degree to which this project team will complete the project successfully.

A well-recorded set of meeting notes will contain this tracking material. If the discussion leader has taken good notes as the sessions proceed, he or she will have most of the material available to distribute soon after the meeting. I recommend that the entire set of notes be provided to each stakeholder for reference and use. The distribution indicates a high level of confidence by the sponsors in the ability of the stakeholders to use the notes well.

Along with the full set of notes for the charter-writing session, the sponsor should provide all stakeholders with a copy of the signed charter. Frequently, the sponsor will have the signed charter copied onto a good grade of paper or simulated parchment and send it to the stakeholders with the meeting notes.

These follow-up actions to the charter writing effort help make the entire experience meaningful to the stakeholders. In addition, stakeholders see that the people who sponsored the partnering are in back of the effort and want it to succeed.

All stakeholders now have in their hands a copy of the signed charter, a full set of meeting notes, plus whatever material they carried away from the charter-writing session. It is time to move to the second major stakeholder effort: writing and implementing a project evaluation system.

12

THE PROJECT PARTNERING EVALUATION SYSTEM

SECTION A: THE BASIC CHARTER

Just as a three-legged stool cannot stand unless all three legs are properly in place, a project partnering system cannot stand unless three critical elements essential to its effective implementation are properly established:

1. A project charter
2. A project evaluation system
3. An issue resolution system

In Chapters 10 and 11 we described how to prepare the first leg of our partnering support system, the charter for the ODS expansion program. In this chapter we begin to unfold the process by which we use the charter to help create the second supporting leg, a project evaluation system. The Olanta project charter described in Chapter 11 and reproduced below consists of a mission statement and a set of project objectives.[1] The Olanta charter was written in one vigorous, action-packed day of study and workshops, signed by all the stakeholders, and is now in effect.

---

[1] The Olanta charter is reproduced in this chapter for reference convenience.

Project Charter for the Olanta Data Systems, Ltd., Expansion Program, Travis, North Dakota, February 24, 19xx

Mission

We, the project team for ODS, commit to design and construct the ODS headquarters profitably as a world-class, architecturally significant, community-based business center safely, on time, and within budget.

Objectives

All stakeholders on the Olanta Data Systems project commit to:

1. Protect Olanta's existing and new facilities to maintain operational continuity within the scope of the contract.
2. Keep project revisions to a minimum. When changes are essential, to issue bulletins promptly, to prepare and submit accurate, fair pricing, and to release the changed work in a timely manner.
3. Be a good on-site and off-site neighbor and to cooperate with all other stakeholders to maintain a clean, safe, secure, and well-organized work site.
4. Actively assist to control and, where possible, reduce project time and cost growth.
5. *Prepare, provide, and process submittals in a timely and fair manner while striving to reduce excessive and unnecessary resubmittals of shop drawings and other documents for which approvals must be provided.
6. Keep paperwork to a minimum while recognizing the importance of essential documentation.
7. Actively participate in preparing, monitoring, updating, and using project network plans and schedules and to strive to construct the project in accordance with current plans and schedules of work.
8. *Prepare, publish, and implement a close-out procedure that will provide other contractors and the owner with a facility delivered on time, in accordance with the desired contract phasing sequence, and so that retention is released in a fair and timely manner.
9. *Prepare, publish, and implement an alternative dispute-resolution system that will assist to avoid litigation and will help resolve problems in an effective, timely, and fair manner at the originating or lowest possible operating level.
10. Communicate effectively and on an as-needed basis at all levels by being available, recognizing the need for quality information, and attending required meetings.
11. *Prepare, publish, and implement a project partnering evaluation system.
12. Pay all properly submitted project invoices promptly.
13. Employ intelligent and timely use of risk-allocation and cost/benefit concepts on the project.
14. Respect and remain open to the needs and ideas of other project stakeholders and staff.
15. Orient, train, and educate all project staff in the principles of effective partnering as defined by this partnering charter.
16. Take pride in our work and encourage others to take pride in theirs.
17. Accept responsibility for our actions.

18. Listen well and try to understand others' concerns.
19. *Prepare and publish an organizational chain of command showing communications flow, responsibilities, and authority.
20. Maintain continuity of key personnel throughout the life of the project.
21. Make timely decisions in all project-related matters.
22. Do it right the first time.
23. Have fun.

---

*Indicates item to be prepared by stakeholders.

SECTION B: PREPARING THE PROJECT PARTNERING EVALUATION SYSTEM

Charter provision 11 states that the stakeholders will maintain partnering effectiveness by preparing, publishing, and implementing a project partnering evaluation system. A project partnering evaluation system usually contains two major elements:

1. A measurement and evaluation system
2. An implementation program for correcting undesirable job trends that threaten achievement of the objectives and the mission contained in the charter

Preparation of a measurement system can be approached in various ways. The main ingredients of the system are the weight or importance of the item being measured and the current condition of the item being measured. A help in getting started is to identify first, the purpose of the measurement system, and second, the results expected from its use.

The entire rationale behind partnering is to establish a set of moral and ethical behavioral guides in noncontract matters that will make the job as successful as possible for all stakeholders on the project (see Chapter 6, Section A, for a detailed description of how to measure construction project success). Therefore, a prime expected result of regularly measuring partnering effectiveness should be a strong attempt by the stakeholders to adhere to the partnering objectives. This effort should result subsequently in reducing the probability of disruptive, costly problems that have the potential for harming the job (see Chapter 11 for samples of Olanta problem statements).

The Olanta charter provisions were specifically built around stakeholders' recommendations that were designed to improve relations and performance on the project. Assigning an order of importance or weight to an objective and then regularly rating the performance quality of the actions needed to achieve that objective should give stakeholders an accurate snapshot of how well the project is proceeding. The ratings should also provide a quick and accurate measurement of what specific items are contributing to the job's good, fair, or poor performance over the evaluation period.

In a nutshell, a periodic partnering evaluation is *a project report card, grading the stakeholders on how well they achieved their project mission and followed the recommendations contained in the charter over the length of the evaluation period.* Then, with intelligent and competent management, stakeholders can convert the results of the analysis to macro and micro management actions that will improve project performance.

With the purpose and results expected from the evaluation identified, the project team can turn its attention to how to assemble the evaluation system. First, when should the evaluation task force begin to do its work? In a one-day partnering charter meeting adequate time is not available to do more than merely provide an overview of the evaluation process. If it is possible to hold a two-day charter meeting, the second day might be devoted to preparing the evaluation system and, possibly, the issue resolution process. However, I have observed that keeping the initial partnering meeting to a one-day session and concentrating that day on writing an excellent charter is the best and most profitable use of the stakeholders' time.

I recommend that the partnering sponsors do not mix the charter preparation period with peripheral topics, regardless of how interesting these topics might be. Write the charter the first day and then follow up with additional conferences over moderately spaced intervals in meetings especially designed to address related partnering topics. Experience has provided several reasons for suggesting this gap between writing the charter and preparing supplementary material to be used in the partnering system.

1. The one-day charter writing session is a long and occasionally stressful series of experiences. By the end of the day, and on signing the charter, most participants are ready to get back to their individual working efforts on the project. Keeping them together too long may damage their understanding and tolerance of each other. Maintenance of both of these at high levels is crucial to successful partnering.

2. Signing the charter is an important and definable intermediate milestone that has been reached on the road to full partnering. Subsequent milestones concern managing, leading, preparing, publishing, implementing, and a myriad of other duties not necessarily done with the same full group who wrote the charter. The charter writers were an assemblage of most of the project ultimate decision makers. They do not all have to be in attendance at the prepare-and-publish working sessions where the evaluation system and the issue resolution system are developed. They must only be given an opportunity to review and comment on each of these as they are developed by other stakeholders acting in a task force mode.

3. In any creative effort, such as writing a well-founded charter, a need is felt somewhere along the way for the participants to think intensely about what they have done. This usually means allowing the work product to

rattle around in the subconscious mind for a while before taking additional actions related to it. The charter has to be mulled over by the subconscious of the stakeholders if they are most effectively to use the charter as a working tool. Allow a few days to elapse, if possible, between writing the charter and building the evaluation and issue resolution systems.

4. The gestation period described in item 3 allows stakeholders' subconscious minds some time to consider which of them are best equipped to prepare the evaluation process. The sponsors of the partnering effort and the other participants can then delegate the task to the qualified stakeholders. In this manner the evaluation system will be produced quickly; it will use the time of the decision makers effectively; and the work product will usually be superior to that resulting from a forced solution arrived at without adequate thought.

Generally, a few of the stakeholders will emerge as leaders in the follow-up efforts required by the charter. If it appears that specific assignments should be made to ensure timely action, the leader and stakeholders might do this during the charter meeting. All follow-up actions, volunteer or free will, committed to during the charter writing should have deadlines set and a timetable established by the leader or the participants.

On the Olanta project, the owner's and the general contractor's representatives have asked Peggy Roethler, project manager for Powers Electric, the electrical contractor, to head up the partnering evaluation task force. Her background includes extensive experience in decision analysis and systems and she is interested and excited by the task. Her boss, Mr. Powers, gives her assignment his blessing, and Mrs. Roethler is ready to begin work. She has been asked by the stakeholders to have a first draft of the evaluation system ready within two weeks. Mrs. Roethler now begins to recruit the stakeholders she wants to work with her on the task force assignment. Usually, there will be enough volunteers to fill the working groups, but the task force leader must be given enough clout by his or her company to assemble a good working group of her choice.

Mrs. Roethler decides that there must be at least one representative from each table of those participating in the charter-writing meeting. Her reasoning is derived from the fact that at the meeting each table group seemed to take on a distinct identity and showed a team cohesiveness that could be continued and encouraged by their participation in the evaluation process. The members Mrs. Roethler selected were:[2]

- Henry Leinenklugle, facilities director, ODS, Ltd., owner; table 1
- John Tarkington, data processing manager, ODS, Ltd., owner; table 2

---

[2] For the company and discipline affiliations of each of the task force members, see the full list in Chapter 10, Section A.

- Mary Charles, computer hardware project manger, Datacomp, Inc., computer systems contractor; table 3
- Ling Metzer, chief designer, Loring and Metzer, architects of record; table 4
- Frank Wilson, Jr., project manager, Frank Wilson and Sons, Inc., mechanical and electrical engineers; table 5
- Marc Smith, project manager, Toonk and Smith, Inc., structural engineers; table 6
- Marla Jinns, project manager, Tiltsen and Greene, general contractors; table 7
- Raymond Strayhorn, project manager, Brown Mechanical, mechanical contractors; table 8
- Mable Tattler, project manager, Efficiency Design, Inc., fixtures, furniture, and equipment contractor; table 10

Mrs. Roethler, who was the representative from table 9 at the charter meeting, calls the task force together for its first session on February 28, the Tuesday following the charter-writing session. They meet at her office and begin to establish the format and mathematics of the evaluation method they will recommend. Mrs. Roethler has set down several questions that are important for the task force to answer during the meetings of the partnering evaluation task force. For this first meeting the questions and the stakeholders' agreed-upon answers were:

Question A: What Is the Purpose of a Partnering Evaluation?   To assist all stakeholders to track job health and provide the project staff with an analysis tool to sustain good performance and take corrective action to correct poor or fair performance.

Question B: What Should Result from the Partnering Evaluation?   A reduction in the probability that disruptive, costly problems having potential for job harm will be experienced. The reduction should improve job morale, increase the potential for true profit, and encourage continuous improvement of stakeholder managerial skills.

Question C: What Data Are Needed to Monitor Project Charter Performance?   Mrs. Roethler says that answering this question is the major job to be done in this first meeting of the Olanta stakeholders.

As mentioned earlier, on the Olanta project the main ingredients to be measured are the relative importance or weight of the charter objectives and the current degree of performance success being realized in achieving each objective. The task force has 23 objectives to which measurements must be applied. As the meeting starts it is obvious that there are differences of opinion among the stakeholders on the task force. Their first heated discussion centers on whether or not to set criteria for measuring the importance of the charter objectives.

If the charter were perfectly written, some members argue, each objective would have the same importance. This perfect charter could have all items assuming the same importance, perhaps a 1 or a 5; or the items could all be said to be of average importance and have a weight of 3 across the board. If all objectives have exactly the same importance, no weight ratings need be assigned. Some professional practitioners feel that the items in the charter should fit this constant-weight model. If they don't, these people believe the item should not be included in the document. Others feel that it is difficult or impossible to write a perfectly balanced charter and that some provisions will always be more important than others. This difference, proponents of the weighted-objective rating believe, can be reflected best in a variable weighting method.

Finally, however, the task force members do agree that weights should be assigned to each objective on a scale of 1 to 5. A weight of 1 means that the objective is of little importance and 5 means that the objective is of great importance. The Olanta evaluation task force stakeholders decide to weight the 23 charter objectives individually. Weights are to be assigned in accordance with the following standards:

Weight	Description
5	Charter objective is of extremely high importance to achieving the mission of the project. If the objective is achieved, its potential contribution to the success of the affected project work is very significant.
4	Charter objective is of above-average importance to achieving the mission of the project. If the objective is achieved, its potential contribution to the success of the affected project work is somewhat over-average but not at the top level of contribution.
3	Charter objective is of average importance to achieving the mission of the project. If the objective is achieved, its potential contribution to the success of the affected project work is at the average for successful similar projects.
2	Charter objective is just below average importance to achieving the mission of the project. If the objective is achieved, its potential contribution to the success of the affected project work is below average but is still of some value to the project.
1	Charter objective is of little or no importance to achieving the mission of the project. If the objective is achieved, its potential value added to the affected project work is minimal and has little impact on overall project success.

Next, Mrs. Roethler leads the task force in setting a method of measuring how well the stakeholders and the project team actually do their work in relation to each charter objective over the evaluation period. Many names can be assigned to this rating. Some call it a value; others call it a performance level or performance quality. The selection of one name over another is made easy by their similarity of meaning. The Olanta team decides to call the gauge of performance during the evaluation period the *performance quality rating,* or PQ. They define performance quality as a measure of how well the project is running. Note that both the weight (W) and the performance quality (PQ) are considered *ratings,* not *rankings.* Each objective of the charter is rated separately, and any weight or performance quality can be the same for multiple objectives.

Mrs. Roethler suggests that the evaluation task force also use a scale of 1 to 5 by which to measure performance quality. The task force agrees and suggests measuring PQ levels at least once per month. They agree to set the detailed procedures and methods of measuring later in the meeting. The task force first addresses how to define performance on the scale of 1 to 5. What does 5 mean, and what does 1 mean? After considerable discussion and some constructive dissension, task force members all agree to the following;

Performance Quality	Description
5	*Best possible performance.* The potential for achieving the objective successfully is very high, due to the excellent performance of the project team and stakeholders over the evaluation period. Their excellence in action has either maintained a previous very high level of value added or has considerably raised a previous lower level of contribution.
4	*Good performance, with the potential for doing better.* The potential for achieving the objective successfully is higher than average, due to the good performance of the project team and the stakeholders over the evaluation period. Their work has either maintained a previous moderate level of contribution or has raised a previous lower level of contribution. There remains room for some performance improvement.
3	*Average performance.* The potential for achieving the objective successfully is average and comes from a moderately competent performance of the project team and the stakeholders over the evaluation period. Their work has not significantly raised lower performance in previous evaluations, nor has it seriously damaged previous moderately

higher levels of contribution. There remains room for considerable performance improvement.

2   *Performance slightly below average and slightly above being unacceptable.* The potential for achieving the objective successfully by this level of performance being continued is below average and comes from a marginal operation of the project team and the stakeholders over the evaluation period. Their work has not significantly raised lower performance levels in previous evaluations and may seriously damage previous higher levels of contribution. There is an important need for sizable performance improvement.

1   *Worst possible performance.* Little, if any, potential exists for achieving the objective successfully by this level of performance. It results from a poor performance of the project team and the stakeholders over the evaluation period. Their work has significantly damaged the likelihood of success and negated previous higher levels of contribution. There is an urgent need for immediate corrective attention and action.

Values assigned can be in intermediate decimals if the stakeholder rating the team's execution of the activity feels the need for this degree of refinement. For instance, if the performance is higher than average but lacks the feel of a truly good job, a stakeholder scoring the performance might it give a rating of 3.5. The interpretation must be that of the stakeholder since it is the way that he or she perceives the work value.

Perception is reality.
                        —From large numbers of experienced people

Next, the task force members are asked to assign each of the charter objectives a weight. The chair of the meeting, Peggy Roethler, can either ask for a consensus of the task force members, or she can have each member write his or her suggested weight values on a sheet of paper and then collect the papers and average the ratings. In this particular instance she has asked for individual ratings on a sheet of paper and has averaged the results. The full group used the averages to begin their discussions. Each task force member spoke to the subject as he or she desired and a final selection was made by consensus of the group.

The weights assigned to each objective are shown in Figure 12.1. The rating form was prepared from a database template using a standard database computer program. The template can also be prepared using most spreadsheet software. If computer equipment is not available, the evaluation form can be prepared and updated manually.

col 1 charter objectives	col 2 weight
01. Protect Olanta's existing and new facilities so as to maintain operational continuity within the scope of the contract.	4.0
02. Keep project revisions to a minimum. When changes are essential, to issue bulletins promptly, to prepare and submit accurate, fair pricing, and to release the work changes in a timely manner.	3.0
03. Be a good on-site and off-site neighbor and to cooperate with all other stakeholders to maintain a clean, safe, secure, and well-organized work site.	3.0
04. Actively assist to control and where possible, reduce project time and cost growth.	3.0
05. Prepare, provide, and process submittals in a timely and fair manner while striving to reduce excessive and unnecessary resubmittals of shop drawings and other documents for which approvals must be provided.	4.0
06. Keep paperwork to an minimum while recognizing the importance of essential documentation.	3.0
07. Actively participate in preparing, monitoring, updating, and using project network plans and schedules, and to strive to construct the project in accordance with current plans and schedules of work.	4.5
08. Prepare, publish, and implement a close out procedure that will provide other contractors and the owner with a facility delivered on time, in accordance with the desired contract phasing sequence, and in a manner so that retention is released in a fair and timely manner.	4.5
09. Prepare, publish and implement an alternative dispute resolution system that will assist to avoid litigation, and will help resolve problems in an effective, timely, and fair manner at the originating or lowest-possible operating level.	4.0
10. Communicate effectively and at on an as-needed basis at all levels by being available, recognizing the need for quality information, and attending required meetings.	4.0
11. Prepare, publish, and implement a project partnering evaluation system.	4.5
12. Pay all properly submitted project invoices promptly.	5.0
13. Employ intelligent and timely use of risk-allocation and cost/benefit concepts on the project.	3.0
14. Respect and remain open to the needs and ideas of other project stakeholders and staff.	3.0
15. Orient, train, and educate all project staff in the principles of effective partnering as defined by this partnering charter.	4.0
16. Take pride in your work and encourage others to take pride in theirs.	3.8
17. Accept responsibility for your actions.	3.5
18. Listen well and try to understand other's concerns.	3.0
19. Prepare and publish an organizational chain of command showing communications flow, responsibilities, and authority.	3.5
20. Maintain continuity of key personnel throughout the life of the project.	4.0
21. Make timely decisions in all project-related matters.	4.0
22. Do it right the first time.	3.5
23. Have fun.	3.5
Average of total	3.7

Figure 12.1 Project partnering evaluation form (listed in charter objective order): weights assigned to objectives.

The chart shows the numbered charter provisions in column 1. Weights assigned to each by the task force are shown in column 2. The average of the total set of 23 weights is shown at the bottom of column 2. This 3.7 average of the total weight represents the stakeholders' feelings of the importance level represented by the entire charter. A weight of 3.0 means that the charter objectives are of average importance and about on the same level as for other successful similar projects. A 4.0 average of the total weight indicates that the charter objectives are of importance but not at the top level of contribution. In the Olanta project the 3.7 rating indicates that the charter objectives are of great importance, particularly when coupled with the other essential ingredients of a successful planning, design, and construction program.

Other important criteria that help measure the degree of success in a construction program and that may not be addressed specifically in the charter are usually programmatic or contractual compared to the noncontract moral issues that are at the heart of a well-written charter. For example, some of these other-than-charter measurements might include:

- Quality of the project program
- Excellence of the construction contract documents
- Financial structure of the project for all participants
- Quality and reputation of the design team
- Promptness of contract awards.

Task force members should remember that the weights assigned will be used for all evaluations until the charter is revisited or the stakeholders agree to revise the weights. Any revision of weights while the project is in progress may endanger accurate trend analysis and should be avoided through careful selection of the assigned weights prior to initiating the evaluation system.

Moving back to measurement of the performance levels, the Olanta task force now has in hand an evaluation scale from 1.0 to 5.0 by which the stakeholders can actually determine what the level of performance is on the project for any given evaluation period. The next step is to establish what level of performance should be considered acceptable, or that would, if maintained, not damage the project seriously but might prevent it from it being an overwhelming success.

For instance, evaluating what level of performance would be acceptable for charter objective 1, the dust-free environment, it is apparent that the Olanta staff would probably consider a level of 5.0 essential to their work. However, buildings are still a field-assembled product marked by glitches, unexpected events, and occasional serious problems of dust, noise, and other disruptive forces not always under the full control of the stakeholders. After considerable discussion, the evaluation task force decided that a performance level of 4.5 would represent an acceptable level of performance. This then becomes the par[3] value against which the project team will be measured over each evaluation period. The weight of charter objective 1 has been set at 4.0. The weight times the performance quality that will be considered an acceptable target is 4.0 multiplied by 4.5, resulting in a weighted rating of w $\times$ pqp = 18.0. This, then, is the numerical standard to be used in measuring project performance over the evaluation period. The results of the first meeting of the partnering evaluation task force are tabulated in Figure 12.2.

Peggy Roethler scheduled the second evaluation task force meeting for the next day. She set a single agenda objective of preparing and writing evaluation procedures for the Olanta project. Questions to be answered are a continuation of the first three outlined in her full agenda of items to be accomplished. The first three were covered in meeting 1 on the previous day. She phrases her remaining points as open questions.

Agenda for Meeting 2 of Olanta Evaluation Task Force, March 1, 19xx (Questions Continued from Meeting 1 Agenda)

D. Who should make the monthly charter performance evaluations?

E. How is information about the project to be collected by those making the evaluation?

F. What format should be used for tabulating, presenting, and evaluating the partnering data?

G. With what frequency should evaluations be made?

H. How should the evaluations be presented?

I. What supplemental information should be provided with the evaluations?

J. What should be done with the evaluation information?

K. What criteria are to be used by the evaluators?

L. How are results to be presented to the project team?

M. To whom should the evaluators report?

The discussion begins by the stakeholders addressing each question in light of the work done on day 1, when questions A, B, and C were answered by members of the task force. The questions for the second day of work include:

Question D: Who Should Make the Monthly Charter Performance Quality Evaluations? Parties responsible for partnering evaluations should, where possible, be stakeholders or those designated directly by a stakeholder to assemble and screen the data. The partnering evaluation task force's responsibility was to develop and recommend a system by which the stakeholders can keep up to date on project progress toward achieving the mission set in the

---

[3] Par performance is defined as the degree of success that must be achieved if the mission and objectives are to be met satisfactorily during the evaluation month.

col 1 charter objectives	col 2 weight (W)	col 3 perf qual per(pqp)	col 4 w x pqp
01. Protect Olanta's existing and new facilities so as to maintain operational continuity within the scope of the contract.	4.0	4.5	18.0
02. Keep project revisions to a minimum. When changes are essential, to issue bulletins promptly, to prepare and submit accurate, fair pricing, and to release the work changes in a timely manner.	3.0	3.5	10.5
03. Be a good on-site and off-site neighbor and to cooperate with all other stakeholders to maintain a clean, safe, secure, and well-organized work site.	3.0	2.5	7.5
04. Actively assist to control and where possible, reduce project time and cost growth.	3.0	3.5	10.5
05. Prepare, provide, and process submittals in a timely and fair manner while striving to reduce excessive and unnecessary resubmittals of shop drawings and other documents for which approvals must be provided.	4.0	4.0	16.0
06. Keep paperwork to an minimum while recognizing the importance of essential documentation.	3.0	3.5	10.5
07. Actively participate in preparing, monitoring, updating, and using project network plans and schedules, and to strive to construct the project in accordance with current plans and schedules of work.	4.5	4.0	18.0
08. Prepare, publish, and implement a close out procedure that will provide other contractors and the owner with a facility delivered on time, in accordance with the desired contract phasing sequence, and in a manner so that retention is released in a fair and timely manner.	4.5	4.0	18.0
09. Prepare, publish and implement an alternative dispute resolution system that will assist to avoid litigation, and will help resolve problems in an effective, timely, and fair manner at the originating or lowest-possible operating level.	4.0	4.0	16.0
10. Communicate effectively and at on an as-needed basis at all levels by being available, recognizing the need for quality information, and attending required meetings.	4.0	3.5	14.0
11. Prepare, publish, and implement a project partnering evaluation system.	4.5	4.0	18.0
12. Pay all properly submitted project invoices promptly.	5.0	4.5	22.5
13. Employ intelligent and timely use of risk-allocation and cost/benefit concepts on the project.	3.0	3.5	10.5
14. Respect and remain open to the needs and ideas of other project stakeholders and staff.	3.0	3.5	10.5
15. Orient, train, and educate all project staff in the principles of effective partnering as defined by this partnering charter.	4.0	3.8	15.2
16. Take pride in your work and encourage others to take pride in theirs.	3.8	4.0	15.2
17. Accept responsibility for your actions.	3.5	4.0	14.0
18. Listen well and try to understand other's concerns.	3.0	3.5	10.5
19. Prepare and publish an organizational chain of command showing communications flow, responsibilities, and authority.	3.5	4.0	14.0
20. Maintain continuity of key personnel throughout the life of the project.	4.0	4.0	16.0
21. Make timely decisions in all project-related matters.	4.0	4.0	16.0
22. Do it right the first time.	3.5	4.0	14.0
23. Have fun.	3.5	3.5	12.3
Average of total	3.7	3.8	14.2

Figure 12.2 Project partnering evaluation form: results of first task force meeting

charter. When the system description has been developed, reviewed, and presented to the full original stakeholder group, the responsibility for maintaining the system shifts to the full stakeholder staff.

The evaluation task force should publish a set of procedures to follow in monitoring the project. The manual should contain recommendations as to what group of people is actually to make the evaluations. In some cases the stakeholders may wish to delegate the evaluation process to a selected few just as the task force was assigned the "prepare and publish" responsibility for the evaluation system. Under other circumstances the stakeholders may wish to maintain the original charter signature group as the monitoring team. The original charter signatories should maintain their heavy involvement up to the point where they are no longer active participants in the project work. Disengagement from the monitoring may occur because the stakeholder's firm has completed its work. Such may be the case when early earthwork or erection of structural steel is completed and the work has been accepted.

Another cause of change in evaluation involvement comes or occurs when the stakeholder has been assigned to other duties within his or her organization. If this occurs and the stakeholder's organization is still active on the job, he or the organization's management staff should select and appoint a replacement to act for the organization in job matters related to the charter.

On the Olanta project the evaluation system task force has recommended that the charter monitoring be done by the full original stakeholder group. They are to meet for breakfast once each month to discuss and analyze the current status of the partnering system components and suggest methods of improving job performance.

Question E: How Is Information about the Project to be Collected by Those Making the Evaluation? The information needed to evaluate the project periodically must be provided by or through the efforts of those most closely associated with the day-to-day project work. As many of the original stakeholders as possible should be encouraged to provide performance quality ratings. They may need to get information from other project staff personnel to supplement their own observations, but the prime responsibility for assembling the evaluation data should always rest with the charter-writing group. If an original stakeholder is no longer able to report accurately on the project condition, he or she should see to it that another competent member of the organization assumes the reporting duties. Continuity and consistency of reporting are critical to an effective evaluation system.

Question F: What Format Should be Used for Tabulating and Evaluating the Partnering Data? Several commonly available graphic methods can be used for entering and calculating partnering data. Most mathematics required in the tabulations are multiplication, addition, and averaging of the totals of columns of figures. If the material is to be prepared manually, any standard typewriter or word processor program is adequate for the work. However, it is tedious work and susceptible to errors in arithmetic.

Other methods use a computer to generate either a database or spreadsheet. The worksheet shown in Figure 12.1 has already been used earlier to enter the charter objectives in column 1. It was also used for recording the agreed-upon weights set at the first Olanta evaluation meeting. These were recorded in column 2 of the worksheet shown in Figure 12.2. Peggy Roethler has decided that this is the format she will recommend the evaluation task force use for the duration of the project. The form that Mrs. Roethler presents for consideration of the stakeholders is shown in Figure 12.3. Notice that column 6 shows all zeros. Column 6 is a computed column showing the product of column 2 multiplied by column 5. Since no values are given in column 5, the computation yields zero values for column 6. A description of the various data columns or fields that she suggests using in this form are described below.

1. *Charter objectives:* taken directly from the written and signed charter.
2. *Weight* (w): the relative importance or weight of each charter objective measured on a scale of 1 for the lowest to 5 for the highest.
3. *Performance quality par* (pqp): the degree of success that must be achieved if the mission and objectives are to be met satisfactorily during the evaluation period.
4. *Performance quality par (pqp) multiplied by the weight (w)* (column 3 multiplied by column 2): the performance needed on a charter objective if the mission and objectives of the charter are to be met satisfactorily modified by the weight or importance of the objective.
5. *Current performance quality (cpq):* the measure of how well the project was running during the evaluation period. A rating of 1 indicates the lowest level and 5 indicates the highest level of performance.
6. *Current performance quality (cpq) multiplied by the weight (w)* (column 5 multiplied by column 2): the actual stakeholder-perceived performance on a charter objective over the evaluation period modified by the weight or importance of the objective.
7. *Previous current performance quality multiplied by the weight* (previous cpq × w): the previous evaluation of current performance quality modified by the objective weight. This column provides an easy comparison of the performance trend over the past month of each charter objective.

At each rating period the individual stakeholders evaluate the current performance quality of the project as they perceive it for each charter objective. The stakeholder then enters the data in column 5 of the evaluation forms shown in Figures 12.4 (worksheet) and 12.5 (stakeholder's completed rating sheet). The stakeholder also lists on the back of the form or on a separate sheet of paper any unresolved project issues that affect his or her work.

col 1 charter objectives	col 2 weight	col 3 perf qual part(pqp)	col 4 w x pqp	col 5 curr qual	col 6 w x cur qual	col 7 prev w x qual
01. Protect Olanta's existing and new facilities so as to maintain operational continuity within the scope of the contract.	4.0	4.5	18.0		0.0	0.0
02. Keep project revisions to a minimum. When changes are essential, to issue bulletins promptly, to prepare and submit accurate, fair pricing, and to release the work changes in a timely manner.	3.0	3.5	10.5		0.0	
03. Be a good on-site and off-site neighbor and to cooperate with all other stakeholders to maintain a clean, safe, secure, and well-organized work site.	3.0	2.5	7.5		0.0	
04. Actively assist to control and where possible, reduce project time and cost growth.	3.0	3.5	10.5		0.0	
05. Prepare, provide, and process submittals in a timely and fair manner while striving to reduce excessive and unnecessary resubmittals of shop drawings and other documents for which approvals must be provided.	4.0	4.0	16.0		0.0	
06. Keep paperwork to a minimum while recognizing the importance of essential documentation.	3.0	3.5	10.5		0.0	
07. Actively participate in preparing, monitoring, updating, and using project network plans and schedules, and to strive to construct the project in accordance with current plans and schedules of work.	4.5	4.0	18.0		0.0	
08. Prepare, publish, and implement a close out procedure that will provide other contractors and the owner with a facility delivered on time, in accordance with the desired contract phasing sequence, and in a manner so that retention is released in a fair and timely manner.	4.5	4.0	18.0		0.0	
09. Prepare, publish and implement an alternative dispute resolution system that will assist to avoid litigation, and will help resolve problems in an effective, timely, and fair manner at the originating or lowest-possible operating level.	4.0	4.0	16.0		0.0	
10. Communicate effectively and at on an as-needed basis at all levels by being available, recognizing the need for quality information, and attending required meetings.	4.0	3.5	14.0		0.0	
11. Prepare, publish, and implement a project partnering evaluation system.	4.5	4.0	18.0		0.0	
12. Pay all properly submitted project invoices promptly.	5.0	4.5	22.5		0.0	
13. Employ intelligent and timely use of risk-allocation and cost/benefit concepts on the project.	3.0	3.5	10.5		0.0	
14. Respect and remain open to the needs and ideas of other project stakeholders and staff.	3.0	3.5	10.5		0.0	
15. Orient, train, and educate all project staff in the principles of effective partnering as defined by this partnering charter.	4.0	3.8	15.2		0.0	
16. Take pride in your work and encourage others to take pride in theirs.	3.8	4.0	15.2		0.0	
17. Accept responsibility for your actions.	3.5	4.0	14.0		0.0	
18. Listen well and try to understand other's concerns.	3.0	3.5	10.5		0.0	
19. Prepare and publish an organizational chain of command showing communications flow, responsibilities, and authority.	3.5	4.0	14.0		0.0	
20. Maintain continuity of key personnel throughout the life of the project.	4.0	4.0	16.0		0.0	
21. Make timely decisions in all project-related matters.	4.0	4.0	16.0		0.0	
22. Do it right the first time.	3.5	4.0	14.0		0.0	
23. Have fun.	3.5	3.5	12.3		0.0	
Average of total	3.7	3.8	14.2		0.0	

Figure 12.3 Project partnering evaluation form for consideration of stakeholders.

col 1 charter objectives	col 2 weight	col 3 perf qual par(pqp)	col 4 w x pqp	col 5 curr qual	col 6 w x curr qual	col 7 prev w x qual
01. Protect Olanta's existing and new facilities so as to maintain operational continuity within the scope of the contract.	4.0	4.5	18.0			0.0
02. Keep project revisions to a minimum. When changes are essential, to issue bulletins promptly, to prepare and submit accurate, fair pricing, and to release the work changes in a timely manner.	3.0	3.5	10.5			0.0
03. Be a good on-site and off-site neighbor and to cooperate with all other stakeholders to maintain a clean, safe, secure, and well-organized work site.	3.0	2.5	7.5			0.0
04. Actively assist to control and where possible, reduce project time and cost growth.	3.0	3.5	10.5			0.0
05. Prepare, provide, and process submittals in a timely and fair manner while striving to reduce excessive and unnecessary resubmittals of shop drawings and other documents for which approvals must be provided.	4.0	4.0	16.0			0.0
06. Keep paperwork to an minimum while recognizing the importance of essential documentation.	3.0	3.5	10.5			0.0
07. Actively participate in preparing, monitoring, updating, and using project network plans and schedules, and to strive to construct the project in accordance with current plans and schedules of work.	4.5	4.0	18.0			0.0
08. Prepare, publish, and implement a close out procedure that will provide other contractors and the owner with a facility delivered on time, in accordance with the desired contract phasing sequence, and in a manner so that retention is released in a fair and timely manner.	4.5	4.0	18.0			0.0
09. Prepare, publish and implement an alternative dispute resolution system that will assist to avoid litigation, and will help resolve problems in an effective, timely, and fair manner at the originating or lowest-possible operating level.	4.0	4.0	16.0			0.0
10. Communicate effectively and at on an as-needed basis at all levels by being available, recognizing the need for quality information, and attending required meetings.	4.0	3.5	14.0			0.0
11. Prepare, publish, and implement a project partnering evaluation system.	4.5	4.0	18.0			0.0
12. Pay all properly submitted project invoices promptly.	5.0	4.5	22.5			0.0
13. Employ intelligent and timely use of risk-allocation and cost/benefit concepts on the project.	3.0	3.5	10.5			0.0
14. Respect and remain open to the needs and ideas of other project stakeholders and staff.	3.0	3.5	10.5			0.0
15. Orient, train, and educate all project staff in the principles of effective partnering as defined by this partnering charter.	4.0	3.8	15.2			0.0
16. Take pride in your work and encourage others to take pride in theirs.	3.8	4.0	15.2			0.0
17. Accept responsibility for your actions.	3.5	4.0	14.0			0.0
18. Listen well and try to understand other's concerns.	3.0	3.5	10.5			0.0
19. Prepare and publish an organizational chain of command showing communications flow, responsibilities, and authority.	3.5	4.0	14.0			0.0
20. Maintain continuity of key personnel throughout the life of the project.	4.0	4.0	16.0			0.0
21. Make timely decisions in all project-related matters.	4.0	4.0	16.0			0.0
22. Do it right the first time.	3.5	4.0	14.0			0.0
23. Have fun.	3.5	3.5	12.3			0.0
By:	3.7	3.8	14.2			0.0

Organization: See other side for listing of any related items

col 1 charter objectives	col 2 weight	col 3 perf qual par(pop)	col 4 w x pop qual	col 5 curr qual	col 6 w x cur qual	col 7 prev w x qual
01. Protect Olanta's existing and new facilities so as to maintain operational continuity within the scope of the contract.	4.0	4.5	18.0	3.8	0.0	0.0
02. Keep project revisions to a minimum. When changes are essential, to issue bulletins promptly, to prepare and submit accurate, fair pricing, and to release the work changes in a timely manner.	3.0	3.5	10.5	3.5	0.0	0.0
03. Be a good on-site and off-site neighbor and to cooperate with all other stakeholders to maintain a clean, safe, secure, and well-organized work site.	3.0	2.5	7.5	3.0	0.0	0.0
04. Actively assist to control and where possible, reduce project time and cost growth.	3.0	3.5	10.5	3.5	0.0	0.0
05. Prepare, provide and process submittals in a timely and fair manner while striving to reduce excessive and unnecessary resubmittals of shop drawings and other documents for which approvals must be provided.	4.0	4.0	16.0	4.2	0.0	0.0
06. Keep paperwork to an minimum while recognizing the importance of essential documentation.	3.0	3.5	10.5	3.8	0.0	0.0
07. Actively participate in preparing, monitoring, updating, and using project network plans and schedules, and to strive to construct the project in accordance with current plans and schedules of work.	4.5	4.0	18.0	4.1	0.0	0.0
08. Prepare, provide,and implement a close out procedure that will provide other contractors and the owner with a facility delivered on time, in accordance with the desired contract phasing sequence, and in a manner so that retention is released in a fair and timely manner.	4.5	4.0	18.0	4.0	0.0	0.0
09. Prepare, publish and implement an alternative dispute resolution system that will assist to avoid litigation, and will help resolve problems in an effective, timely, and fair manner at the originating or lowest-possible operating level.	4.0	4.0	16.0	4.0	0.0	0.0
10. Communicate effectively and at on an as-needed basis at all levels by being available, recognizing the need for quality information, and attending required meetings.	4.0	3.5	14.0	3.6	0.0	0.0
11. Prepare, publish, and implement a project partnering evaluation system.	4.5	4.0	18.0	5.0	0.0	0.0
12. Pay all properly submitted project invoices promptly.	5.0	4.5	22.5	4.5	0.0	0.0
13. Employ intelligent and timely use of risk-allocation and cost/benefit concepts on the project.	3.0	3.5	10.5	3.5	0.0	0.0
14. Respect and remain open to the needs and ideas of other project stakeholders and staff.	3.0	3.5	10.5	3.9	0.0	0.0
15. Orient, train, and educate all project staff in the principles of effective partnering as defined by this partnering charter.	4.0	3.8	15.2	3.3	0.0	0.0
16. Take pride in your work and encourage others to take pride in theirs.	3.8	4.0	15.2	3.8	0.0	0.0
17. Accept responsibility for your actions.	3.5	4.0	14.0	4.0	0.0	0.0
18. Listen well and try to understand other's concerns.	3.0	3.5	10.5	3.5	0.0	0.0
19. Prepare and publish an organizational chain of command showing communications flow, responsibilities, and authority.	3.5	4.0	14.0	3.0	0.0	0.0
20. Maintain continuity of key personnel throughout the life of the project.	4.0	4.0	16.0	4.0	0.0	0.0
21. Make timely decisions in all project-related matters.	4.0	4.0	16.0	4.0	0.0	0.0
22. Do it right the first time.	3.5	4.0	14.0	4.0	0.0	0.0
23. Have fun.	3.5	3.5	12.3	3.0	0.0	0.0
Average of total	3.7	3.8	14.2			0.0

By: Henry Leinenklugle   Organization: ODS   See other side for listing of unresolved issues

Figure 12.5 Completed stakeholder evaluation worksheet.

The stakeholder sends the performance ratings and the unresolved issue list to the chair of the charter monitoring task force. Usually, a deadline for the response is set to allow the chair to prepare the worksheet for the next regular charter monitoring meeting. The chair computes the averages from all the responses, puts the information into a database format, and prepares a computed spreadsheet (Figure 12.6). A list of stakeholder-mentioned issues to be resolved is also prepared by the chair of the charter monitoring task force, and the spreadsheet and the issue resolution list are sent to the stakeholders by the chair for their study and discussion prior to the monthly evaluation meeting.

Question G: With What Frequency Should Evaluations be Made? Frequency of evaluations is dependent on many variables, such as the project length, cost, complexity, delivery system, location, claim-prone potential, and other impacts of a similar nature.[4] On most projects the evaluations will be spaced from once per month to once each quarter. Stakeholders on the Olanta project decided that their evaluations should be made monthly. This choice was made to fit the evaluations into a time frame similar to that of the monthly pay request submittals. The stakeholders further recommended that the monthly evaluation be requested as a required attachment to the pay request.

I recommend that like project monitoring of job progress, the partnering evaluations be held often enough that trends can easily be spotted. The period, however, should not be so long as to hinder properly timed action to correct undesirable trends on the project. Overall, I believe that the most effective timing of project partnering evaluations is in regular meetings held from once a month to once every two months. The evaluation sheet should be turned in to the chair of the evaluation task force at least one week before the date of the next meeting. This gives the task force leader adequate time to enter the information into the project evaluation template and distribute the completed evaluation package to stakeholders. The stakeholders should be required to analyze the results and prepare their comments before the next meeting.

Evaluation meetings should be held at an on- or off-site conference space. Meetings might be coupled with a breakfast or lunch served after the discussion session. Usually, meetings of this nature are best held early in the day, when stakeholders are most alert.

Question H: How Should the Evaluations be Presented? At each evaluation meeting the stakeholders review the material, discuss each of the 23 charter objectives, comment on the ratings, and select those that need special attention from the list. Deviations either up or down of the current weighted performance (column 6) from the par-weighted performance figures (column 4) should be watched carefully. The monthly evaluation meeting offers a good

---

[4] See Chapter 9 for a list of factors that influence the format of a partnering meeting. Many of these items also affect or influence the optimum frequency of the partnering evaluations.

charter objectives	weight	perf qual par (pqp)	w x qual par (pqp)	curr qual	w x curr qual	prev w x qual
01. Protect Olanta's existing and new facilities so as to maintain operational continuity within the scope of the contract.	4.0	4.5	18.0	3.8	15.2	
02. Keep project revisions to a minimum. When changes are essential, to issue bulletins promptly, to prepare and submit accurate, fair pricing, and to release the work changes in a timely manner.	3.0	3.5	10.5	4.0	12.0	
03. Be a good on-site and off-site neighbor and to cooperate with all other stakeholders to maintain a clean, safe, secure, and well-organized work site.	3.0	2.5	7.5	3.8	11.4	
04. Actively assist to control and where possible, reduce project time and cost growth.	3.0	3.5	10.5	3.0	9.0	
05. Prepare, provide, and process submittals in a timely and fair manner while striving to reduce excessive and unnecessary resubmittals of shop drawings and other documents for which approvals must be provided.	4.0	4.0	16.0	4.0	16.0	
06. Keep paperwork to an minimum while recognizing the importance of essential documentation.	3.0	3.5	10.5	3.9	11.7	
07. Actively participate in preparing, monitoring, updating, and using project network plans and schedules, and to strive to construct the project in accordance with current plans and schedules of work.	4.5	4.0	18.0	3.8	17.1	
08. Prepare, publish, and implement a close out procedure that will provide other contractors and the owner with a facility delivered on time, in accordance with the desired contract phasing sequence, and in a manner so that retention is released in a fair and timely manner.	4.5	4.0	18.0	4.0	18.0	
09. Prepare, publish and implement an alternative dispute resolution system that will assist to avoid litigation, and will help resolve problems in an effective, timely, and fair manner at the originating or lowest-possible operating level.	4.0	4.0	16.0	4.0	16.0	
10. Communicate effectively and at on an as-needed basis at all levels by being available, recognizing the need for quality information, and attending required meetings.	4.0	3.5	14.0	3.7	14.8	
11. Prepare, publish, and implement a project partnering evaluation system.	4.5	4.0	18.0	4.5	20.3	
12. Pay all properly submitted project invoices promptly.	5.0	4.5	22.5	4.5	22.5	
13. Employ intelligent and timely use of risk-allocation and cost/benefit concepts on the project.	3.0	3.5	10.5	3.0	9.0	
14. Respect and remain open to the needs and ideas of other project stakeholders and staff.	3.0	3.5	10.5	3.9	11.7	
15. Orient, train, and educate all project staff in the principles of effective partnering as defined by this partnering charter.	4.0	3.8	15.2	3.0	12.0	
16. Take pride in your work and encourage others to take pride in theirs.	3.8	4.0	15.2	3.5	13.3	
17. Accept responsibility for your actions.	3.5	4.0	14.0	4.2	14.7	
18. Listen well and try to understand other's concerns.	3.0	3.5	10.5	3.4	10.2	
19. Prepare and publish an organizational chain of command showing communications flow, responsibilities, and authority.	3.5	4.0	14.0	3.1	10.9	
20. Maintain continuity of key personnel throughout the life of the project.	4.0	4.0	16.0	4.0	16.0	
21. Make timely decisions in all project-related matters.	4.0	4.0	16.0	3.8	15.2	
22. Do it right the first time.	3.5	4.0	14.0	3.6	12.6	
23. Have fun.	3.5	3.5	12.3	3.0	10.5	
Average of total	3.7	3.8	14.2	3.7	13.9	

By: All stakeholders   Organization: See other side for listing of unresolved issues

Figure 12.6 Stakeholder evaluation worksheet: averages from all responses.

neutral ground to stress the importance of correcting subpar performance where needed. It also allows stakeholders to obtain advice from the other stakeholders to help the project team accomplish this within the spirit of the charter.

Question I: What Supplemental Information Should be Provided with the Evaluations? The evaluation task force has suggested that stakeholders list the unresolved issues on their evaluation sheets and return the issue list to the chair with the evaluations. The chair can then tabulate the unresolved issues and the entire group of stakeholders can focus their combined attention on the problems. If a stakeholder does not list an unresolved matter or a disruptive issue, it is assumed there are none that the stakeholder wishes the group to discuss. If there are undisclosed problems or issues, the evaluation of the ratings is very likely to disclose the unresolved issue. If there are issues in the process of being considered and resolved, these too can be brought up as agenda items in the evaluation session.

One of the major values of an evaluation or monitoring is that the process tends to force those involved to look ahead. Monitoring project progress by the use of network models and bar charts is a similar process. The look-ahead insight given by tracking trends in project schedules is invaluable. Similarly, the look-ahead potential in the partnering evaluation provides a great benefit to the stakeholders who are doing the rating of the partnering performance.

Question J: What Should be Done with the Evaluation Information? The material collected each evaluation period should regularly be summarized, distributed, and analyzed by the stakeholders active in the evaluation process. The simplest summary is a list of the charter-weighted performance ratings by month containing current and previous ratings. This number is found in columns 3 through 6 of Figure 12.7. For a graphic display the ratings can be plotted with the rating shown on the vertical axis and months shown on the horizontal axis.

More detailed analyses of the partnering system can be had by following a similar process and tabulating the $w \times cpq$ values for each of the 23 charter objectives. This could be done by setting dates on the horizontal axis and showing the $w \times cpq$ value for each objective for each month. This format could also be converted to a graphic series of curves, one set for each objective showing the rating for each month. There is a point where the analysis potential for improvement begins to diminish because of the large amount of data being shown. Each project and each group of stakeholders is different and the tabular and graphic tools of use to them may vary widely.

On a recent project one of the workshop teams suggested that the evaluation committee publish a look-ahead newsletter for the project team. In the newsletter the evaluation committee would provide narrative and graphic descriptions of progress on the project. It might also forecast what problems and opportunities are seen in the near and distant future for improving job perfor-

col 1 charter objectives	col 2 par	col 3 period #1	col 4 period #2	col 5 period #3	col 6 period #4
01. Protect Olanta's existing and new facilities so as to maintain operational continuity within the scope of the contract.	18.0	15.2	16.4	17.0	18.5
02. Keep project revisions to a minimum. When changes are essential, to issue bulletins promptly, to prepare and submit accurate, fair pricing, and to release the work changes in a timely manner.	10.5	12.0	11.7	10.3	9.5
03. Be a good on-site and off-site neighbor and to cooperate with all other stakeholders to maintain a clean, safe, secure, and well-organized work site.	7.5	11.4	11.4	11.0	9.0
04. Actively assist to control and where possible, reduce project time and cost growth.	10.5	9.0	9.6	10.0	10.5
05. Prepare, provide, and process submittals in a timely and fair manner while striving to reduce excessive and unnecessary resubmittals of shop drawings and other documents for which approvals must be provided.	16.0	16.0	15.2	14.0	14.5
06. Keep paperwork to an minimum while recognizing the importance of essential documentation.	10.5	11.7	11.7	10.0	9.5
07. Actively participate in preparing, monitoring, updating, and using project network plans and schedules, and to strive to construct the project in accordance with current plans and schedules of work.	18.0	17.1	18.0	19.0	19.5
08. Prepare, publish, and implement a close out procedure that will provide other contractors and the owner with a facility delivered on time, in accordance with the desired contract phasing sequence, and in a manner so that retention is released in a fair and timely manner.	18.0	18.0	18.0	17.0	15.0
09. Prepare, publish and implement an alternative dispute resolution system that will assist to avoid litigation, and will help resolve problems in an effective, timely, and fair manner at the originating or lowest-possible operating level.	16.0	16.0	15.2	14.5	14.8
10. Communicate effectively and at on an as-needed basis at all levels by being available, recognizing the need for quality information, and attending required meetings.	14.0	14.8	15.2	13.0	12.0
11. Prepare, publish, and implement a project partnering evaluation system.	18.0	20.3	20.3	21.0	20.5
12. Pay all properly submitted project invoices promptly.	22.5	22.5	22.5	18.0	18.5
13. Employ intelligent and timely use of risk-allocation and cost/benefit concepts on the project.	10.5	9.0	9.3	9.0	10.2
14. Respect and remain open to the needs and ideas of other project stakeholders and staff.	10.5	11.7	11.4	10.0	10.5
15. Orient, train, and educate all project staff in the principles of effective partnering as defined by this partnering charter.	15.2	12.0	14.0	14.2	14.8
16. Take pride in your work and encourage others to take pride in theirs.	15.2	13.3	16.3	14.4	15.5
17. Accept responsibility for your actions.	14.0	14.7	12.6	10.5	13.1
18. Listen well and try to understand other's concerns.	10.5	10.2	10.8	10.0	9.0
19. Prepare and publish an organizational chain of command showing communications flow, responsibilities, and authority.	14.0	10.9	14.0	15.0	14.2
20. Maintain continuity of key personnel throughout the life of the project.	16.0	16.0	16.0	15.0	13.0
21. Make timely decisions in all project-related matters	16.0	15.2	16.0	15.0	14.0
22. Do it right the first time	14.0	12.6	13.3	14.0	12.0
23. Have fun.	12.3	10.5	12.6	10.0	9.0
Average of total	14.2	13.9	14.4	13.6	13.4

Figure 12.7 Weight × current performance quality evaluations for periods 1 through 4.

mance in specific disciplines and areas of work. A well-prepared newsletter could be used as a very effective communications tool to improve involvement of staff that might not be directly involved in partnering.

Question K: What Criteria Are to be Used by the Evaluators?   Normally, the performance quality criteria given earlier in this chapter are sufficient to make accurate assessments of the project condition. However, as more and more partnering evaluations are made, the stakeholders may wish to improve specific performance-level descriptions. This means that the evaluators will be able to better interpret trends and shifts in project elements as the job proceeds. They may find that they get better performance measures by giving each charter objective its own rating system. For instance, two samples of individual charter objective rating criteria are given below.

- *Charter objective 1:* Protect Olanta's existing and new facilities in order to maintain operational continuity within the scope of the contract.

Rating	Description
5	*Highest:* No breaches in temporary protection and no complaints from personnel occupying spaces adjacent to construction operations.
3	*Average:* A few easily fixed breaches in temporary protection and some easily addressed and quickly resolved complaints from personnel occupying spaces adjacent to construction operations.
1	*Lowest:* Continual serious breaches in temporary protection and frequent, serious complaints from personnel occupying spaces adjacent to construction operations.

- *Charter objective 2:* Keep project revisions to a minimum. When changes are essential, to issue bulletins promptly, to prepare and submit accurate, fair pricing, and to release the changed work in a timely manner.

Rating	Description
5	*Highest:* fewer than three bulletins issued, totaling under $5000. Those issued were estimated, submitted, approved, and released within an average of 5 working days.
3	*Average:* fewer than seven bulletins issued, totaling less than $10,000. Those issued were estimated, submitted, approved, and released within an average of 10 working days.
1	*Lowest:* more than ten bulletins issued, totaling over $20,000. Those issued were estimated, submitted, approved, and released within an average of 15 working days or more.

Question L: How Are Results to be Presented to the Project Team?   The results of the stakeholder's monthly partnering evaluation meetings serve a useful purpose within the stakeholder group. They can serve an even broader purpose by making them a part of the regular construction meetings. Attendance at construction meetings is usually by the project personnel concerned mainly with operational aspects of the project. These are elements of the job that are shown or specified in the construction documents and required by the legal contracts of each of the parties with others.

The partnering provisions are designed to make it easier, more profitable, and more gratifying to work at the operational levels. A review of what is going well or not so well in the many soft areas of partnering could very well help a foreman better understand what he can do to improve performance on the project. For instance, suppose that midway through the project the partnering evaluations for the past three months have shown a steady deterioration in the quality of the worksite condition.[5] The $w \times cpq$ values (the weighted performance quality) have been 6.5, 6.0, and 5.5 for the three-month period. Par is 7.5. Chances are that the same trend has been noticed by the foreman for the site utilities contractor and perhaps a few others around the table.

A quick discussion of the partnering evaluation findings for this element of the job reveals that the field crews have been working for the three months with a new owner inspector who is repeatedly demanding that already installed work be torn out and replaced for what appears to be trivial reasons. However, the money and time being spent appeasing this owner representative is getting serious. The field managers are reluctant to make too much noise about it, and therefore the problem has been festering. The partnering evaluations confirm that the actions of the inspector are affecting other stakeholders. Combining the observations of the field crews and the commitment of the project staff to the charter concepts makes addressing the problem a matter to be initiated immediately.

Question M: To Whom Should the Evaluators Report?   Stakeholders, all of whom have committed to and signed the project partnering charter, are those to whom the evaluation should be most significant. The project team managers and support staff will also have a great interest in the condition of the project and the trends for each of the charter objectives. Reporting on the partnering condition should be kept simple, and for the most part, limited to publishing the monthly evaluation sheets to those interested. This will allow the stakeholders to discuss the results intelligently in their monthly monitoring meetings, and along with the other recipients of the monitoring or evaluation data sheets, draw their own conclusions and take their own corrective or

---

[5] Charter objective 3 of the Olanta charter reads: "Be a good on-site and off-site neighbor and cooperate with all other stakeholders to maintain a clean, safe, secure, and well-organized work site.

sustaining action. The managers with higher management responsibilities will be acting at the macro level, while the middle and lower operational managers will be concentrating on limited macro and large applications of micro management in the week-to-week and day-to-day management of the project.

The main reporting path for the partnering data from the charter monitoring task force, however, should be to the stakeholders and related project managers. Further distribution, vertically or horizontally, of the information can then be made at the discretion of the prime recipients of the monitoring data. There may be some special provisions for distribution to those important decision makers and managers who are not stakeholders but who have a strong interest in the results of the charter monitoring and the health of the job. These might include the owner, planner, designer, contractor, or regulatory project participants who will be affected by the actions of the stakeholders. Some of these might not have been on board when the charter was written but now have a management interest in the project. Such extra nonstakeholder distribution should be at the discretion of the stakeholders.

After the charter evaluation task force had its second meeting, Mrs. Roethler prepared a set of recommended procedures and guidelines for conducting the charter monitoring evaluations. The guidelines were reviewed and approved by the task force for submission to the stakeholders. The document submitted is reproduced below.

Recommended Procedures and Guidelines for Monitoring and Evaluating Project Partnering Performance for the Olanta Data Systems, Ltd. Facilities Expansion Program, March 6, 19xx

A. The purpose of regularly monitoring and evaluating project partnering performance is to provide stakeholders with a management tool by which they can sustain good and excellent project performance and take effective action to improve poor or fair project performance.

B. The charter to be used for monitoring and evaluating the Olanta Data Systems, Ltd., partnering performance is that written and signed by the ODS stakeholders on February 24, 19xx.

C. The charter evaluation committee is to be composed of all stakeholders who were signatories to the Olanta project partnering charter dated February 24, 19xx. For the full list, refer to the roster of those attending, contained in the meeting notes. If any of the original signatories is unable to attend a meeting, they will be responsible for appointing an alternate from their organization to attend the meeting in their place.

D. Information to be used for monitoring and evaluating project partnering performance will be as discussed and adopted in the charter evaluation task force meeting of March 1, 19xx. The data fields to be used include:
  1. Charter objectives
  2. Charter objective weight (w)
  3. Performance quality par (pqp)
  4. Performance quality par multiplied by the weight (pqp $\times$ w)

5. Current performance quality (cpq)
6. Current performance quality multiplied by the weight (cpq × w)
7. Previous current performance quality multiplied by the weight (previous cpq multiplied by w)

For details of the information contained in these fields please refer to the March 1, 19xx meeting notes attached.

The measure of current performance quality (cpq, column 6) is based on the rating levels developed and adopted in the partnering evaluation task force meeting of February 28, 19xx in response to question C. For details of the rating system, see the meeting notes attached.

E. Charter monitoring meetings are to be held once each month from 7:00 to 10:00 A.M. at the Snowbird Inn, 3233 Carlsbad Road, Travis, North Dakota. Breakfast will be served from 7:00 to 7:30 A.M. The evaluation meeting will be from 7:30 to 10:00 A.M. If the discussion ends earlier, adjournment will be earlier. Under special conditions the evaluation session may be required to extend beyond 10:00 A.M. However, it will be extended only by permission of the stakeholders attending.

F. The evaluation meeting will be chaired by a stakeholder to be selected by the members of the evaluation group. The chair will serve for three successive monthly meetings and will be responsible for:
   • Setting and publishing the meeting agenda
   • Preparing and publishing the current partnering performance evaluation sheet
   • Chairing the evaluation meeting
   • Preparing and publishing the evaluation meeting notes
   • Updating the evaluation data sheet from stakeholder responses each month
   The chair of the initial four meetings will be Marla Jinns, stakeholder and project manager for Tiltsen and Greene.

G. Project partnering evaluation worksheets (see Exhibit B) will be provided by the current chair to each member of the evaluation group one week before the monthly evaluation meeting. The stakeholders will fill in their evaluation of the current performance quality in column 5 and return the worksheet to the current chair for tabulation and ultimate distribution to the evaluation group members. The tabulated worksheets should be returned at least five calendar days prior to the next evaluation meeting.

H. Where there are insufficient data to make a judgment of the performance on any given charter objective, the stakeholder should assign the item the par performance rating. This will allow subsequent stakeholder evaluations a base from which to measure a decline or improvement in performance for the charter item.

I. The current chair will calculate the average charter objective performance ratings for each charter objective and return the updated rating sheet with columns 5, 6, and 7 inputted and calculated. This document will provide the basic material for the monthly evaluation meeting.

J. If there are unresolved issues that are considered disruptive to project health, these should be described on the back of the rating sheets before returning the sheet to the current chair.

K. Minutes of the charter monitoring task force will be taken by a scribe to be appointed by the current chair of the task force. The minutes should be brief, terse, and objective. Contents of the minutes of the meeting should contain:

1. Date and number of the charter monitoring meeting
2. Those attending
3. Location of meeting
4. Time of start and of finish
5. Current tabulation of stakeholder performance quality ratings for past month
6. List of outstanding issues
7. Summary of actions taken or to be taken on outstanding issues
8. Brief analysis of project condition *in relation to charter objectives*
9. List of those to whom the meeting minutes are to be given

L. These guidelines should be reviewed each six months and revised as may be appropriate and felt necessary by the stakeholders. If the charter is revisited and revised, the guidelines should also be revised as the stakeholders desire. The monitoring procedure, once begun, should produce a consistently based set of data throughout the current charter period.

Case Study 12.1: Olanta Project Evaluation 1 as of March 24, 19xx

The guidelines above were agreed to informally by the stakeholders on March 10 at the regular job meeting, and Mrs. Roethler was given instructions to go ahead with the monthly charter monitorings. The first evaluation is to be held on Tuesday, April 4.

Mrs. Roethler immediately prepared the data-entry sheet shown in Figure 12.4 and sent it to each of the stakeholders. She asked the stakeholders to rate the project for each of the charter objectives for the period from the start of construction field work on February 28 through March 27, the end of the pay period for March. She further requested that the ratings be returned to her no later than the evening of March 27 so that she could tabulate the material for the meeting on April 4.

The schedule of data preparation is as follows:

February 28	Start of field work on project.
March 10	Stakeholders approved charter evaluation guidelines.
March 13	Stakeholders receive performance rating form.
March 27	Stakeholders return performance rating by end of day.
March 31	Mrs. Roethler distributes computed evaluations to stakeholders.
April 4	First meeting of stakeholders to monitor charter.

SECTION C: MONITORING CHARTER PERFORMANCE

The condition of the project as of March 27, 19xx is as follows (see Chapter 10, Section A, for a description of the Olanta project and the project team):

- Temporary protection of several penetrations into the existing building has repeatedly been torn to shreds by high winds. The owner and the general contractor have had some mild and easily resolved differences of opinion about extra costs for the replacement work. Usually, the general has paid for the replacement.
- The new building has been laid out, some foundation caissons have been drilled and filled, and construction of walls and caisson caps is actively in process. Foundations are about 20 percent complete. All anchor bolts have been fabricated and delivered. Some problems are being encountered in timely delivery of embeds in the foundations, but these are being worked on continuously by the mechanical and electrical engineer and the contractors affected. Field work is on schedule with the short-term network and bar chart issued by the general contractor.
- No bulletins for revisions have been issued and the architect and owner say that none are expected in the near future.
- Some grumbling from the residential neighbors about the truck noise and the dirt on the roads into the site has been heard by the excavation contractor. Both the owner and the contractors are trying to resolve the complaints, and are making progress.
- Separation of Olanta employee parking and construction parking has been a problem, but temporary construction parking areas are being installed and should be available by early April.
- Guaranteed maximum and hard-money costs for the project are still being held by all project stakeholders.
- All submittals to date have been provided, reviewed, approved, and returned promptly. The architect, engineers, owner, and the general have met and prepared a set of procedures to follow on setting priorities and processing submittals.
- The first month of the project has produced a very modest total of 15 letters, including transmittals, from all stakeholders. The volume is so low there is some concern by the architect that not enough record keeping is being maintained. However, all others are delighted with the low-cost administrative efforts of the stakeholders.
- The general contractor has met three times with the owner, the design team, and all major subcontractors to prepare a plan and schedule for the new building. Issue 1 of the network model and bar chart schedule will be issued on April 10.
- The owner has requested the general contractor and the architect to prepare a set of recommendations for closing out contracts as the subcontract work is complete and the work is accepted by the owner or other contractors. There is some reluctance to get into this matter so early, but Olanta management insists, pointing out that there were at least 10 mentions of close-out as a problem during the charter-writing session.

- The alternative dispute resolution task force will meet next week to start preparing recommendations on establishing and implementing an alternative dispute resolution procedure for the project. To date there have been no serious conflicts on the project, and it appears that most of the problems that do occur are being worked out at the operating levels.
- Communications among the stakeholders has been excellent over the past month. Communications between stakeholders and other project participants have failed occasionally, and there seems to be a breakdown in informing nonstakeholders about the charter and other partnering concepts. This is particularly true in the owner's organization, where some of the lower-management facilities personnel seem to feel that the partnering agreement is a method for the contractors to get out of fulfilling their contract obligations.
- The project partnering evaluation system has been put in place and is operating.
- This first pay period is a rather small draw. However, all subcontractors have been very conscientious about getting their requests for payment to the general contractor, and the general contractor's request has been submitted and a quick payment by Olanta is expected.
- No major substitutions or alternates have been submitted or suggested. The architect, engineers, owner, and the general contractor are presently doing a costing study of the second phase of the project, but this will have little effect on current project field work.
- The entire field and office staff for the stakeholders on the project have proven easy to reach and to get along with during this first month of the project. There has been some minor confusion about the organizational, authority, and responsibility relationships on the job. These have been observed by Mr. Leinenklugle of Olanta's staff, and he is taking steps with the major stakeholders to clarify who's who in the management structure of the project and in the stakeholder's organizations. Some reluctance to participate in this has been noted among the subcontractors.
- All stakeholders have maintained high morale and good attitudes, and all have professed to be proud of their work in preparing and signing the charter. In particular, the members of the charter evaluation task force are very pleased and impressed with how well Mrs. Roethler did her homework and led them so effortlessly through a relatively complex process.
- The project staff seems to be proud of being on a partnered project. Most of the nonstakeholder group still do not know much about the system, but they sense that they are part of a somewhat elite design and construction program that is leading the way into better ways of getting the work done properly.
- Project staff morale is moderately high. Also, Olanta employees seem to be enjoying their close-in view of construction on their new building.

It is the morning of March 27, 19xx. Mr. Henry Leinenklugle, the Olanta facilities director, is studying his partnering evaluation worksheet, getting ready to fill it out and return it to Mrs. Roethler. Mr. Leinenklugle is in charge of facilities construction and maintenance for ODS and has spent considerable time on the job during the past month. He completes his information entry and reviews the ratings that he has given. His opinion of the current status of the project relative to the charter provisions is shown in column 5 of Figure 12.5.

Marla Jinns, the chair of the partnering evaluation committee, receives Mr. Leinenklugle's evaluation along with one from each of the other signatories of the charter by the late morning of March 27, 19xx. She averages each of the responses for each charter line item, and enters the averages in the database template she is using for the project.[6] The results of the data entry and the computations are shown in Figure 12.6, the charter status report. This document is then distributed to the stakeholders along with a list of the outstanding issues to be resolved and a meeting agenda for their analysis and use in the charter evaluation meeting on April 4, 19xx.

Notice that the values in column 7, previous w × qual, are empty since this is the first report. In subsequent reports this column will be filled in with the previous month's data in column 6. Including this information gives the stakeholders information about the job trend for each charter objective over the past month's evaluation period.

As the project proceeds over the next four months, the stakeholders and Marla Jinns prepare a similar evaluation monitoring near the end of the pay period for each month. It has become an accepted policy for each stakeholder to submit the evaluation sheets at the same time that he or she submits monthly requests for payment. The owner stakeholders have also been submitting their evaluations at the same time.

The tabular results of charter objective evaluations for each of the first four-month periods of the job are shown in Figure 12.7. Charter evaluation period 1 begins with the weighted total performance average of 14.2 and moves down to the average of 13.9 at the end of the monthly rating period. This is repeated for the other periods, showing successive total weighted performance ratings of 14.4, 13.6, and 13.4. The total charter average for each of the four periods shows a trend downward. The trend is a gradual move with a decline from the average at the end of period 4 of about 0.8 point or 5.6 percent, measured against the 14.2 average. This decline is not too alarming, although it may be a little early in the project to encounter two successive months of performance drop, as shown in periods 3 and 4.

---

[6] Mrs. Roethler has prepared a database entry form which has the charter objective listed in the left-hand column followed by a separate column or field for each stakeholder. She also has entered a computed field which she specifies to contain the averages of the stakeholders' entries. She enters the data from the individual stakeholder sheets in the proper column, and the averages are automatically computed in the average field. She then enters this average data in the meeting 1 charter report form.

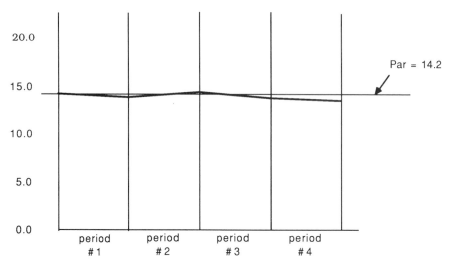

Figure 12.8   Partnering evaluation full charter rating, periods 1 through 4.

The trend may be seen more easily in a graphic plot of the trend. The total Olanta charter performance using the grand total averages for all 23 objectives is shown in the graph in Figure 12.8. The deviation from the total average can be seen to be relatively small. Caution should be used in reading the tabular and graphic translations. The lowest weighted performance par rating is 7.5 for charter objective 3. The weight assigned to objective 3 was slightly lower than the overall average, but the par performance considered acceptable

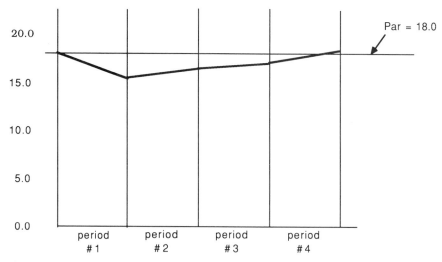

Figure 12.9   Partnering evaluation for charter objective 1: Protect Olanta's existing and new facilities so as to maintain operational continuity within the scope of the contract.

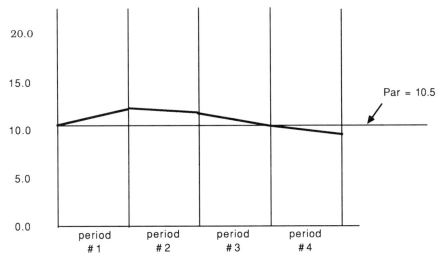

Figure 12.10   Partnering evaluation for charter objective 2: Keep project revisions to a minimum. When changes are essential, to issue bulletins promptly, to prepare and submit accurate, fair pricing, and to release the work changes in a timely manner.

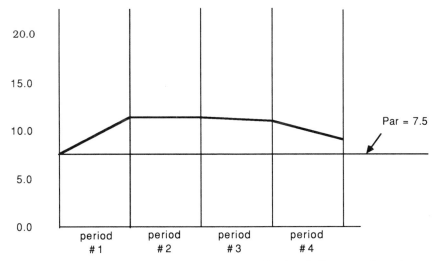

Figure 12.11   Partnering evaluation for charter objective 3: Be a good on-site and off-site neighbor and to cooperate with all other stakeholders to maintain a clean, safe, secure, and well-organized work site.

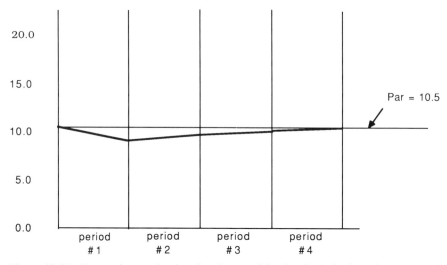

Figure 12.12   Partnering evaluation for charter objective 4: Actively assist to control and, where possible, reduce project time and cost growth.

was only 2.5, considerably lower than for most of the other objectives. The highest weighted performance par rating was 22.5 for charter objective 12. Drops in weighted performance ratings for high-par-value objectives are generally more significant and have greater impact on the total job than do drops in lower-par-rated objectives.

Further refinement in measurement analysis can be obtained if each of the 23 charter objectives is plotted on the same scale for the four-month period. Plots for charter objectives 1, 2, 3, 4, and 5 are shown in Figures 12.9, 12.10, 12.11, 12.12, and 12.13. If an overview of the plots is desired, the individual charts can be combined in a collage of all the charts plus the plot of the total charter performance measure. An example of such a collection showing six charts to a sheet is shown in Figure 12.14.

The charter monitoring process must be carried out carefully, taking into account some of the distorting influences that may affect the evaluations. One of the most critical of these impacts is any one-time but especially disturbing occurrence that may have happened shortly before the evaluation is made. For instance, on the Olanta project, Sylvia Goldsmitty, the Olanta office manager, did not get the word about a weekend power shutdown. The outage had been agreed to a week ago by Alfred Joiner, Powers Electric's field superintendent, John Tarkington, Olanta's data processing manager, and Henry Leinenklugle, Olanta's facilities manager. The outage lost a full day of data input for Ms. Goldsmitty.

She is still very angry at the three people who planned the outage without telling her and has just finished taking Mr. Joiner to task. She had already told Mr. Tarkington and Henry Leinenklugle what she thought of them.

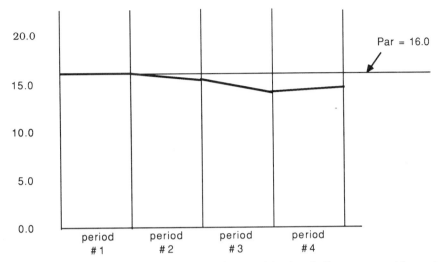

Figure 12.13 Partnering evaluation for charter objective 5: Prepare, provide, and process submittals in a timely and fair manner while striving to reduce excessive and unnecessary resubmittals of shop drawings and other documents for which approvals must be provided.

Ms. Goldsmitty is now back at her desk making out her monthly partnering evaluation sheet. Her performance ratings of the objectives dealing with operational continuity, being a good neighbor, and communicating effectively are about to be lowered a point or so because of her recent poor experience. The objects of her indignation realize the reasons for it but are still upset and a little angry themselves. They too, will have to provide their charter evaluations, and these will undoubtedly be influenced by the short-term discomfort all are feeling.

The net result of this event is that charter objectives 1, 3, and 10 get a lower rating than the par value they were going to get. This can have an unexplained and distorting impact on these objectives if the reasons for the lowered rating are not discussed. Sylvia Goldsmitty adds a note on the back of the rating sheet that requests a short discussion about future electrical and other types of disruptions to the facilities that may affect her work and that of her staff.

There are other impacts that can distort findings temporarily. They include such items as inadequate knowledge of the current job status, faulty communications systems that provide inaccurate information, dislike of people and an attempt to get even, misunderstandings about policies and decisions, and current interpersonal conflicts. Detection and moderation of temporary influences that overbias evaluations is a responsibility of each stakeholder.

The partnering charter evaluation system must mesh closely with the alternative dispute resolution system that is to be implemented on the project.

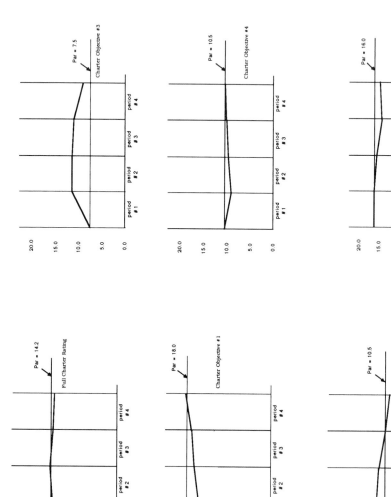

Figure 12.14 Summary of partnering evaluation charter objectives.

Good evaluations and proper analysis of the results can often help identify potential issues that should be resolved early. The charter is leg 1 in our three-legged stool of partnering. The periodic charter evaluation system provides leg 2. Leg 3 is provided by developing an ever-ready alternative dispute resolution system to deal with problems identified in the charter evaluation process.

13

PREPARING AND IMPLEMENTING THE PROJECT PARTNERING ISSUE RESOLUTION SYSTEM

SECTION A: BASIC ISSUE RESOLUTION CONSIDERATIONS

The partnering charter for Olanta Data Systems, Ltd., is now in place, the charter has been written and signed, and an evaluation process has been defined (see Chapter 12 for description of the partnering charter evaluation system). Project performance is already being measured against the charter through a weighted-performance system that requires the stakeholders to provide their perceptions and opinions of job health each month. Stakeholders are also expected to identify any outstanding issues that they feel might cause problems on the job if not resolved promptly. These are to be noted on the charter evaluation worksheets prepared by the stakeholders each month. In this manner those most responsible for job performance and ultimate project success are kept current on actual and perceived project problems.

Although valuable, calling attention to an outstanding issue does not necessarily resolve the issue. It might help force the project team to find a solution, but there must be an active, driving force that provides the route by which those involved can solve problems without booting them up the management ladder or outside the project staff immediately involved. Stakeholders must put a system in place that encourages resolution within the group where the issue first surfaces.

The Olanta charter addresses this matter when it specifically charges the stakeholders to prepare an issue resolution system (see Chapter 11, Section D, for the Olanta charter). Charter objective 9 states:

All stakeholders on the Olanta Data Systems project commit to prepare, publish, and implement an alternative dispute-resolution system that will assist to avoid litigation and will help resolve problems in an effective, timely, and fair manner at the originating or lowest possible operating level.

The stakeholders fulfill this charter objective by establishing an issue resolution task force to study, write, and implement an issue resolution system.

We discussed alternative dispute resolution briefly in Chapters 6 and 7. These earlier discussions concentrated on describing the various methods that can be used to resolve conflict issues without resorting to third-party binding solutions such as arbitration or litigation. In the discussion we also graphically depicted the steps or route of issue and dispute resolution. These steps, less the binding techniques, are shown in Figure 13.1.

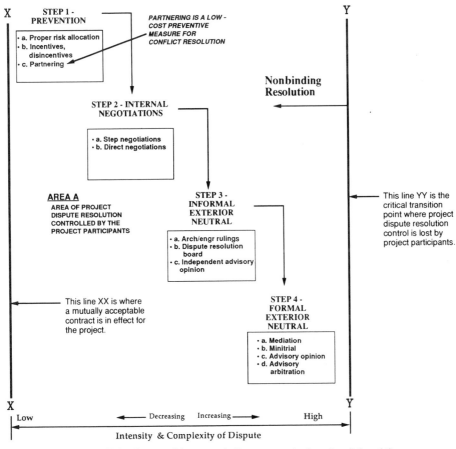

Figure 13.1 Steps of issue and dispute resolution detail level 2.

The process of resolution becomes increasingly complex as an issue moves from the upper left corner system of prevention (step 1), through internal negotiations (Step 2), the use of an informal exterior neutral (step 3), and finally, bringing in a formal exterior neutral (step 4). Issue resolution through prevention (step 1, Figure 13.1) is a simple and inexpensive conflict management and avoidance measure taken very early in the project life. It is at this time that prevention techniques of issue resolution are usually included in the contract documents and are used most effectively to resolve potentially destructive conflict.

Partnering, and incentives and disincentives are two issue-prevention systems that can be defined easily in the formal contract documents. By placing them there the entire project team is made fully aware of the preferred methods to be used to help the job through to completion with a minimum potential for major legal entanglements. The third issue prevention technique, proper risk allocation, is basically a management technique. Risk allocation is difficult to specify and should be identified within the stakeholder's organizations as an essential partnering tool to be used with intelligence and skill by all managers on a project.

As we have seen in earlier chapters, partnering provides a set of informal, morally binding agreements that guide the stakeholders in their day-to-day noncontract activities. It is a proven system that has benefited nearly all those who have used it. If stakeholders adhere to all or most of the charter recommendations generated in the partnering documents, the number of problems remaining unresolved would be diminished considerably. Partnering should be specified in the general requirements of the project if it is the intent of the sponsors and the owner to engage in a partnering effort. A suggested format for a project partnering specification is shown below.

Construction Project Partnering Specification for New Facility Olanta Data Systems, Ltd., Facilities Expansion and Renovation

Olanta Data Systems, Ltd., and its design and construction consultants will encourage, support, and implement a project partnering system on their expansion program. Olanta intends to do this with the full participation of its staff, the design firms, the contractors, and their subcontractors. Partnering is a performance system designed to achieve an optimal relationship among all parties to a construction contract. Further, it is a method of conducting business in the planning, design, and construction profession without unnecessary, excessive, or disruptive external party involvement.

The partnering system is structured to draw on the strengths of each participating organization to identify and achieve mutually profitable objectives. The Olanta partnering system will consist of three main actions: preparing a partnering charter, establishing and implementing a partnering evaluation technique, and establishing and implementing an issue resolution procedure.

Team members will be required to participate in establishing these three elements of the partnering system in conjunction with Olanta Data Systems, Ltd., and its consultants. Olanta anticipates that within 14 calendar days of issuing a notice to proceed with construction, Olanta, Ltd., its consultants, and the prime contractors on the project, with their subcontractors, will participate in a one-day meeting during which they will write a partnering charter in a team effort.

The partnering charter is the basic manual for operating a partnering system. It includes, at a minimum, the mission of the project and the objectives of the project team. In addition, it outlines in broad terms the project evaluation methods to be used and the dispute resolution process to be applied to conflict issues as they arise on the job.

Olanta further expects that within 14 calendar days after the partnering charter meeting a partnering evaluation task force will be appointed by mutual agreement among the partnering charter participants. This task force will establish and publish a partnering effectiveness evaluation method. The partnering evaluation method will set guidelines for periodic measurement of project performance against the mission and objectives set out in the charter.

Also within 14 calendar days after the partnering charter meeting a mutually selected issue resolution task force will be appointed from the partnering charter participants. This task force will establish and publish an issue resolution procedure encouraging the use of alternative dispute resolution (ADR) techniques. Alternative resolution methods are voluntary and designed to help resolve conflicts quickly, satisfactorily, and as near as possible to the originating level of the conflict.

As a part of their expected contract performance, all parties will be expected to participate in the preparation and maintenance of the charter, the periodic evaluations, and the issue resolution process. Outside costs for effectuating the partnership will be mutually agreed to by all parties.

The second preventive step, an incentive–disincentive payment system, is used in construction to pay a bonus or incentive to a contract party for performing its work in a superior manner to that specified. The bonus may relate to cost, time, quality, safety, or other such measurable component of the total job performance. If the standards set are not achieved by a measurable point on the project, the disincentive clause is triggered and the contract party is penalized for less-than-satisfactory performance on the project.

Incentives–disincentives tend to affect the project staff in a manner that is considered by some as threatening. Others consider it a stimulant to do the work better within the contract specified quality. It is important to understand that incentives and disincentives do not necessarily resolve issues. They encourage, by the promise of reward, prompt resolution of those problems that endanger high levels of job performance.

The incentive–disincentive system carries several caveats—one of the most important being that an exact definition must be provided about where incentives and disincentives begin. Parties to the contract must know precisely when the project, the phase of work, or the element being measured has been completed in accordance with the contract-defined incentive–disincentive conditions. The importance of definition can be seen in two specific examples of incentive–disincentive systems, one used in a workable application on a highway project, the other used in an unworkable application to a building construction project that potentially could have resulted in disaster for all stakeholders.[1]

Case Study 13.1: The Timely Expressway Project

Franklin Paving and Timely Road Builders were awarded a joint venture contract to pave eight miles of the George Cohan urban expressway, all to be done without traffic being allowed on the under-construction portion of the roadway. The incentive was that the State Department of Transportation would pay Franklin-Timely $10,000 for each full calendar day by which they completed the northbound section of the George Cohan roadway ahead of the target completion of 11:00 P.M. on August 3.

Franklin-Timely was also to be paid $10,000 for each full calendar day by which they completed the southbound section of the George Cohan roadway ahead of the target completion time of 11:00 P.M. on November 30. Fractions of days were to be prorated over a 24-hour period. Work on the southbound section was not to begin until the northbound section of the George Cohan was open to full traffic load as defined by the Highway Department.

If Franklin-Timely failed to complete the northbound and southbound freeway sections by the specified date, a one-week grace period went into effect during which no bonus was paid nor was any penalty imposed. If the project was not completed within the 7 calendar days of grace after the agreed-on completion dates, a disincentive or penalty system kicked in, and the Franklin-Timely fixed contract price was reduced by $10,000 per calendar day measured from the end of the 7-day grace period.

The system worked beautifully, with Franklin-Timely completing both sections of the George Cohan well ahead of time. The Highway Department gladly paid Franklin-Timely nearly $600,000 dollars in incentives. The public, the Highway Department, and Franklin-Timely were all pleased with the results. Conflict issues within the project were resolved promptly since all subcontractors and vendors shared in the payments of the bonus. Application of this prevention technique certainly paid handsome returns on a modest investment in high-grade planning, management, and issue resolution.

An incentive–disincentive program that had considerably different results than the George Cohan Expressway delivery system was experienced by an-

---

[1] The case studies for the George Cohan Expressway and for the Girard Home Furnishings store are taken from actual projects. Names have been altered to conceal the actual identity of people, organizations, and places.

other project team during their remodeling of the Girard Home Furnishings store, a high quality-furniture store in Epworth, Colorado.

Case Study 13.2: The Girard Home Furnishings Store Remodeling

Girard had negotiated a fixed-sum contract with Columbia Construction Company for the entire remodeling of 18 different departments in the store and construction of a 40,000-square-foot two-story addition on the north end of the existing facility. Mr. Lawrence Girard, president of Girard Home Furnishings, had heard of the incentive–disincentive program from a friend of his at the county road commission. The friend told him how well it had worked on a recent repaving program and suggested its use for the Girard project. Mr. Girard was impressed and felt it might be a good way of stimulating above-average performance on his project. He, his architect, and his department heads set what they all thought were realistic completion dates for each phase of the construction program. They then tied this into an incentive–disincentive system where completing the phase ahead of schedule for each remodeling completion gave Columbia a bonus of $500 per calendar day. Completing later than the schedule dates cost Columbia $500 per calendar day.

Columbia reluctantly agreed to the system and the project began. The first remodeling of 3000 square feet of dining room furnishings was 80% complete when the department head, Ms. Kingman, said she wanted to move into the nearly complete space while Columbia finished its remodeling of that department. The date was a full week before the target completion date specified. Columbia decided not to press for their incentive credit until they completed all work in the dining room furnishings department.

The early occupancy of the first department by Ms. Kingman interfered considerably with completing the work in the remaining 20% of the area, and Columbia was two weeks late in finishing the remodeling. However, the partial occupancy of dining room furnishings had permitted an early start on the home library furniture department and remodeling there was proceeding very well, with completion anticipated by Columbia about three weeks ahead of the incentive–disincentive date for that department.

Bob Beaufort, the project manager for Columbia, foresaw that Columbia was either going to go bankrupt on the job or was going to make more money in bonuses than the Girards could afford to pay. The problem, of course, was that accurately determining when any given remodeling was complete became, under field conditions, almost impossible. The future appeared to hold nothing but disagreements over starting and completion dates achieved, interferences with ongoing work by departments returning to their spaces early, and arguments with store staff over the working conditions they and Columbia had to endure.

One day, there was a particularly stormy job meeting in which two department heads were questioned about changes they had suggested that could delay construction by two weeks. The delay potentially could cost Girard nearly $5000 under the incentive–disincentive program and involve a difficult claim negotiation with Columbia. Mr. Girard invited Bob Beaufort to lunch that day and asked what to do about the difficulty of setting accurate schedules for the remodeling. Mr. Beaufort suggested first writing a change order eliminating the incentive–disincentive contract provision. He said the clause had already cost Girard nearly $20,000. He quickly added that Columbia had no intention of trying to collect the money since Girard was a valued client. However, Mr. Beaufort was concerned that the arguments and counterclaims might forever destroy the good relations that were now endangered. Mr. Girard agreed, the clause was removed from the contract, and Mr. Beaufort, the architect, and the department heads cooperated to tailor a workable schedule for the remaining remodeling in accordance with the current conditions, and without the cloud of finger-pointing that had begun to harm the project.

The third preventive dispute resolution method suggested in Figure 13.1 is proper risk allocation. Risk allocation has been addressed in detail in Chapter 4, Section A. The discussion there should provide the manager enough basic information to allow him or her to establish an early risk allocation procedure that will be effective in reducing the probability of destructive conflict during the project. A valuable risk management reminder to the stakeholders is contained in the Olanta partnering charter objective 13, which states that *all stakeholders on the Olanta Data Systems project commit to employ intelligent and timely use of risk allocation and cost/benefit concepts on the project* (see Chapter 11, Section D, for the Olanta charter). Like all ADR methods, risk allocation is most effective when used in conjunction with partnering.

It is easy to see from the examples above that the use of prevention techniques—properly specified in a timely manner and intelligently implemented—can provide a powerful incentive to reducing job problems from before or at the start of the project. Once actual project contract work has begun, the second and succeeding steps of the resolution process are usually called into play. Normally, these are in effect only when potentially damaging issues are encountered.

The second step in alternative dispute resolution shown in Figure 13.1 is internal negotiation techniques. These include two approaches, step negotiations and direct negotiations (a description of internal negotiation methods is given in Chapter 6, Section C). Parties involved in a dispute and desiring to use the internal negotiation process should apply it as quickly as possible after the dispute arises and should strive to obtain consensus among the disputing participants.

The first of the techniques, step negotiation, is a method by which the participants at the issue-origination level first conduct internal efforts to resolve the dispute among themselves. This system can be seen at work almost any day of the week on most design or construction projects. The following situation illustrates how the step negotiation process is applied to problem resolution in the field.

Case Study 13.3: The Problem of the Damaging In-Wall Work

Glenn Lamb is the electrical foreman on a small office building, Tom Templeton is the drywall foreman, and Arnold Wolf is the plumbing foreman. The structure is weathertight, and gypsumboard on interior walls is being hung by Tom concurrent with the in-wall plumbing and electrical work being installed between the studs by Glenn and Arnold. Tom has learned by long experience that the in-wall work trades, particularly plumbers and electricians, often damage unbraced metal studs.

The problem is that the in-wall tradesmen find it most convenient to move from room to room through the studs if possible. They also like to get at both sides of the wall without having to leave the immediate area of their current work. Workers moving back and forth between the studs, coupled with the occasional dislocation of the studs by conduit and piping installation work, often damage the studs inadvertently. Usually, this damage requires costly reworking of the stud work before the gypsumboard can be hung properly.

To help alleviate this situation, Tom has been installing one side of the drywall partitions close behind setting his metal studs. The one side of the board provides bracing for the studs but also blocks passage of tradesmen through the wall. This forces the electrician and plumber to work from one side of the wall only and they have to move from room to room through the door openings instead of through the studs to work on the other side of the wall. Both of these trade practices tend to increase mechanical and electrical installation time, and consequently, the cost of the plumbing and electrical work quickly increases, as does the time of installation.

One morning, Tom, Arnold, and Glenn had it with the constant arguing and physical and financial damage being done to their work efforts. They all remember that they were a participant in a partnering charter preparation at the beginning of the job. They also remember that in the charter they signed they agreed they would attempt to resolve job-related problems in an effective, timely, and fair manner at the lowest-possible operating level. So Glenn suggests they stop their bickering and see what can be done now to prevent deterioration of their work efforts with each other. It is a difficult project and there is still nearly 10 months of work to do before the three are complete with their work.

They have found a quiet corner of the project, and over hot coffee are testing the waters of conflict to see what can be done. Tom says he does not mind the plumbers and electricians working in the studs without either side of the board being hung if they exercise reasonable care in their work.

Above all, he says he wishes they would keep their own movement through the studs to a minimum. So far he has spent nearly $1500 on this job alone repairing dislocated and bent studs. His boss is beginning to think that he does not know how to run the work.

After a few minutes of thought, Arnold says that he understands Tom's position and hates to see him unfairly criticized by his superiors. Arnold adds that most of their work can be done from one side and he certainly can tell his crews not to push their way through the studs now that he understands the magnitude of the problem faced by Tom. Glenn says substantially the same thing as Arnold and adds that if the studs could be left freestanding to allow him to work from one side of the wall without interference, that he certainly can instruct his lead men and journeymen to keep their damage to the studs at a minimum and under control. He adds that he will make an over-and-above effort each day to track any damage to studs and to negotiate payment for these damages with Tom.

Arnold seconds this commitment and accepts it as good alternative to what Tom is doing now. So the three agree to survey each day's work and to agree at the end of the day on what damage was caused, who caused it, and how much, if anything, repairs to the wall framing would cost. The three agree further that this informal agreement is within their individual authority and responsibility range so they do not have to boot it up to their next level of management to get approval for its implementation. They finish their coffee and get back to work with the issue resolved to each party's satisfaction and with a far lower potential for escalation of the issue than was the case before the negotiating coffee break.

If the dispute over in-wall work described in the case study above had not been resolved by Tom, Gregg, and Arnold in their informal meeting, the step negotiation approach would be the next technique to be employed. Internal step negotiations would require the three to move the dispute up a notch in the field management organization to their respective superintendents. This might have been the case if any of the three foremen did not want to go along with the suggested solution or if he did not have the authority to trade off costs of any magnitude to repair damage without the approval of his superior, the superintendent.

Occasionally, problems like those that of Tom, Gregg, and Arnold cannot be resolved at the originating level nor by their respective superintendents at the next level up. In such a case the matter is progressively raised to successive levels of management until a resolution or a stalemate is reached. If no resolution is reached at the highest level felt appropriate, the matter is moved into the next step of the alternative dispute resolution process. The incentive to solve project problems quickly is high when those at the originating level are formally expected to resolve these difficulties among themselves. They know that each time they go to their boss for him or her to solve a lower-level problem, the lower-level supervisor's reputation for managing is damaged.

The second method within step 2, the internal resolution process, is the direct negotiation technique. Here the conflict is taken directly to the management level that is considered best able to resolve it without movement to higher management levels. Use of the direct negotiation method can be seen in the story of the delayed payment.

Case Study 13.4: The Case of the Delayed Payment

In a design conference, Larry Strong, the architect of record on the Brownell Manufacturing facility, an agricultural feed plant in central Texas, was asked by the structural engineer, Red Jones, why fee payments to him were so slow lately. Red added that he now had two months of unpaid invoices and was visiting the bank for short-term loans more than he liked. Frank Peal, the mechanical and electrical consultant, echoed Red's comments. Both Red and Frank, who were consultants to Larry Strong, had thought it was Larry who was holding up their payments, and it seemed to them it was time they resolved the issue and eliminated the potential for serious conflict.

Larry Strong said that he had not been paid for two months either and that he was certainly open to suggestions as to how to shake the money loose. The design team's request for payment was always made to Kim Bligh, the owner's project manager. She was not at the design conference, so after the meeting Larry, Red, and Frank made a conference call to Ms. Bligh. She was very surprised that the three had not been paid. She added that she had been forwarding all design invoices to the accounting department, but to no one in particular.

"This is serious enough to warrant quick action" said Larry to Kim. "I suggest we get some clarification on payment practices from Sam Jones." Sam was the vice president of facilities and Kim's boss. Kim said that she had just been working five months at Brownell, and had talked only infrequently to Mr. Jones about invoicing. Frankly, she said "I'm a little afraid to go that high in the company and possibly get in trouble with accounting and perhaps with Mr. Jones." Red and Larry told Kim that getting to see Sam Jones was no problem—he was a good professional and personal friend of both of them. However, they had not gone directly to him about their payment because they did not want to get Kim in trouble on her new job.

Kim said that she would appreciate some support and would be glad to set up a resolution conference with the design team, Mr. Jones, and herself. She immediately called the vice president and explained the situation and the reasons why she and the others felt they would like his help with the payment process. He was pleased to see her take the initiative in this matter and agreed to meet with them.

At the meeting Sam Jones outlined the steps the design team and Kim were to take in processing payments. Kim was to receive them on the 29th of each month. On the first of the succeeding month she was to submit them directly to Bernice Lant, Brownell's controller. Sam Jones called Mrs. Lant into the meeting and outlined to her what procedure he was requesting that they all follow. Mrs. Lant agreed that the process should work well.

Larry and his consultants were very pleased, and Kim learned a little more about her job and how to do it better. Sam Jones was impressed that Kim came to him in an effort to avoid serious problems and perhaps even a claim against the young job. As the meeting drew to a close, Larry asked Mrs. Lant, "How can we get our current invoices paid? They were approved by Kim and sent to accounting." Mrs. Lant called down to accounting and found that the invoices had been received but had been lost in the paper flow from accounting to the controller. She asked Larry and his consultants to resubmit their invoices to Kim. She told Kim to bring the invoices directly to her and added that the two of them would walk the invoices through the approval and payment process together. Payment would be within one day of resubmittal.

So Larry, Red, and Frank finally got paid, the project invoicing process was improved, Kim gained a new friend in Mrs. Lant while learning more about the Brownell Company, and Sam Jones had gained an increased respect for Kim. He was pleased that she had the gumption and good sense to nip a potentially serious problem in the bud and get it resolved promptly and fairly.

The case of Brownell Manufacturing is an illustration of how to use direct internal negotiation effectively. Lower-management participants were led into a meeting directly with the ultimate decision maker, Mr. Jones. He quickly made the judgments needed to implement a system that was badly needed on the job. In addition, a flawed payment process was revised and some good precedents were set that should carry over into the construction phase. Internal negotiations require that strong efforts be made to resolve issues quickly before they are escalated by external efforts to force a binding solution. It is usually a low-cost resolution method, and if the parties concerned truly want a settlement with minimum difficulty and at the lowest cost, it is a very effective method to use.

If a settlement cannot be reached by either step or direct negotiation, the resolution process is moved to step 3, the use of an informal exterior neutral.[2] The key to effective use of systems involving neutral parties is to obtain a qualified, unbiased expert who can grasp the problem quickly, perceive the alternates available, make sound decisions that are fair, and clearly state a

---

[2] A description of the informal exterior neutral method is given in Chapter 6, Section C. Also see Appendix A for a definition of neutral and standing neutral.

recommended course of action. To use this system of resolution a neutral is selected and is presented with the key information about the project, the issues, and the people involved. He or she then recommends a course of action that has a high potential to produce resolution results consistent with the facts as presented. The neutral is normally paid a fee, although there are some professional societies who provide neutral services without charge as a public service to planning, design, and construction professionals.

A neutral who is recommended by a professional society is usually required to complete a specified number of training hours in dispute resolution. The neutral, however, also needs to be qualified by background, education, experience, temperament, and lack of bias to act as a professional advisor in issue resolution situations. Those neutrals who are available and on call from a pool of available, qualified neutrals are known as standing neutrals. For instance, a Michigan Society of Professional Engineers task force on standing neutrals in construction has recently begun efforts to encourage the use of standing neutrals in conflict resolution. The task force has developed suggested language, shown below, to incorporate into the contract documents if the parties to the contract desire to use the neutral concept. Standing neutrals under the Michigan plan are presently required to receive 8 hours of training in dispute resolution.

Suggested Contract Language for Incorporating the Standing Neutral Concept in Construction Specifications

In an effort to resolve conflicts that may arise during design or construction of the project, and as a condition precedent to the initiation of any action, litigation, or formal arbitration between the parties, the owner, the designer, and the contractor agree that all disputes among them arising out of, or related to, this agreement shall be first submitted to nonbinding mediation according to the procedures of the Michigan Society of Professional Engineers Standing Neutral Program, unless the parties mutually agree otherwise. The owner, the designer, and the contractor further agree to include a similar mediation provision in all agreements with independent consultants, contractors, and subcontractors retained for the project, and to require all such consultants, contractors, and subcontractors to include similar mediation provisions for all of their subcontractors and subconsultants so retained.

Although the Michigan specification calls for use of nonbinding mediation, standing neutrals can assist in any of the step 2, 3, and 4 issue resolution techniques.

The step 3 informal exterior neutral process shown in Figure 13.1 consists of three techniques: the architect/engineer ruling, the dispute resolution board, and the independent advisory opinion (see Chapter 6 Section C, for a description of these techniques under step 3). In the past the architect/engineer ruling method was frequently found in the construction specifications and was used occasionally. However, as pointed out in the discussion in Chapter 6, this resource is seldom called upon today, for many reasons. The main objection to using the technique is that the architects and engineers of record are nearly always stakeholders in the project. In disputed issues they may be hard-pressed to maintain an unbiased position, particularly if they and their firm are affected by the outcome.

The second method, the dispute resolution board, is a system that shows great promise for helping to settle conflicts in a truly effective manner. As with all standing neutral systems, the technique depends on the availability of competent and truly unbiased persons having a recognized planning, design, and construction background. The dispute resolution board is normally most useful in cases of disputes that occur during construction, whereas many of the other alternative dispute resolution systems can be used during the planning and design periods as well as during construction.

In this system the dispute resolution board is usually appointed prior to or concurrent with the start of active work on any given project phase to which it is to be applied. They remain available for duty whenever a specific contested issue arises, although the board can also be asked to meet regularly to review project progress and analyze project evaluations and documentation to see if any signs of future job difficulties are visible.

The third method in step 3 is the independent advisory opinion. This resolution method makes use of a qualified neutral to provide an opinion of what outcomes can be expected if certain courses of action, including doing nothing, are followed. The advisory opinion route varies slightly in emphasis since it produces an opinion of the results of a proposed course of action rather than actually producing a resolution action. The information exchange meeting with the neutral is unstructured and informal. No major presentation need be prepared. Instead, the parties only need to be equipped to discuss the facts surrounding the dispute.

There are no established procedural rules and meetings are held in the manner best suited to the neutral's desires. The neutral's opinion can be written or oral and is given during the meeting or shortly thereafter to the participants for their review, study, thought, and action. The value of the informal neutral opinion is that if rendered by a qualified neutral party, it can often provide an insight into what may be expected by the parties that they did not know or understand previously. For instance, the neutral's conflict observations could provide guidelines not well known or understood, such as:

- In binding arbitration or litigation the judgment or settlement will generally be from 20 to 30% of the claim amount.
- The amount of a claim will almost always increase substantially once it enters binding arbitration or litigation. This is due primarily to the additional legal fees required, the heavy demands on the time of the partici-

pants, the costs for expert witnesses, and the expense of assembling formal presentation material.

- The time a major claim may take to be settled through binding techniques can be from 1 to 10 years for litigation and from 2 to 5 months for arbitration.

- The longer the time of settlement, and the more nonproject parties are involved in the settlement of the dispute, the more difficult it will be to obtain people to provide firsthand information about the difficulty. Most will be busy elsewhere, will not want to be involved in a legal problem, will have moved elsewhere, or are no longer actively involved with the disputants. Some will have died.

- As less and less individual and personal knowledge of the problem situation becomes available, the amount of needed backup documentation prepared by noninvolved technicians increases. Document collection, indexing, review, and analysis are very expensive. Information available firsthand may cost as little as $40 per hour to record if done at the time the information is first available. A similar analysis might cost as much as $120 per hour and require much more time if done later after the original paper trail has been lost and firsthand knowledge of the job problem has disappeared.

The last of the steps to be taken in alternative dispute resolution is step 4, the use of a formal exterior neutral (see Chapter 6, Section C, for other information on formal exterior neutral methods). The four techniques in step 4 include mediation, minitrial, advisory opinion, and advisory arbitration. The major difference between these methods and the informal methods is that considerably more preparation is necessary when employing them. In step 4 methods the parties are pressing for an actual ruling which, although not binding, can be used for backup and evidence in the event the dispute goes to binding arbitration or litigation.

Usually, formal exterior neutral techniques are used in specific instances where the issue has been escalated into a major dispute, with the disputants clearly identified and the elements of the dispute reasonably well documented. I do not recommend that the more sophisticated and relatively expensive methods shown in step 4 be included in an issue resolution process. Instead, they should be kept in reserve to be used as needed and where their special features make them a last resort resolution possibility before crossing line YY into the binding resolution area.

SECTION B: PREPARING THE PROJECT PARTNERING ISSUE RESOLUTION SYSTEM

To trace through the actual preparation of an issue resolution policy I have used the Olanta project and its charter as a starting point for our description (see Chapter 12 Section A, for the contents of the Olanta charter). Olanta charter objective 9 says that "all stakeholders on the Olanta Data Systems project commit to prepare, publish, and implement an alternative dispute-resolution system that will assist to avoid litigation, and will help resolve problems in an effective, timely, and fair manner at the originating or lowest possible operating level."

In a telephone survey by Karl Largo, Olanta's vice president of operations, shortly after the charter was written and signed, most stakeholders agreed that Lincoln Brown, President of Brown Mechanical, was a logical choice to chair the issue resolution task force. Mr. Brown was very active in writing the charter, has had a wealth of experience as an arbitrator and mediator, and has been a student and advocate of alternative dispute resolution for several years. He is also listed as a member of several groups of standing neutrals.

Mr. Brown has accepted the temporary assignment and requested that the following people from the Olanta project team be placed on the task force:

- Joyce Hallmark, public relations manager, Olanta Data Systems, Ltd., owner
- Travis Loring, president and chief operating officer, Loring and Metzer, architects
- Tobin Strendel, vice president, Strendel Geotechnical
- Charles Flyer, field superintendent, Tiltsen and Greene, general contractor
- Gerald Powers, president, Powers Electric, electrical contractor
- Harry Wolfson, project manager for installation, Efficiency Design, Inc., fixtures, furniture, and equipment contractor

Mr. Brown immediately sent his task force members a memo containing a suggested agenda for their first meeting. It included a statement of potential objectives of the task force, along with a list of several points that might be of help in writing the Olanta issue resolution policy. The memo is reproduced below:

Date: March 1, 19xx

To: All issue resolution task force members, Olanta Data Systems, Ltd., project

From: Lincoln Brown, task force chairman

Re: Agenda, objectives, and guidelines for meeting 1 discussion

Meeting date: March 7, 19xx

Meeting time: 10:00 A.M. to 2:00 P.M.; lunch will be served

Location: Offices of Brown Mechanical, 3303 Second Avenue, Travis, North Dakota

A. *Suggested agenda*
   1. Set objectives of task force.
   2. Review points of importance in alternative dispute resolution.
   3. Prepare rough draft of issue resolution policy.
B. *Suggested objectives of task force.* These are to be discussed and rewritten at meeting.
   1. By March 21, to complete the final draft of a proposed issue resolution policy for submission to all stakeholders for critique and comments.
   2. To have an issue resolution system functioning within two weeks after approval by the stakeholder group.
C. *Items of possible interest in preparing and implementing the Olanta issue resolution policy.* The following are considerations to be taken into account in selecting and implementing an issue resolution policy.[3] They are in no special order, nor are they intended as fixed points of reference. I suggest that we use the items as ideas to stimulate our thought processes about issue resolution.
   1. The method selected to resolve issues must encourage and assist participants to get to the root of a conflict quickly, to determine the causes of the dispute, to identify the actual people involved, and to assess the circumstances surrounding the conflict.
   2. Issue resolution methods selected must emphasize seeking firsthand information when collecting information about a problem that has surfaced. It must not depend solely on gossip, grapevine data, or hearsay to provide a basis of resolution.
   3. When helping to resolve issues, the stakeholders should seek out the parties most affected and most knowledgeable to use for determining the root facts about the conflict.
   4. Determine how serious the conflict is before taking action. It may be that the best solution is one of allowing time for the parties involved to resolve the matter among themselves.
   5. Get rid of as many as possible of the biases you have about the people involved in the situation.
   6. Get rid of as many biases as possible about the organizations involved.
   7. Base your resolution decisions on what you see people are doing and on what you know people are like, rather than on what you think people are thinking and what you think they are like.
   8. Encourage that individual and organizational disputes be settled at the originating level wherever possible.

[3] These items have been mentioned previously in some of the case studies in Chapter 3. However, their critical nature is worthwhile studying in different situations since the lessons they convey about resolving disputes is valuable in establishing an issue resolution system.

9. Do not allow a dispute to be pushed into the informal communication, conversation, and grapevine systems.
10. Stop and reconsider before you talk loosely about someone else's mistakes and deficiencies. They may be neither.
11. Always think twice, or more, about airing problems outside the participant arena.
12. Determine what the roles and responsibilities for all levels of the stakeholders are in the issue resolution process.
13. Follow the most ancient of good advice—if you can't say something positive about someone don't say anything. Loose and casual criticism does not belong in a professional business world.
14. Promptness and fairness are two major ingredients of the issue resolution process.
15. The stakeholders and management of the project must give the people who are expected to resolve disputes at the originating level the authority and support they need to act.
16. Poorly delegated issue resolution policies may seriously weaken the partnering system. Stakeholders must be responsible for, and assume authority for, the creation and use of partnering techniques for the partnering system to be successful.

I am pleased that you accepted this assignment to assist in a very important partnering function. It will be a valuable and contributive effort on the Olanta project.

Signed: Lincoln Brown

At the first meeting of all task force members on March 7, Mr. Brown had the members reintroduce themselves and then asked for suggestions on what the task force objectives should be. The two put forth in the meeting announcement by Lincoln Brown were accepted as generally suitable for inclusion in the final issue resolution report to the stakeholders. However, all those present felt the timetable was too tight for them to meet at this busy early stage of the project. After a brief discussion, the task force adopted the following schedule for its work:

March 14   Issue resolution task force meeting. Outline dispute resolution policy and prepare preliminary outline of resolution process to be followed.

March 21   Issue resolution task force meeting 3, if needed. Task force review and approve stakeholder draft of the issue resolution system.

March 24   Task force submits recommended issue resolution system draft to all stakeholders.

March 30   All stakeholders meet to discuss, revise, and approve the issue resolution system draft.
April 3   Issue resolution system released to full project team at special job meeting.
April 10   Issue resolution system functioning.

Since no other suggestions for objectives were offered at this time, the stakeholders began immediately to determine what the content of the issue resolution policy and system should be. Joyce Hallmark said that recently she had been collecting information on issue resolution in the construction industry. She added that she had found several brief, well-written policy statements that seemed to express the principles of issue resolution extremely well. Ms. Hallmark made copies of several of the statements and distributed them to the stakeholders at the meeting. The task force agreed that many of them were good model statements of intent for issue resolution and began to build on the concepts expressed in their discussions. By noon the group had prepared the rough outline below for the structure of their issue resolution system.

**Dispute Resolution Process Outline, Olanta Data Systems, Ltd.,
Travis, North Dakota**

A. *Statement of intent.* All stakeholders will attempt to resolve issues that show signs of escalating into destructive disputes by taking prompt action at the originating level. This will be achieved by open, honest, and businesslike communication, by attacking the problem and not the person, and by appropriately assigning and accepting risk. When the issue cannot be resolved at the originating level, the matter will be referred to successively higher levels of management for consideration. If, at any point, it is obvious that a settlement cannot be reached among the project stakeholders and their staff, it will be submitted to an impartial third party for recommendations and settlement. Every appropriate channel of alternative dispute resolution action must be explored before considering binding solutions such as arbitration or litigation.
B. *Outline of steps to be taken in producing a fair and well-considered solution to the problem.* Succeeding steps are to be taken only if the previous step fails.
   1. Attempt to settle at the originating level among the disputants.
   2. Attempt to settle at the job level using peer review of the problem.
   3. Attempt to settle at successively higher levels of job management.
   4. Attempt to settle at successively higher levels of organizational management.
   5. Call in a third-party neutral or neutrals for advisory opinions as to what will happen if a solution is not found.
   6. Call in a third-party neutral or neutrals for settlement recommendations.

7. Consider resolution by crossing line YY of Figure 13.1 (the boundary between nonbinding resolution and binding resolution) and bring disputants back in for a second effort.
8. Take formal steps to resolve the issue by binding methods.

It is now 2:00 P.M. and the group adjourns with the assignment to bring to the next meeting, on March 21, a full description of the specific methods to be followed in each of the eight steps above so as to achieve a workable procedure within each step. On March 21, the issue resolution task force reassembled to take up the content of each step in the dispute resolution process. The first point in the discussion is brought up by Mr. Flyer, field superintendent for Tiltsen and Greene, the general contractor: How do the stakeholders know when an issue exists? This leads into the question of the difference between a trivial argument that is part of the day-to-day process of communicating and managing on the job, and a serious problem that has high potential for stimulating destructive conflict.

Mr. Brown suggests that an issue needing stakeholder attention be formally noted when the issue is submitted on the monthly partnering evaluation sheet prepared by each stakeholder. The charter monitoring system was approved on March 10 and all stakeholders have a copy of the procedures written by the charter evaluation task force (see Chapter 12, Section B). Section J of the evaluation procedures states that *"if there are unresolved issues that are considered disruptive to project health these should be described on the back of the rating sheets before returning the sheet to the current chair."*

"The question then," poses Mr. Flyer, "is what do we recommend be done in the time period between when the issue surfaces and when it can be put on the evaluation sheet? I suggest that we make each stakeholder manager responsible for immediately applying step 1 preventive measures if the dispute is more than just informal and conversational. To do this will require some training at the foreman and lead person levels, but I'm convinced such training will pay high dividends if done right and with the key leaders in that group." Mr. Flyer adds, "I'll personally set up a training program if some of you can help me prepare the course outline."

Others on the task force were very interested in experimenting with some issue resolution training at the lower management level. Travis Loring, the architect of record, said that his professional liability carrier had been after him for months to get his production staff better acquainted with dispute resolution, and this was exactly the right time to do it.

All agreed and Joyce Hallmark of the owner's office said that she had received some formal dispute and issue resolution training in college, along with attending two later seminars on the subject in her employment with ODS. She offered to prepare a suggested course curriculum and to help teach the course with Mr. Flyer and Travis Loring if they were willing. The group recommended that the training program be put in the draft of the issue resolution system description under the heading "Stakeholder Internal Training Program in Issue and Dispute Resolution." Ms. Hallmark, Mr. Flyer, and Mr. Loring are to have this program ready for review and approval at the next meeting of the task force, on March 24.

The remainder of task force meeting 2 was spent in writing descriptions of progressive issue resolution steps 1 through 8. Mr. Brown had the results of the conference typed and was able to give the task force stakeholders a hard copy of the step descriptions before they left his office. At meeting 3 the group was to merge the ideas and material prepared to date into the stakeholder review draft for formal adoption in late March.

Meeting 3 of the issue resolution task force started at 7:30 A.M. with breakfast and an intensive work session which the participants hoped would be complete by noon or shortly after. Task force members showed up promptly and began their work on the stakeholder approval draft. By 11:30 P.M. they had produced a rough draft ready for typing, their final approval, and distribution to the full stakeholder group. The task force adjourned for lunch and then came back to make its final check of the document. By 1:30 P.M. the members had completed their review, signed off on the issue resolution system description, and were on the way back to their respective jobs.

The final draft of the issue resolution system was a conglomerate of ideas from the task force, from materials used in researching the subject, and from the experiences of the task force members. The final stakeholder recommendation issue draft is reproduced below.

Recommended Issue Resolution System, Olanta Data Systems, Ltd., Facilities Expansion Program, March 21, 19xx

A. *Statement of Olanta stakeholder intent re issue resolution for the Olanta project.*
It is the objective of the Olanta project stakeholders first and foremost to avoid unnecessary disputes and conflict on the job. All stakeholders will attempt to resolve issues that show signs of escalating into destructive disputes by taking prompt action at the originating level. This will be achieved by open, honest, and businesslike communications, by considering the problem and not the person, by looking at the other side's point of view, and by appropriately assigning and accepting risk.

In all cases, individuals involved should be businesslike and not resort to personal attack. The principles outlined in the partnering charter mission and charter should be followed at all times in resolving differences. In seeking resolution of an issue, involved parties will attempt to:
- Thoroughly understand the issues
- Maintain an appreciation for other points of view
- Communicate thoughts promptly, openly, and clearly
- Clearly document the issue resolution

Any issue presented should be clearly defined and alternative solutions suggested. A log of unresolved issues will be maintained from meeting to meeting of the partnering evaluation task force.

When the issue cannot be or is not resolved at the originating level the matter will be referred to successively higher levels of management for consideration. If resolution cannot be reached at the job site, the principals of the involved firms or agencies should attempt to reach resolution through informal discussion. If at any point it is obvious that a settlement cannot be reached by the project stakeholders and their staffs, the dispute will be submitted to an impartial design and construction professional third party for recommendations and settlement.

Upon third-party resolution of disputes, a written résumé of the issues and the details surrounding its resolution should be prepared by the parties involved and provided to the stakeholders from those organizations involved. Every channel of resolution action must be reviewed and considered before moving to binding solutions such as arbitration or litigation.

This issue resolution system will be reviewed every six months by the current members of the issue resolution task force or their selected successors. The review will be for the purpose of evaluating the effectiveness of the resolution process and to offer an opportunity to revise the steps to resolution as may be appropriate. Strong efforts should be made to keep major revisions of the policy to a minimum so as not to lose the original intent of the resolution system if such basic intent remains valid.

B. *Outline of steps to be taken in producing a fair and well-considered solution to the problem.* Succeeding steps are to be taken only if the preceding step fails.

1. Attempt to settle at the originating level among the disputants. No project issue should be allowed to remain unresolved for more than one week without those involved seeking a solution and settlement. If the parties involved in the original issue cannot agree on a resolution within one week, the matter will be listed as an unresolved issue and be recorded as an open issue in the stakeholder charter performance monitoring submittal.

   The immediate superiors of those involved in the issue shall be responsible for encouraging resolution if appropriate. If no resolution is reached within the week, the immediate superiors are responsible for seeing that the issue is listed in their company stakeholder's monthly charter performance evaluation. At the end of one week, if a resolution of the issue is not forthcoming, the immediate superiors of those involved are authorized *at their discretion* to take step 2: appoint a peer review panel of three to further review the matter.[4] If the superiors do not wish to appoint a peer panel to review the matter they should move immediately to step 3.

2. Attempt to settle at the job level using peer panel review of the problem. Once the item appears in the charter evaluation form as an unresolved issue, the stakeholders will, if appropriate in their opinion, convene a peer review group to hear the facts in the unresolved issue. Those on the peer panel should be at a similar or higher management level to those most directly involved in the dispute. The peer panel, when used, will be selected by the

---

[4] Not every project is suited to the use of the peer panel. Where poor relations, bad feelings, or strong trade loyalties exist, the people involved may be too biased to allow the method to be effective. Management and the stakeholders should determine each issue on an individual basis.

stakeholders present at the job meeting closest to the expiration of the one-week settlement deadline.

The facts in the case will be presented by the disputants in an informal job setting. The peer panel will make a nonbinding recommendation that expresses its opinion about what the ultimate outcome will be if the issue is not settled. The peer panel will also be encouraged to express opinions about how the issue could be resolved. Peer panel discussions should be kept informal, short, and focused on the key points in the issue, not on the nature of the people involved.

3. Attempt to settle at successively higher levels of job management. If the matter is not resolved at the originating level or by the use of the peer panel within two weeks, the issue shall be deemed to be a dispute and should be booted to the next level of management within the project. The matter will continue to be discussed at successive levels of management for a maximum of one week per management level until a resolution or an impasse is reached. The succession of management layers through which an issue or dispute will be moved is to be determined by the stakeholders for those organizations most directly involved in the dispute. The open issue will be reported in the charter monitoring report until it is resolved. Each reporting will include a description of the management level at which the issue or dispute is currently located.

4. Attempt to settle at successively higher levels of organizational management. If and when the dispute reaches an impasse or has passed the highest management level of the project without a solution, it may be transferred by mutual stakeholder consent to higher levels of organizational management within the stakeholder groups.

    (Note from the issue resolution task force: This move up to higher organizational levels is *not recommended*. It is included here only to indicate that such action is an alternative. The reason for not recommending the course of action is that by the time the issue has moved from its origin through step 3 with no settlement at the project level, the chance of a higher-level organizational settlement is slim.)

5. Refer the matter to the impartial third party. If an impasse is reached or all project management levels have been consulted without a solution, the stakeholders will call in an impartial third-party neutral or neutrals to provide an advisory opinion as to what will happen if a solution is not found. Within one week of the neutral being notified or appointed, each party will submit a brief written description of the issues, facts, and conclusions supporting their positions. The neutral will meet with the participants in the matter and each participant will be allowed a maximum of one-half hour to present his or her side of the issue or dispute. This presentation should be based on the description submitted to the neutral before the presentation. At the close of the discussion the participants and the affected stakeholders should be given another opportunity to resolve the dispute.

6. If the dispute is not resolved in one week, the stakeholders call in a third party neutral or neutrals for settlement recommendations. The neutral will again meet with the disputing parties and the parties, under the leadership of the neutral, will have two days to reach a mutually agreeable solution. If an agreement cannot be reached by both parties, the neutral shall adjourn the

negotiations and issue a written recommendation and the backup for the recommendation to the disputants and their stakeholders no later than one week after conclusion of the step 6 negotiations. The neutral party will be paid for by both sides equally.
7. Make one final attempt to negotiate a resolution before crossing line YY of Figure 13.1 (the boundary between nonbinding resolution and binding resolution).
8. Take formal steps to resolve the issue by binding methods.
C. *Outline for stakeholder Internal Training Program in Issue and Dispute Resolution*
   1. *Program purpose.* The purpose of the program is to train lead personnel, foremen and women, and superintendents in the basics of conflict and issue resolution. The course is designed to build on the individual's industry knowledge and objectivity to allow intelligent participation in dispute resolution situations.
   2. *General program information.* The recommended course as proposed by the issue resolution task force is based on six hours of training held in three meetings, one per week. A suggested course outline and information is available upon request from:
      • Joyce Hallmark, public relations manager, Olanta Data Systems, Ltd., or
      • Charles Flyer, field superintendent, Tiltsen and Greene
      Stakeholders may sponsor their own in-house classes in conflict and issue resolution as they desire. The owner, Olanta, Ltd., is presently holding in-house training sessions for its project staff. A new class begins each week and all design and construction personnel working on the Olanta project are eligible to enroll for sessions.
   3. *Subjects addressed in the course outline:*
      • Basic assumptions about people and conflict
      • The nature and sources of destructive conflict
      • Elements of effective management of destructive and positive conflict
      • Procedures commonly used to resolve disputes and problems
      • Alternative dispute resolution (ADR)
      • Experiences and applications in dispute resolution
      • Risk and conflict in design and construction
      • Reasons why disputes are not resolved promptly and fairly

The issue resolution system above was issued to the stakeholders on March 24 for their study and comments. On March 30, the stakeholders met and approved the issue resolution policy as written by the task force. Some of the stakeholders had reservations about including steps 2 and 4. However, Mr. Brown, the chairman of the task force, pointed out that the issue resolution policy was to be reviewed each six months. He added that some elements of these steps had been questioned by individual task force members but ultimately were approved for inclusion by the current task force. If in six months there are solid reasons for revising or removing any steps, that course of action will be reviewed as appropriate. With these assurances the stakeholder group gave the go-ahead to issue the policy at a special job meeting to be held on April 3. It is expected that the issue resolution system will be functioning and in full use by April 10.

The Olanta partnering mission and charter have been signed and issued, the partnering charter monitoring process is in use, and implementation of the issue resolution system is about to begin. Completion of these three elements puts all the legs of our three-legged stool into place and provides the Olanta Data Systems, Ltd., project team a stable base upon which successful project partnering effort depends.

14

USING THE PARTNERING SYSTEM TO IMPROVE PERFORMANCE

Once the charter is written, the charter monitoring system is in place, and the issue resolution process is approved and published, we have put most of the basic tools in place ready to use. As in any results-oriented program, the true test of the preparation work is how well it works when real men and women use the system to guide their actions in removing the obstacles to excellent performance.

> The test of a man's or woman's breeding is how they behave in a quarrel.
>
> —G. B. Shaw

Always to be remembered is that partnering system components are built of words that should be easily convertible to actions. They should require a minimum of questions as to their intent and meaning within the system. Evaluating the performance of the project staff by measuring perceived or actual work results against charter provision expectations is much like monitoring project progress against a critical path model of the work. A recommended system of charter performance measurement is described in detail in Chapter 12, Sections B and C. This regular evaluation system is generally accepted as an excellent communication tool for the total project team. Less well known is the potential contribution of accurate measurement to improved execution of work in the project stakeholder organizations.

The key to effective use of performance measurement is knowing what to do with the findings. Monitoring is a diagnostic tool, not an end in itself. Performance measurement is a management-by-exception (MX) process that sounds an alarm to the manager when problems have appeared or are about to appear. The alarms are silent when there are no problems. An MX system helps identify a problem as an exception; it permits the effective manager to manage the exception while leaving the smoothly running operations to continue without management meddling.

Good management of planning, design, and construction depends on knowing what it is that needs managing. Perceiving and isolating job problems is often very complex and requires application of talent, experience, training, and education. In some projects, particularly where complex interactions between people, equipment, and materials exist, a sensing, or instinct of what is wrong is often of immense supplemental help, particularly if it comes from the mature manager. Management tools commonly used in planning, design, and construction, such as charters, schedules, models, and the like, should be considered support elements to help a person make better use his of her brains, experience, intelligence, and training.

Partnering is proving to be an excellent technique to help locate and solve both project and organizational problems. One reason is that the large number of problem statements that have been identified in charter workshops indicates that a what-to-do approach to project problems will be needed almost as soon as the project is put into work. In this chapter I have delineated several case studies that illustrate the problem identification and resolution features of partnering as they may be used to improve individual and organizational performance. The improvement may help correct a deficiency in the project's management; it might assist to resolve a current or potential threat to the project's health, be used to sustain and improve project execution, or it may be a catalyst that encourages raising the stakeholder organization's potential for good work.

Below are listed some of the most often-mentioned causes of problems encountered on design and construction programs. They offer a rich source of material from which to simulate problems and solutions for difficulties that have actually been met in the field and in the office. Some problem statements were found to be answers to the question "What is it that others do that cause you problems?" and others were answers to "What is it that you do that causes problems for others?" Note that these are the major subjects addressed in the Olanta charter workshops 1 and 2 described in Chapter 11, Section C. Whether they are caused by others or by us is immaterial to this discussion. They occur—and they cause trouble.

To prepare the material below, 2855 specific problem statements from 23 charter meetings were tabulated and coded. The eight most frequently mentioned problem categories were used to build the subject discussion. Figure 14.1 shows a histogram of the frequency of total mentions of the eight problem types. One or more of the problem statements is illustrated in each case study. Case studies have been specially written to demonstrate how the problem, its identification, and the partnering system might be used to improve individual and organizational performance. A sampling[1] of specific problems mentioned

---

[1] Problem statements have been taken from actual partnering documents. Some have been paraphrased to clarify their meaning.

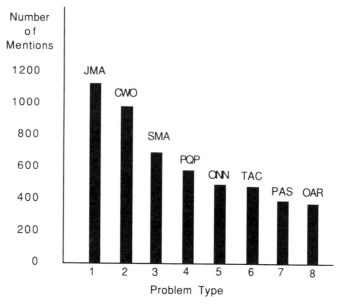

Figure 14.1   Eight most frequently mentioned design and construction problems from a total of 2855. JMA, job management, 1146 mentions; CWO, communicating with others, 984 mentions; SMA, staff morale and attitudes, 684 mentions; PQP, personnel quality and problems, 593 mentions; ONN, being a good on-site neighbor, 475 mentions; TAC, timely action, 467 mentions; PAS, planning and scheduling, 396 mentions; OAR, organization, authority and responsibility, 371 mentions.

in each category is provided under each of the eight categories to impart to you an overall flavor of the problem category. The number of times the problem category appears in the master list is given both as a number of appearances and as a percentage of the 2855 problem descriptions.

- *Problem 1: Job management* (appears 1146 times, 40% of total)

Sample Problem Statements
- Passing the buck
- Requiring unreasonable personnel assignments
- Failing to meet project deadlines
- Inadequately supervising subcontractors
- Preparing and issuing unrealistic schedules
- Not providing survey control lines
- Not providing adequate material storage space
- Contractors failing to clean up site at the end of a project

- Asking owner staff to cover for omissions in project documents
- Using tunnel vision approach to project—failing to realize that others are involved or affected besides our group
- Prime contractor not paying subcontractors and suppliers promptly
- Poor site management
- Failing to resolve conflicts with other subcontractors
- Unrealistic owner scheduling of utility tie-ins and outages
- Dictating how to do the work rather than what is to be done
- *Problem 2: Communicating with others* (appears 984 times, 35% of total)

Sample Problem Statements

- Given too short a notice for project meetings and equipment demonstrations
- Avoiding the agreed-upon chain of authority and responsibility
- Slow turnaround by subcontractors on quotes
- Making schedule and specification changes without informing other participants
- Not following agreed-upon project communication channels
- Not providing sufficient notification of material deliveries
- Not bringing problems to the proper and responsible persons
- Lack of timely information flow to consultant
- Multiple sources of owner direction
- Failure to respond to warranty obligations[2]
- Failure to notify those affected of utility outages
- Not identifying who is responsible for abatement in existing facility
- *Problem 3: Staff morale and attitudes* (appears 684 times, 24% of total)

Sample Problem Statements

- Lack of respect for other partner's space, materials, and work in place
- Carrying over previous personality conflicts
- People missing job meetings
- Criticism of design without suggested alternatives
- Improper passing of general conditions responsibility to subcontractors
- Slow punch list and deficiency correction

---

[2] A warranty is a legally enforceable assurance of the duration of satisfactory performance or quality of a product, a piece of equipment, or of work performed. Often, the warranty period begins when the installation is turned over to the owner.

- Being available
- Lack of design details and direction from architect/engineer
- Proper resolution of punch lists
- Perfunctory treatment of critical path plan and schedule
- Indifference to need for decisions
- Slow on closing out job
- Not returning phone calls in a timely manner
- *Problem 4: Personnel quality and problems* (appears 593 times, 21% of total)

Sample Problem Statements

- People who attempt to take advantage of or intimidate others
- Contractor personnel not checking shop drawings
- Contractor personnel not familiar with contract documents
- Letting designer and owner decisions slip at the expense of estimate or construction time.
- Failing to follow up on what was agreed to
- Failing to recognize the urgency of the problem
- Failing to understand trade contractor's details
- Failing to prepare proper shop drawing details
- Preparing inaccurate estimates
- Making incomplete responses
- Preparing incomplete submittals and providing them late
- Lack of understanding of other disciplines' requirements
- Slow shop drawings and submittal preparation
- Summer help doing testing
- Team members failing to prepare for meetings
- Not verifying that existing utilities are shut down
- Weak and late code research
- *Problem 5: Being a good on-site neighbor* (appears 475 times, 17% of total)

Sample Problem Statements

- Bad housekeeping
- Breakage by others
- Poor cleanup of own materials
- Dumpsters not on site
- Construction noise in early morning affecting neighbors
- Damaging installed products and materials
- Damaging other contractors' work without notifying them

- Electrical contractor not providing advance notice of outages
- Placing equipment in others' way
- Contractors failing to recognize the impact of construction work on public
- Failure to protect completed work
- Design team members being inaccessible or unavailable
- Exhibiting an "It's not my job" attitude
- Lack of respect for owner's property
- Storing materials in others' ways
- Poor consideration of priorities of other disciplines
- Not attending meetings
- Running up walls without notifying other contractors
- Taking up more space than required
- Using too much room on site
- Desire to get "our stuff" in tight spaces first
- *Problem 6: Timely action* (appears 467 times, 16% of total)

Sample Problem Statements

- Providing late answers to questions.
- Issuing change orders too late, after work has started
- Delaying start date
- Delaying work
- Fail to get submittals in on time
- Incomplete and late submittals of estimates
- Getting late information responses from other trades
- Getting late information responses from consultants
- Delays to laboratory's preparing and transmitting reports
- Late notification of required inspections
- Owner late in providing equipment information and equipment
- Not allowing enough time for other team members to react
- Not identifying and discussing problems early
- Not on time at job site
- Not sharing information in a timely fashion
- Too short notice on scheduling of subcontractors on job
- Slow paperwork response time
- Slow response and processing of change orders
- Slow response to payment applications
- Untimely addressing of problems and their solutions

- Inadequate notification of required shutdowns
- Delays to acceptance of work
- Untimely responses to field changes
- Untimely correction of construction errors
- *Problem 7: Planning and scheduling* (appears 396 times, 14% of total)

Sample Problem Statements

- Not providing good project planning
- Lack of consideration for what nature can do to a job site
- Lack of communication on short-term scheduling
- Poor planning that results in crises management by owner
- Lack of commitment to schedule by contractor
- Not ordering materials in a timely manner
- Untimely delivery of supplies and equipment
- Poor schedule for submittals that requires too many at one time
- Schedules established without input from affected parties
- Failure to obtain utility input when preparing construction schedule
- Late or untimely selection of colors/materials
- Doing work out of sequence
- Owner not providing definitive program and schedule of their work
- Not setting design due dates for schematics and design development work
- Bid shopping resulting in delays and damaging schedule impacts
- *Problem 8: Organization, authority, and responsibility,* (appears 371 times, 13% of total)

Sample Problem Statements

- Lack of clearly defined chain of command
- Not taking responsibility for actions
- Lack of clear identification of decision makers
- Lack of clear path for resolution of field problems
- Lack of clear responsibility for cleanup
- Lack of clear responsibility for providing material-lay-down areas
- Unclear definition of project mission and budget
- Design team pushing design responsibility to trade contractors
- Lack of field engineers having authority to make decisions
- Lack of usable organization plan and chain of command
- Poor time management: inefficient meetings, tardiness

- Poor definition of the method of processing progress payments
- Lack of authoritative owner project manager

Charter objectives are derived from problem statements such as those listed above which are identified in the charter workshops. If the problem statements are translated properly into charter objectives, they should provide an accurate preview of possible problems and give the stakeholders some ideas on how to avoid them. An accurate preview conveyed to competent managers will often encourage early resolution of some of the most common project problems. If the problems most frequently found to be troublesome in a planning, design, and construction project can be anticipated and prevented, the charter and the partnering system will have been well worth the effort. This is a basic reason for using partnering.

A few case studies related directly to some of the most prevalent problem areas might help provoke thought as to how extended uses of partnering workshop results can be used to improve management practices in construction-related organizations.

Problem 1: Job Management (Appears 1146 Times, 40% of Total)   The most frequently mentioned problem category is job management. Its high number of mentions, 1146 or 40% of the total, indicates that it is an area of potential improvement that should be explored continually to determine how to make job management work better for the total company good. Knowledge of the process that produces good design and construction results is a critical ingredient of effective design and construction project management. It is a process—knowledge gained through technical education, participation in successful (and not so successful) design and construction efforts, and experience adequate to predict accurately what might happen given a certain course of action.

The quality of job management is also related to the education[3] and training[4] that the emerging manager receives. Education without training may produce a manager who believes that the educated manager can successfully apply his or her skills to any system that requires managing to work effectively. Training without education tends to produce a manager whose confidence in the learned–doing skills can be coupled with minimal education to be a successful manager. Above all, the management of planning, design, and construction programs requires a heavy potion of sound leadership in its managers. The debate rages as to whether or not good managers must be good leaders and whether good leaders must be good managers. We find all combinations in generic construction. Few, however, will argue against the thesis that if the

---

[3] Education is the teaching and learning process by which the principles of doing things are conveyed to the learner.

[4] Training is the teaching and learning process by which specific, explicit methods and systems of doing something, usually by rote, are conveyed to the learner.

two skills are found together in a person, the probability of that person being a successful manager is very high.

The Thompsonville Learning Experience is a case study in which the qualities of education, training, and leadership are illustrated and merged to produce an example of what can happen when synergy is at work in intelligent and understanding people.

Case Study 14.1: The Thompsonville Learning Experience

Thompsonville voters have just approved a long-awaited school bond issue. The school district staff has started early planning work on a $20,000,000 elementary, middle, and high school remodeling and new construction program. Excitement is running high in Thompsonville among teachers, parents of school children, and those involved in the planning, design, and construction profession.

Of major interest is that there have been some changes on the school board since the bond issue was first proposed. One significant event was the election of a successful local businessman, Al Lockland, to the board. Mr. Lockland is known to advocate very tight controls on all school-related expenditures. He has little or no planning, design, or construction experience but does exert considerable influence in Thompsonville. People tend to follow his leadership. At Mr. Lockland's suggestion the board voted to save the remainder of the project programming fees by having the school staff write the project program instead of retaining the architect and engineer to prepare it. Once written, they plan to have their consulting architect/engineer, Smith and McConnell, review and critique the program document.

Smith and McConnell (S & M) was selected prior to the bond issue to help the school district set the program parameters and preliminary target costs. These were used to determine the amount of the bond issue to be voted on. S & M had assumed that their firm would be selected to write the full program if the bond issue passed. They also had been tentatively selected by the school district's staff as the architect/engineer of record. With this dual responsibility implied, Martin Smith, one of the partners, took on the project manager job for S & M to ensure that it received top management and design attention. He is understandably puzzled by the school district's decision to attempt writing the program in-house.

Mr. Smith is an experienced and highly regarded design professional. He knows that the path to a successful project sometimes takes unexpected twist and turns. S & M has had a long and very good relation with Thompsonville schools, and Mr. Smith feels that he will probably have to question the programming decision at an appropriate time in the near future. However, he and his partner, Lynn McConnell, have decided they will make every effort now to work in the interest of the Thompsonville School District and to help carry out the program as anticipated in the successful money-raising campaign.

The school board has appointed Carl Burton, the district's financial officer, as its program writer and project manager. Mr. Burton is a conscientious and capable financial officer but knows very little about design or construction. He and Martin Smith get along very well and have been through many smaller projects together. In discussing how to proceed with the project, Martin has been retained by Carl to help him with the program on an hourly fee basis and to review the material as it progresses. There is no word from the school board whether or not S & M will be the design team finally selected.

As the project proceeds, Carl Burton quickly saw he had neither the time, the experience, nor the knowledge to write the complex program appropriate for a job of this size. Mr. Burton brought the matter to the board and after considerable discussion and some disagreement among the board members, the board finally agreed to consider a different course of action—this only after a very harassed Carl Burton said that he could not possibly do the programming job he had been assigned without professional help. He had been consulting with Martin Smith on an informal basis over an occasional lunch or breakfast to save as much as possible on the programming costs. When he decided to shed the programmer assignment, Carl Burton recommended that Martin be given the program responsibility on a fee to be negotiated. At the board's request, Carl agreed to continue his work with S & M at a reduced level of involvement.

Ultimately, the full program was written by Martin Smith, who was also selected as the architect of record; S & M prepared the contract documents on schedule, and these were issued for competitive bids to several local general contractors and specialty contractors. The Thompsonville School Board was heavily pressured by Mr. Lockland to award separate prime contracts for general, mechanical, electrical, food service, and controls work. These separate primes were ultimately to be assigned to the general contractor.

The low-bid general, Maxima Construction, has a doubtful reputation for both integrity and competence. Its track record on payment practices and job management is especially poor. This is well known among construction professionals in the Thompsonville area. The low-bid mechanical, electrical, and food services contractors are all good contractors who were dismayed to see that Maxima was the low general. When the bids were opened, all three would have refused the job except for the fact that they would have had to forfeit their bid bond. Such action would have been a mark against their record. Also, they were looking to Martin Smith and Carl Burton to guide the project through to completion. The controls contractor was satisfied with his contract and had no objection to being assigned to Maxima.

The attitudes of the successful contractors and the nature of the project delivery system being used tended to make this a moderately high potential claim-prone job (see Chapter 5 for a description of claim-prone construction projects). Despite the shaky start, the school expansion program seemed to get off on the right foot. This was in part because S & M and Carl Burton strongly recommended that the school district establish a project partnering system on the job. Mr. Burton had continued being the official school board representative on the project. In this situation he commanded considerable respect because of his reputation for fairness, balanced judgment, and good management of the school district's financial affairs.

Nobody on the project had been involved as a stakeholder in a partnering system previously. However, most of the project participants had heard of project partnering and were enthusiastic about trying it. Even the superintendent and project manager for Maxima were excited about the partnering effort. They had been telling their company management that it was time to change the local industry's perception of Maxima as a less-than-reliable contractor to one more favorable to the future careers of key staff members. They were also enthused about receiving help to mold a smoothly working project team from the good design and contractor group assembled for the project. The field superintendent, Phil Marconi, and the project manager, Lou Hillston, are competent people and have been with Maxima about five years. They both have a string of successful projects in other communities with the Maxima organization.

The partnering charter writing came off well, with nearly 35 stakeholders participating. All signed the charter, and many stakeholders volunteered to participate in the task force work to write a partnering evaluation system and an issue resolution system. The charter monitoring task force has written and implemented its evaluation system. The monitoring system has been received cautiously but positively by the full project team. Monitorings are once per month and reported on by the stakeholders to the entire project management team.

Five of the charter objectives that were of special significance to the stakeholders on the job are listed below. Achievement of these objectives is important to project success since they represent some of the most important management actions to be measured by monitoring the charter. They are also concerned with some of the historically weak leadership and directive abilities of two principal players on the project, the school district management, and the general contractor's ownership and management.

The objective par weight (w) and par performance (p) assigned by the stakeholders is noted following each objective. The product of the two gives the acceptable target level of performance, designated as the par rating (see Chapter 12, Section B, for a detailed description of the various elements of the partnering evaluation system).

- *Objective 2.* Prepare and maintain with all stakeholders an accurate project critical path network and a short-term schedule of work; $w = 4.2$, $p = 3.5$, par rating = 14.7.

- *Objective 3.* Make prompt, full payment of all properly submitted pay requests; w = 4.0, p = 4.5, par rating = 18.0.
- *Objective 8.* Prepare, publish, and adhere to a chart showing channels of communication, responsibility, and authority; w = 3.5, p = 3.8, par rating = 13.3.
- *Objective 13.* Prepare, publish, and implement procedures to be followed for resolving disputed issues and problems promptly and at the originating level where possible; w = 4.0, p = 4.0, par rating = 16.0.
- *Objective 14.* Be a good on-site and off-site neighbor; w = 3.0, p = 3.5, par rating = 10.5.

The first month's rating of partnering performance (Table 14.1) shows the result for these five items. The reason for the importance of the rating trends for these five charter objectives is that they relate to potential deficiencies in the stakeholder organizations, particularly Maxima's. The calculated par weight multiplied by the actual performance rating each month is shown in column 2.

After the first month's ratings were tabulated, Martin Smith, the architect, Carl Burton, and Mr. Lockland of the Thompsonville school district met in the school cafeteria for lunch. Mr. Lockland had seen the ratings and was questioning some of the results since he had submitted a performance rating of his own. "The stakeholder ratings," he said, "were all out of kilter with his." Being an intelligent and well-intentioned man, he was curious as to where his thinking and measuring were different than those of the stakeholders. Martin Smith said: "Let's look at the five objectives that Carl and I are specially interested in. They represent, in our opinion, a good measure of some of the project staff weaknesses that must be strengthened for the project to be successfully completed."

The first objective they discussed was charter provision 2: *Prepare and maintain with all stakeholders an accurate project critical path network and a short-term schedule of work.* Par performance was set at 14.7; the first month's rating came in at 10.0, considerably below what should be expected

TABLE 14.1   Charter Ratings for Thompsonville School

(1)	(2)	(3)	(4)	(5)	(6)	(7)
				Month		
Objective	Par	1	2	3	4	5
2. Planning and scheduling	14.7	10.0	—	—	—	—
3. Payments	18.0	11.3	—	—	—	—
8. Organization, communication, authority	13.3	10.5	—	—	—	—
13. Issue resolution	16.0	09.5	—	—	—	—
14. Be a good on-site and off-site neighbor	10.5	10.0	—	—	—	—

during so critical a planning period in the job. Mr. Lockland said that he had expected a higher rating also. He also related that Maxima's field superintendent, Phil Marconi, had told him on a field visit that the president of Maxima wants to keep scheduling expenditures to an absolute minimum since it is basically an overhead expense. The resulting lack of attention by the general contractor is what has caused the rest of the stakeholders to have given the vital scheduling process a low performance rating.

The second objective reviewed by the group was charter provision 3: *Make prompt, full payment of all properly submitted pay requests.* Mr. Lockland said he thought the single monthly payment to date had been made with reasonable promptness. He had rated this at 17.0, just under the par of 18.0. His lower-than-par rating, as he admitted, was given because he had tried to lower the request amount from that approved by Martin Smith last month, and the delay in considering the reduction caused the payment item to miss the agenda deadline. Carl Burton said to Mr. Lockland that payments are the lifeline of any construction project and that tampering with the contract-stipulated process would bring quick negative response. Martin Smith said this was certainly true and that the stakeholder rating of 11.3 was an exceptionally low report card grade for the owner.

Mr. Lockland somewhat unwillingly agreed that he would not interfere again with the payment process. He correctly deduced that the field and design staff were perfectly competent to evaluate the accuracy of the draw. Shopping for delays to payment in order to produce added interest for the owner was, as Mr. Smith pointed out, one certain quick way to lose the support of the construction team.

The next charter objective to be discussed was charter provision 3: *Prepare, publish, and adhere to a chart showing channels of communication, responsibility, and authority.* The par rating was 13.3, the stakeholder rating was 10.5, again, as with the preceding two items, considerably lower than it should be. The problem, said Martin Smith, was that Maxima did not want to expose its management and authority structure to the other stakeholders on the job. "Why, then," asked Carl Burton, "did they agree to it and support it as charter provision?" Mr. Lockland added a bit of wisdom from his business experience: "Maxima management just doesn't want their own internal staff to find out how little clout the staff actually has. Perhaps an informal discussion with Maxima's president and two or three of the school board's members might help shake some kind of chart or statement loose. We'll tackle this issue and report back to you, Martin, on the result. We've got to get that rating up."

Up for discussion next was charter provision 13: *Prepare, publish, and implement procedures to be followed for resolving disputed issues and problems promptly, at the originating level where possible.* Carl Burton related that the procedures had been published and were in effect, but the very first issue that arose was the early slow payment situation. In this matter, which the contractors called an issue, the school, and perhaps Smith and

McConnell also, were not interested in following the "resolving at the originating level promptly" precept in their work. The result was, predictably, a very low rating of 9.5 compared to a par of 16.0.

The luncheon partners next touched on the fifth objective—14: *Be a good on-site and off site-neighbor.* Mr. Lockland asked how the on-site and off-site neighbor relations were being maintained. Martin Smith said that the residents in the area were beginning to complain about the dirt being dragged from the site by the excavator. He said the specifications required Maxima to clean up but that they had been somewhat derelict. He added that he would talk to Lou Hillston, Maxima's project manager, this afternoon about doing a better job of off-site cleanup. Otherwise, the rating seems to indicate that most stakeholders believe the trades are keeping their own areas clean and passable, that site conditions for parking and access are being well maintained, and that layout of the site went well with no major discrepancies encountered. The key to ongoing good performance will be in how effectively the existing school users will be protected from construction work as project foundations are completed.

The total charter par target is 17.5, and the first month's performance rating for all charter objectives was 14.0. This is not even an average performance, and all three admit that it must be improved if the project is going to be considered successful. In summary, Mr. Lockland says, their informal discussion about the five charter objectives selected has brought out the following points.

1. Maxima, the subcontractors, the design team, and the owner must immediately raise their performance level in planning and scheduling. Mr. Smith will discuss methods of doing this at the site tomorrow with Phil Marconi and Lou Hillston of Maxima.

2. Mr. Lockland, with guidance from Mr. Burton and Martin Smith, will take steps to streamline and improve the payment process. If Mr. Lockland wants to be involved in reviewing pay requests, he will not interfere or delay payment of legitimate billings properly submitted.

3. Irrespective of the internal organizational structure that is used by the stakeholders, the owner, the architect/engineer, and all interacting contractors are entitled to know what relations they are expected to maintain with other stakeholders. Mr. Lockland and two or three others from the board will meet informally with Maxima's president and ask him to set an example and take the leadership in getting a project organization chart that will accurately depict the interrelations of people on the project team.

4. All three agreed that they would have to shift more support to the issue resolution policy, particularly since the owner seemed to have been one of those apparently showing little interest in resolving the very first issue of importance to the stakeholders, a delay in processing

payment requests. Future payments are to be expedited to the best of the architect's, engineer's, and school board's abilities.

5. Being a good on-site and off-site neighbor was rated fairly well. However, there will be an ongoing need for cleaning roads, maintaining security, providing night lighting that will not be a nuisance to the neighbors, and each of the other items that can endanger job performance. Mr. Lockland said as they left the cafeteria that if there were no objections, he might organize a be-a-good-neighbor task force from the school board to monitor neighborhood attitudes toward the project. This, he feels, might bring the school users into a closer relation to the project and improve local relationships.

Over the next four months S & M and the school board watched the charter evaluations carefully. The monthly evaluations for the five objectives of greatest concern to them are shown in Table 14.2.

As the project moved ahead there appeared to be a perceptible improvement in management of the project. Planning and scheduling improved and at the end of month 5, the master network plan has been prepared, and four-week short-term schedules are being issued by Maxima at each weekly job meeting. Subcontractors are still not providing all the input they could, but Maxima's superintendent, Phil Marconi, did convince his executive staff that the school board was very serious about good planning and scheduling. Maxima has hired an outstanding planning and scheduling consultant to work with the field superintendent and the project manager. Lou Hillston, Maxima's project manager, feels that by evaluation period 6, the rating will have reached or be greater than the par of 14.7.

Payment practices from the owner and from Maxima have improved but still have a long way to go. Maxima has glue on its fingers when it comes to paying the subcontractors. However, the improvement in owner payment turnaround has removed one of Maxima's excuses for late payment. In addition, Mr. Lockland has had some long conversations with

TABLE 14.2 Monthly Evaluations of Chapter Objectives

(1)	(2)	(3)	(4)	(5)	(6)	(7)
				Month		
Objective	Par	1	2	3	4	5
2. Planning and scheduling	14.7	10.0	11.2	12.3	11.5	12.8
3. Payments	18.0	11.3	12.5	14.0	15.0	17.0
8. Organization, communication, authority	13.3	10.5	13.2	14.2	13.0	14.0
13. Issue resolution	16.0	09.5	12.2	13.3	15.0	14.0
14. Be a good on-site and off-site neighbor	10.5	10.0	11.0	10.5	12.0	12.5

Maxima's president about his future work on Thompsonville projects. Management of money has evened out since and the improved project performance shows the positive effect of this on job management.

Organizational relationships have been clearly defined now and most job personnel know much more about the chain of command, the authority and responsibility patterns, and the communications channels they have to use to be most effective. The performance ratings on charter objective 8—*Prepare, publish, and adhere to a chart showing channels of communication, responsibility, and authority*—have risen from 10.5 in the first month to 14.0 in the fifth month, measured against a par of 13.3. This is good news and shows considerable improvement in job management.

Issue resolution problems have not appeared since the first conflict about delayed payments. There are still some rumblings about some of the owner's strong-arm methods in reviewing requests for payment. Mr. Lockland, though, has become a crusader in getting prompter action on pay requests.

The last of the five management-oriented charter provisions was being a good neighbor. The good-neighbor task force assembled from the school board has done its job well and ratings have risen 2.5 points in five months. This is 2.0 full points above par. The effort to improve off-site relations carried over into the on-site work, and currently everyone seems happy with how the site is being managed.

Problem 2: Communicating with Others (Appears 984 Times, 35% of Total) Effective communications of all types and at all levels play a crucial role in the generic construction industry. The difficulty encountered when trying to improve communications is that the analysis of what is wrong and the solutions rarely consider the problem in adequate detail to make communications better. Communications technology today has reached a point where the machines are so dazzling and exciting that many feel that the mere presence of a fax machine or the use of fiber optics or capturing of signals by a communications dish will right all the wrongs of our communications dilemma. Not so. Effective communications among and between people on design and construction projects need careful substantive[5] attention in their development and use. The machines we use to communicate merely allow us to increase and accelerate information transfer. They do not automatically improve the results we obtain from the interchange. Case study 14.2 illustrates how the need for communication within a company can change as the organization changes. Effective communication results from human systems working in synch with technological systems to produce a product that people can and want to use as tools to improve their performance.

---

[5] Used here to mean belonging to the real nature or essential part of something.

Case Study 14.2: Larson Development's Communication Overkill

Frank Larson is a very successful developer. He also is a man who believes in operating his business on an everyone-should-know[6] basis. This system was put into use at Larson Development when the total staff included Frank, his wife, one secretary, and a part-time real estate broker. The four of them met twice a week for about an hour to bring each other up to date on what was happening on their projects and holdings. These meetings were absolutely essential to the four working productively without expensive overlaps and destructive conflicts. Construction volume in these early days was about $1 million annually.

In a short time Larson Development grew to its present level of 40 employees and a real estate management portfolio valued at nearly $50 million. The annual construction volume has increased to about $15 million. However, most of the employees are still expected to maintain an everyone-should-know communications system on their projects and in their daily work. Exceptions are made for confidential leads and on early financing and syndicating actions.

Larson is about to begin construction of a $15 million office campus in a nearby community. Tom Davis, a construction project manager with 15 years of solid experience, has been hired to run the project for Larson. He has been very busy in his first five weeks moving the new job into design and producing the first construction document package for site preparation and construction of the foundations, superstructure, and the exterior skin of the building. Tom has recommended to Frank Larson that the project team plan and put into effect a partnering system to be implemented at various phase points in the job (see Chapter 8, Section B, and Chapter 10, Section A, for discussions of the relation of project phases to partnering efforts). Mr. Larson, an astute people-oriented person, has been studying the partnering concept for several months and agrees that the job should be partnered.

Tom Davis is very happy at Larson Development. He has found it to be a highly professional firm with a good reputation that is staffed with some exceptionally high-caliber people. The only irritant in his first five weeks has been the constant information requests from Mr. and Mrs. Larson, the bookkeeper, the financial analyst, and some of the salespeople and tenant coordinators, all of whom have asked him to keep them posted weekly on the project's progress. Tom is a man who is accustomed to being a good team player while retaining the prerogative to move ahead in the direction that he feels is best without always having to share his reasons with those outside the immediate action arena. Tom realizes that this course

---

[6] An organizational communications system based on the managerial belief that if everyone knows what all or most people in the organization are doing and working on, the organization's overall output quality will be superior.

of action is not always correct or best. However, he has fallen into an independent action mode and understands that he may have to modify his management style in his work at Larson.

During project mobilization and move onto the project site, Tom and Mr. Larson announced that partnering was to be used on the project. A week later the charter-writing session was held at a hotel near the site. By this time a general contract had been awarded to Berke Construction and all of the major sitework, foundation, superstucture, and close-in subcontractors had been placed under contract directly with Berke. The owner-provided furnishings contract was awarded by Larson directly to an interiors design–build firm.

Many of the participants and stakeholders selected for the project have worked with Larson Development in the past. Many compliments and congratulations have been passed along to Tom about his new position with the firm. All who know him have indicated their pleasure at having him on the job for Larson. In general, field work on the new building seems to be off to a good start.

During the charter problem identification workshops, Tom noted a heavy emphasis by the stakeholders on potential communications difficulties in the project. At first he thought these were the usual indistinct, general, and sometimes fuzzy statements about generic communications. As the charter writing proceeded, however, Tom began to see emerging in the session a specific theme of communication alerts from the project team. Typical of potential problems the stakeholders mentioned were:

- Not defining and following agree-upon lines of authority and responsibility
- Several parties presuming to speak for the owner, the design team, and the prime contractor
- Too many project spokespersons
- Making schedule and specification changes without informing the participants affected

Other problem comments were made which strongly indicated that project stakeholders wanted project communication, authority, and lines of responsibility established clearly on the site. They also wanted to reduce or eliminate any reporting channels to off-site personnel in any of the stakeholder offices. The entire matter of communicating within and outside the stakeholder project team was addressed in the charter objective that states:

5. Stakeholders will attempt to communicate effectively and on an as-needed basis. Stakeholders commit to be available, to recognize the need for high-quality information provided in a timely manner, and to identify clearly

the project chain of command, showing communications flow, responsibilities, and authority.

Mr. Larson, who had attended the charter meeting for the full day, was chatting with Tom at the social hour after the charter had been signed. He asked Tom, "What was the significance of charter provision 5?" He continued: "Today, some of the stakeholders who know me particularly well asked if I really intended to communicate more effectively on this project." Tom tactfully pointed out that from his viewpoint the charter objective was designed to help streamline communications among the stakeholders and cut down on the amount of casual project-related information that had in the past been sent or received by those not directly involved in the job. Frank Larson said that he would be interested in seeing how well the stakeholders adhered to the need-to-know policy since he had always thought that most people liked to know all about their work and the work of others.

In the first charter monitoring the stakeholder's evaluation gave charter provision 5, the communications and chain of command provision, a weighted performance rating of 12.0 against a weighted performance par of 14.8. The second month the communications rating dropped to 10.9 and some grumbling was heard about slow payments, cross communications from the leasing people, and conflicting directions being given by Larson's office staff to some field personnel. Mr. Larson asked Tom to have breakfast with him and asked why some of the weighted performance ratings were declining. Tom briefly outlined the reasons and his recommendations to turn the ratings around. He said there seemed to be three main reasons, based on discussions at the evaluation sessions:

1. Larson's accounting staff insists on reviewing each general and subcontractor request for payment directly with the contractors. This meant that some of the subcontractors had to justify their payment requests once with the general contractor, once with him, and once with the accounting group. It was requiring a lot of expensive time and reworking to accomplish a relatively simple job.

   Tom's recommendation: Eliminate the accounting group's review of subcontractor's payment applications. Tom can make his review of the general contractor's and the subcontractor's request at one time and bring them to the office for accounting to check. Or they can come to the field to do the check. He estimates that Larson can trim at least one week in processing all payments, can considerably enhance its reputation with all the contractors, and might even begin to get better pricing from the contractors on future jobs.

2. The Larson office staff has resisted (formally and informally) all efforts by Tom to prepare a project organization, responsibility, and communications flowchart. The office staff has also been slow to

provide information to Tom that would allow a stakeholder task force to complete an organizational matrix required by the charter. This chart is to show responsibilities to and from all job management personnel. Lack of information from Larson for the task force has interfered with stakeholder efforts to complete the required matrix.

Tom's recommendation: That Mr. Larson and Tom prepare and distribute an organization chart as it applies to this project only. In addition, that Mr. Larson provide Tom with additional responsibility, authority, and communications information that can be considered when preparing the stakeholder's task force responsibility and communications matrix (see Case Study 14.8 for an example of a communications matrix).

3. Many documents, including critical requests for information from the stakeholders, have gone astray in the Larson information system. Apparently, over the years the firms that have worked on Larson's smaller projects have merely sent all documents to the main office, addressing them to Mr. Larson. He then gave them to his secretary for distribution. This project, however, is much larger than any previous Larson project. The need for prompt, well-directed transmission of spoken and printed directions is far more urgent now than in the past. In addition, there are many more faces, some very new, in the office. These people do not know much about how things were done in the past and have not been told how to handle and route designer and contractor communications.

Tom's recommendation: That project-related communications flowing into and out of the project be directed through his field office. Frank Larson and Tom should meet immediately with key members of the Larson support staff and prepare a description and instructions as to how all project-related information is to be routed and distributed on the job.

Several other points were discussed at breakfast, many aimed at strengthening Larson Development's efforts to improve communications on the project. However, Mr. Larson and Tom Davis took immediate steps to plug the dike where many of the complaints were originating.

At the third charter monitoring session the communications rating on charter objective 5 dropped slightly, from 10.9 to 10.6. Mr. Larson by this time saw that the communication techniques he had needed and encouraged in his smaller company had not worked nearly as well in his growing organization. He also saw that any reduced overhead and increased development profitability was closely related to the ease with which information could be obtained and transmitted to the people best equipped to take action.

Meanwhile, Tom Davis took considerable heat from some of the office staff as he gradually redirected project-related communications from an unstructured flow to a very highly directed flow. He was somewhat used to this type of criticism however, and as his recommendations were implemented and worked without hurting anyone, his stature increased both within the company and on the job. Good results began to show up in the fourth charter monitoring, when charter objective 5 got a rating of 12.0, up from 10.6 the month before. The fifth rating, made a month later, showed further improvement, to 14.0, only 0.8 point below par. Of course, not all of the steady improvement could be credited to Larson Development and Tom Davis. Their actions had merely stimulated a general move toward improving project-wide communications through gradual company improvements.

In the Larson Development example, the thoughtful, openminded consideration given to the partnering processes by Frank Larson and Tom Davis, particularly to the partnering evaluation ratings, produced two results. The first was an improvement in project performance that benefited all stakeholders. Second, Larson's efforts helped to streamline the payment process and began to identify the true responsibilities and authority of the field and office staff relative to this specific project. As improvements continue to be made they should help make the firm more effective in their work, reduce overhead, and improve profitability. These results will benefit everyone employed in the Larson organization.

Problem 3: Staff Morale and Attitudes (Appears 684 Times, 24% of Total) Planning, design, and construction people are just like other people. Most of us want to be respected, want our talents recognized, like to receive deserved (and sometimes, undeserved) praise, need help and encouragement in times of trouble, and want our work efforts appreciated. A situation encountered in Dallas, Texas, several years ago illustrates the importance of an appreciation of good work efforts. A well-respected general contractor was completing work on a new 900-room hotel. Hotel furnishings were being put in the guest rooms and as an intermediate step were being stored in vacant areas of the hotel. One of the temporary storage areas was on the second floor in a large open space. Both rough and finish building trades were working in the area and by some quirk of fate, the masonry contractor, who had earlier left out a wall to allow construction trade access to a floor area, had come back in a work lull elsewhere to finish work on that floor. The trade access was no longer needed, so the masonry crew completed the wall across the open space.

Unfortunately, the interior furnishings crew was not watching and did not see the block wall being completed. The wall was up and complete in one day and the next day the furnishings contractor was down at the mason contractor's field office demanding that they remove the new wall. The furnishing contractor had been blocked from the freight elevators by the newly erected wall. There was no immediate solution except for the general contractor and the masonry subcontractor to remove a part of the wall they had just erected. Ample assurance was given the mason contractor by the general that he would be paid for the wall rework.

When the masonry foreman was told to take the wall down, he flatly refused. His rationale was that if higher job management did not know what was going on at that space, why should he have to correct their mistake by taking down his well-built block wall, put up in record time and built to his high-quality level. He refused to demolish the wall or to put a new opening in the wall as a matter of principle.

Extreme?—perhaps, but certainly understandable from the viewpoint of a craftsman who didn't need a lot of practice building and demolishing masonry walls—particularly when it resulted, in his perception, from a dumb mistake by someone else. In his opinion, his work effort was not appreciated, and his reaction, though strong, exhibits how any good worker might feel if what he or she has done can be undone so easily.

This type of problem involves staff morale and attitudes. Partnering meetings bring out morale and attitude statements that without resolution can seriously harm work quality and production on the job site. This can be seen by the fact that such problems received the third most mentions of the 45 problem classifications and 2855 problem statements. The real management challenge faced on a project is what to do about morale and attitude problems once they are identified. Earlier in this chapter I listed some selected problem statements that were made in charter meetings. How does the manager bring about performance improvement by better understanding and overcoming these types of problems in the office and on the site?

Case Study 14.3: The Bullseye Project

Bullseye Recreational Products, Inc., is a large assembler of motorized recreational equipment. The product line includes off-the-road vehicles, snowmobiles, jet water skis, and an extensive line of hard and soft accessories such as lightweight outdoor shelters, boots, backpacks, all-weather jackets, carrying attachments for sports equipment, and many others. Most of their products are assembled from parts that others manufacture in all parts of the United States. The company and its employees are very proud of the fact that to the best of their knowledge, 85% of the parts used in the products they sell are produced in one or more of the 50 states comprising the United States.

Bullseye is currently building a new plant in its home town of Oxford, Wisconsin. It is to be a 200,000-square-foot addition to the present facility. Design documents are complete, a general contract has been awarded to Stone Construction Company, a good local general contractor, and Stone has given purchase orders to about 10 local subcontractors for work on the new plant. There still remain about five small contracts to be awarded as the job moves into the field.

Art Perry is the field superintendent on the project for Stone. He has been with Stone 30 years, first working there as a laborer during the summer months in high school. Art is a loyal, competent, conscientious man who gets along well with most people. He has good leadership skills and tries hard to study and learn new working management techniques. His exposure to the management literature available at the Oxford library has lead Art to recommend to Mr. Charles Stone, owner of Stone Construction, that the company institute an in-house management training program for its new generation of field and office managers. Art has been so busy that he has put the program on hold for a while. Mr. Stone has many other things on his mind and has not followed up on the idea.

Art notes that the mechanical contractor on the Bullseye project has assigned Jerry Stressland, a foreman-superintendent, to the new job. Jerry had been on one of Stone's jobs about 10 years ago. He and Art had had a very noisy and acrimonious argument one day that resulted in both men being strongly criticized by their superiors. Jerry was actually moved to another job by his firm because of the altercation. This created hard feelings strong enough to cause the two to avoid each other deliberately during the intervening 10 years.

Art, as part of his ongoing education, has been studying the history of partnering in construction. He is impressed by the track record produced by its use in the Corps of Engineers and on several heavy construction jobs out west. As a result of his study, he has requested that Stone participate with the owner, architect, engineer, and subcontractors in a partnering effort on the job. Mr. Stone is agreeable, as is Barry Eychel, owner of Bullseye.

The major subcontractor managers on the job seem to be pleased by the opportunity to try partnering, although the old antagonisms still show in Jerry Stressland's attitude. He and Art have had some mild discussions about the Bullseye job, and Art is anxious to heal the old wounds, which are still a bit raw from their decade-past conflict. Jerry does not seem quite as anxious as Art to rebuild their relationship. He is somewhat neutral toward partnering, but not hostile.

The decision to proceed with partnering has been made, the charter-writing meeting is in progress, and stakeholders are currently working in the problem identification workshops. The participants have completed identifying and presenting to each other the problems others cause them. One of the stakeholder teams mentioned two problem items that were somewhat related to the Jerry-and-Art conflict:

- Lack of respect for other partner's space, materials and work in place.
- Poor attitudes and previous personality conflicts.

Art began to think over his 10-year-old confrontation with Jerry. It was caused when the two men were working on a very tight site. Art was at that time a newly appointed superintendent for Stone. He had been told by the owner of the project that Jerry's trailer was incorrectly located in an area that the contract specifications clearly said was reserved for owner parking on the job site.

Jerry, the mechanical foreman on the job, when approached by Art said that he knew nothing about the restriction since he did not have the general requirements for construction parking in his bid set. Jerry was right—the specification section had been accidentally left out of the mechanical specifications. However, Art felt he had no alternative but to insist that Jerry move to a far corner of the site. Jerry felt betrayed by Art and by Stone both, and considered that he had been abandoned when he needed their help. The result was a steadily eroding relationship that lead to the severe criticism leveled against both of them by their bosses.

As the workshop discussions of potential problems on the Bullseye job continued, Art noticed several other items that he had been somewhat indifferent to in his own management career. He had tended to shift job cleanup costs from Stone down to the subcontractors all too often. In fact, on the Bullseye job he had already transferred more cleanup from mobilizing and moving on the site to his subcontractors than was warranted. Bullseye was very fussy about site cleanliness, and the image presented to the public was very important, particularly to its Wisconsin neighbors. Costs for cleanup had been higher than Art wanted to spend this early, so he had spread the extra cost among his subcontractors, and not too fairly.

Others in the room also began to think about the matter of their attitudes and morale as the presentations and discussions continued. Mr. Eychel, owner of Bullseye, was thinking as the meeting proceeded that he had to tighten up on his decision making and provide project design and construction answers more promptly than he had been doing over the past several months. Frank Fischer, the architectural project manager, said to himself as the matter of slow shop drawing review and approvals was brought up that he was spending more time on shop drawings and being far more critical than was warranted. Frank had two or three recent unfortunate experiences with contractors trying to push him too hard on submittals. He admitted to himself he was becoming more nit-pickey than was warranted—maybe this was a time of change.

Jerry Stressland was thinking about the 10-year-old argument with Art Perry and about the Bullseye job and about his present situation. He remembered his boss telling him yesterday that the argument he had with Art 10 years ago was past. Just before Jerry left for the partnering session, the boss added: "Jerry you're where you are now, and you can't be anywhere else. Don't always be reliving history."

And so it went. By the end of the long charter day many of the stakeholders had some second and even third thoughts about past experiences and present problems that were presumably caused by others' real or imagined attitudes and remarks. The open discussions of problems that might be encountered and the injection of outside opinions of the importance of some of these problems had a sobering but relieving effect on many of the stakeholders.

A few days after the charter meeting Art Perry decided to approach Jerry Stressland and try to patch up their differences. He did and his attempt at a reconciliation impressed Jerry. It was a good move, it worked to a limited extent, and it was welcomed by Jerry. There were still some rough edges to smooth, but the old resentment began fading fast in the knowledge that today's work must be done by people who are willing to live in today's world, not yesterday's.

A side effect of the partnering meeting was that Mr. Stone was very impressed by Art's performance in the charter meeting. After the meeting he told Art that the partnering concept might provide the base of a good in-house management training program for new field and office managers. Art volunteered to work on the project, and his offer was accepted immediately by Mr. Stone.

So the Bullseye project was up and running. Some past emnities were resolved, the owner's decision-making processes were improved, shop drawing processing got better and quicker, and Art is off on a highly visible training effort. Partnering helped provide the invisible catalyst for change because it helped stakeholders see where they could change for the better.

Problem 4: Personnel Quality and Problems (Appears 593 Times, 21% of Total)   Quality is an elusive element to define in any person's performance. Usually, results must speak for the level of excellence. Other measurements are apt to be subjective and to contain emotional overtones that reduce their worth as reliable barometers of value added. Partnering contributes to the overall good of a project by exposing all stakeholders to the scrutiny of their subordinates, peers, and superiors without encouraging any level of observer to become overly judgmental. A well-conducted charter meeting first mixes individuals within disciplines and then gradually combines them into a total group. An important benefit that the less-than-competent planner, designer, or constructor gains from the intense interaction among stakeholders is an opportunity to improve.

Partnering does not benefit the stakeholders by actually improving their work performance. Instead, it benefits each person by demonstrating what good can come from achieving higher competence levels. This is accomplished by all responsible project practitioners being exposed to the most competent people on the job. If we assume, as is usually the case, that the charter meeting is attended by most of the ultimate decision makers, all attending are exposed to the management influence of those leading the project. Learning is then accomplished by benchmarking against these leaders. Of course, excellence does not always rub off automatically, and it may be too much to expect all men and women who attend a charter meeting to improve instantly. However, the opportunity to improve is given, and this is often a sufficient motivator to act as a needed catalyst for improvement.

Case Study 14.4: The Maturing of John Fenton and William Tole

Langtree Robotic's president, Thomas Langtree, has for years visualized an office and industrial park development that contains related industries—businesses that might benefit from being near each other. He has decided to implement his idea by purchasing 1100 acres of undeveloped land on the outskirts of Redford City, a commercial- and industrial-based community of about 200,000 people. Mr. Langtree selected a site very close to a projected new regional airport, one that will be served by a new freeway, U.S. 18, now being designed by the highway department. The airport is planned to be operational in four years. The freeway is to be open to traffic in two years.

Mr. Langtree intends to build a 150,000-square-foot state-of-the-art manufacturing plant as the initial facility on the site. It will have adjoining design offices, laboratories, and testing and demonstration facilities containing another 100,000 square feet of floor space on two levels. The total proforma cost set for the Langtree phase one facility is about $30 million. Mr. Langtree has inquired of his business and professional associates about architects, engineers, and contractors who might be equipped to work on this prototype. He has finally settled on an planner, architect/engineer, and contractor team that he feels will produce the facility he wants.

Wilson Allen, Ltd., is the planner; Langley Evans Associates is the architect and engineer; and Rhodes Construction Company has been selected as the construction advisor. Langley Evans has appointed John Fenton as project manager. Mr. Fenton is a reasonably good designer and production architect—much accustomed to working with contractors on hard-money lump-sum jobs. Rhodes Construction has selected Bill Tole as its project manager. Mr. Tole recently joined the firm as a project manager and will be in charge of all work for Rhodes, through planning, design, and construction. His background is basically in industrial manufacturing buildings, although he is knowledgeable about office and research and development facilities.

Mr. Langtree has given Wilson Allen the go-ahead to start on site plan studies and to work with the architect/engineer and the construction advisor to write a program for the job. As these program meetings proceed, three facts begin to emerge:

1. The concept of an integrated business park of related enterprises is understood by the planner. However, John Fenton, the architect, has expressed his feeling that working drawings for Langtree Robotics must be started immediately if his firm is to finish the contract documents in a timely fashion. Mr. Langtree expects considerably more planning and total program effort from the team before moving into schematics and design development work.

2. The construction advisor, Bill Tole, is pushing hard for construction documents to be started so that he can begin early procurement on

the project. He appears to be marshaling support among the Langtree Robotics staff for following this course of action. They are anxious to get into the new facility

3. Wilson Allen, the planning firm project manager, has not exhibited the leadership qualities that Mr. Langtree had anticipated. The vacuum created is being filled in part by John Fenton, but especially by William Tole.

The unintended shift in control of the project planning and programming to production-oriented team members is viewed with some alarm by Mr. Langtree. He sees his vision of a prototype business park threatened by hasty action to get a project built. His view is that he is willing to allow time and to spend the money to produce a product consistent with his long-range ideas. Efforts to convey this to the project team in meetings have been ineffective. One reason is that the meetings are dominated by Mr. Allen. Another reason is that Mr. Langtree is not an expert in planning, design, and construction. He is depending on his project team to do this work. However, he sees now that they do not understand what he wants and he is not certain they know how to produce it even if he could explain it well enough.

He arranges a dinner with Larry Rhodes, president of Rhodes Construction, to make a final effort to clear things up. At the meeting Mr. Langtree lays it on the line. He admits that he has probably not explained the project concept well enough to those doing the work; he is deeply concerned that he made a mistake having Wilson Allen, the planner, so heavily involved in writing the program; and he is worried that the present personnel on the job do not have the abilities to accomplish what he expects from them.

Larry Rhodes is not startled by what he is hearing since he sensed the loss of direction from regular reports he receives from Bill Tole. Larry also keeps in contact with Langley Evans, the architectural principal for the project. Langley is worried about his manager, John Fenton, pushing too hard and too fast. Larry suggests that the current project team assemble and outline where they are, what their mission is, what they can do about achieving it, and then decide on a course of action. In essence he is suggesting that they convene an informal partnering meeting to turn the project around by examining their problems and creating some solutions. Mr. Langtree is delighted. He wants his job to succeed and he feels the people now involved can do the job. He just does not know how they will do it based on their present actions.

The partnering session is called, the entire project team is there, including the Langtree Robotics operations and manufacturing staff, the principals, and project managers from the planners, architect/engineer, and the construction advisor, together with Mr. Langtree. During the workshops the problems and their solutions begin to surface. Some of the facts that emerged included:

1. Mr. Langtree's concept of an interrelated industrial park was not known by some and not understood by others. In essence, the basic idea had disappeared as a planning guideline.

2. Neither Wilson Allen, the planner, nor John Fenton are experienced at program writing. They have been proceeding as if their parameters were set primarily by the site boundaries and what the Langtree operations and manufacturing staff had told them.

3. The pro forma analysis expected from the efforts of the planner, architect/engineer, and constructor had not even been discussed in their meetings. Further, none of them had the experience or knowledge to prepare a pro forma cost analysis to set design parameters.

4. The Robotics staff believed that Mr. Langtree was totally aware of the work going on and had given the project team permission to move immediately into working drawings and final estimating.

5. No mission statement had ever been produced by the project team.

6. The talents of the project team were not being applied fully to work they knew they could do. Similarly, some of the things expected of them, they freely admitted, they were not qualified to do, such as writing the program, preparing a pro forma analysis, and several other critical early elements.

The list of problem statements was long and embarrassing. However, the solutions were quick to come. All of the firms wanted to stay on the project, the Robotics staff were excited about the concept and wanted to participate, and aside from a few bent noses among the participants, all agreed that a turnabout in management of the project was essential. Things to be done included:

1. Langley Evans, the architectural/engineering firm principal and a reasonably good program writer, would immediately take over the project and see that a tailormade program was written.

2. Bill Tole would shift his immediate attention to providing cost estimates at the scale and time frame consistent with the accuracy of the information provided by the architect/engineer.

3. Wilson Allen would immediately set a phased site development budget with Mr. Langtree and Robotics financial advisors. He would then work to that schedule of expenditures to produce a phased site plan construction program.

4. Bill Tole would work with Wilson Allen to price out the proposed site work plan and set cost controls to meet the projected site work budget.

5. Langley Evans would retain a civil engineering firm to take the phased site plan and translate it into site working drawings.

6. Langley Evans and John Fenton would identify and schedule all front end work and design package production work needed to start the job in the field within the next five months.

There were many other action elements that were decided on in the meeting. Some of them appeared in a charter that all of them signed. Other, more production-oriented activities, were noted to be done within the next week by the project staff. By the intelligent and adaptive use of partnering, a potential design and construction disaster might be avoided. Might? It is too early to determine if the people on the job are competent or not. Presumably they are. In that case the job could well be turned around and be a huge success. If not, the people involved learned a hard lesson about people quality and problems.

Problem 5: Being a Good On-Site Neighbor (Appears 475 Times, 17% of Total)   Many of the good-neighbor problems identified in partnering meetings concern bad manners and poor management. Experience with partnering results in respect to being a good on-site neighbor is not conclusive since measuring results must often be done in a relative rather than an absolute sense. However, reports back from partnered jobs indicates that the respect shown others, the avoidance of damage to others' work, the general attitudes of stakeholders toward each other, and the improvement in practices that make the work site a better place to be generally improve if the charter addresses them specifically.

To see why, we must go back to the concept of what people are like and what they want in order to see how their attitudes are improved if they give full attention to living by the charter. As we have said earlier, most people are honest, concerned, desirous of challenge, need attention, and welcome help in times of turmoil. I recently was helping write a charter on a very large office building addition and the stakeholders were struggling to write a section in the charter about how they should behave toward each other. This matter had only a distant relation to extra costs, delays to the job, bad weather, pending change orders, and other more common problems. Yet those 48 highly paid people spent an inordinate amount of time struggling with how to express how to be a good neighbor.

One of the stakeholder finally suggested "Why don't we just say—treat other stakeholders as you would have them treat you?" There was a moment of dead silence in the room and then a ground swell of agreement that clearly said—that's it, that's the charter objective that says what we were looking for! It's as simple as that sometimes.

Case Study 14.5: The Good Neighbors to the Vacationers

The Faraday Resort Village on the outskirts of Faraday, Michigan, was expanding. Not just a few rooms and a new sauna, but an honest-to-goodness full-scale resort expansion of 200 new rooms and all the related amenities. The owner, Toona Dole, was a good businesswoman and she was always especially careful to take care of her resort guests so that they came back time and time again. The expansion was a great event for her. It was her first major construction program since she had assumed control of the family-owned business 10 years ago. She had retained a youthful but enthusiastic local design firm, Wiggins and Loi to prepare the contract documents for the expansion. Laura Wiggins, the architect/engineer of record, was about ready to begin preparing full working drawings and specifications. All design development work had been approved and Miss Wiggins had recommended that as the construction documents were being prepared, construction should overlap. She estimated that foundations could start in about a month. This delivery system would require awarding an early construction agreement to a general contractor and some of the early subcontractors

While talking to Laura, Toona suddenly remembered a visit she had paid to an early construction conference for work to be done on a hotel and restaurant expansion in the nearby town of Dunbaer. The firms participating in that project had assigned their best superintendents, foremen, and tradespeople to the work because the owner (a good friend of Lorna's) paid well and promptly for good work. The meeting was for the purpose of writing a partnering charter, and the charter turned out to be a key element in keeping the existing facility in business during the expansion.

Good housekeeping was a major objective built into the partnering charter by the project participants. The contractor's constant concern for customers using the facilities and for maintaining the ongoing business operations while the facilities were being renovated and built all kept a steady flow of business coming during construction. All of the elements Toona Dole felt strongly about in her business life seemed to all be present in the partnering system she had seen at work on the Dunbaer project.

So when the opportunity to partner on her new project appeared, she jumped at the chance. First, she negotiated a contract with the general contractor who had worked on her friend's project. This, she explained to him, was because he had done such a good job of construction and of community and guest relations in Dunbaer. Next, she requested that he use the same subcontractors he had used in Dunbaer if it was at all possible. Lorna felt that having worked together and performed so well that they might just have the same chemistry going on her job. And finally, she requested that the general contractor establish and implement a partnering system on the Faraday expansion. Toona had some ambitious plans for the future of her Faraday resort and she wanted her financing team and business partners to see how her projects were initiated, planned, designed, and built. What better way than to have them sit through a partnering session and see and hear the people who were actually going to design and build the project?

The partnering session went well with the general contractor, his few early subcontractors, some of the long-lead-time equipment suppliers, the design team, and Toona and her business guests. The stakeholders wrote a good charter, stressing the need for over-and-above care for the users of the Faraday Resort. The stakeholders were invited to a social hour after the charter meeting, a dinner on Toona, and a free night's lodging for the stakeholders and their families at the Resort Village. To make a long story short but happy, the job turned out well, was built around the existing buildings, and guests hardly knew a construction program was in work. The work came in on time and right on budget. Already the other projects that Toona had in mind are on Laura's drawing board, and the entire construction team for Faraday are gearing up for a good season of construction.

The moral commitments that people make as part of their business contracts do work. They impress clients who do business the same way. For the Faraday jobs, the entire successful affair started because a bunch of construction people agreed with their client that people on vacation should be allowed to enjoy their vacation.

Problem 6: Timely Action (Appears 467 Times, 16% of Total)   Timely action is a catch-all phrase that must be reduced to specifics if its benefits are to be fully realized. Fully understanding what timely action means and how to take it is absolutely essential to good management of generic construction. Timely action can mean early, as needed, or late action, and it can also involve the duration of the action. Sometimes a delay in doing something can be considered timely action. For instance, if the design team needs three more days to devise a solution to a nasty traffic problem in a company parking lot and their study cannot be used until the contractor completes its cost estimate of the site grading and mass excavation in two weeks, timely action may consist of allowing the design team the additional time to improve their work product.

The application of network modeling and critical path analysis is an excellent method to help make better time-related decisions. A good critical path diagram allows the manager to see immediately how much discretionary management or float time is available to each activity. Float time is the number of time units available after subtracting the early start from the late start of an activity. It is the amount of time over and above the estimated duration of an activity that the manager actually has in which to complete the action. Proper use of float time can be considered taking timely action if it adds value to the action.

The current phase of a job along the line of action is also a critical determinant in determining if an action is timely or not. (See Chapter 2, Section A, for a discussion of the line of action. Also see Chapter 8, Section B, for a detailed description of the elements of the line of action.) If the project team is attempting to improve the quality of a to-be-built building, the members are best advised to determine how that quality is to be achieved while the project is in the program or design period, where the expenditures for experimenting, analyzing, testing, and other small-scale simulations are relatively small. Testing the wearing characteristics of a new laminated flooring material is best done on a small scale in the program or early design period. Testing the material by actually installing it in the building may risk sizable amounts of money in a gamble that could fail at an untimely moment. The case of the untested pavers may best illustrate the concept of timeliness in the construction process.

Case Study 14.6: The Untimely Paver Action

Tom Whitney, Bob Tulk, and Larry Allen were conferring about the landscaping, site work, and sidewalks that were to surround the new performing arts center in Marias Bay, New York. Marias Bay was a far-north community with a climate more suited to polar bears than music lovers. The town had somehow generated a reputation for presenting unique and well-done musical presentations from jazz to opera. Its musical efforts were well rewarded when the citizens of Marias Bay voted in a large revenue bond issue to build a performing arts center. Tom, Bob, and Larry were part of the architectural engineering team from TBL Associates, the designer of record, which had been working over a year on the project and was now getting the final working drawings and specifications ready for issue.

At the beginning of the program and design work, the Marias Bay Performing Arts Authority, the client, had decided to use partnering on the project. The authority board had held a partnering meeting at the beginning of the schematic design phase and had invited the program writer, the authority project manager, the project leaders from the architect and engineering offices, and the operating staff of the present music hall, which was being replaced by the new performing arts center. At this early partnering meeting the stakeholders decided to do all their value engineering up front during design and they recommended that a construction advisor be retained for the value engineering work. The authority, with the help of the design team, selected a good local contractor who had been active in value engineering in the region for about 15 years. Bryan Construction Engineers, the firm that was selected, has a staff of registered professionals working full time with owners, architects, and engineers. Their job: to add value to projects by intelligent, thorough, up-front cost/value analyses.

One of the first value engineering areas studied was the exposed surfaces of the project. Marias Bay winters were long and harsh. Because of this the Bryan staff recommended many different materials, giving the authority and TBL Associates a wide range of choices. TBL had acted on most of these, but for some reason, the design team had missed making a final selection of site sidewalk pavers. There was nearly a mile of exterior walks on the performing arts site and Bryan had recommended three paver types from which the final selection should be made. All three, in their opinion, would easily survive the summer rains, the freezing and thawing temperatures, and the havoc wreaked by snow plows and brushes after the heavy winter snowfalls.

TBL's late selection was the first of the untimely paver actions on the project. The next was the final forced selection of one of the three that turned out during the bidding period to have a manufacturing and delivery lead time for production that would bring them to the job, under normal conditions, at least two months after they were needed. So a paver selection decision was deferred, and TBL and the owner issued a prebid addendum including exterior pavers as an allowance. Hard-money bids were received, a single general contractor was selected, he submitted his subcontractor list, it was approved, subcontracts were awarded (except for pavers), and the job began.

The authority held a construction project partnering meeting as field work began, and among the problem statements was the paver reminder from the general contractor—we need a paver selection now! This time TBL responded and an intensive crash-selection process ensued. Tom, Bob, and Larry met with the general contractor project manager, who reviewed Bryan's earlier recommendations, concurred with the choices, and added that he knew the second choice paver would be available by the time they were needed. He had just come from a job 300 miles farther north where that very tile had been used.

Tom, Bob, and Larry got on the phone with the superintendent and confirmed that the tile was available. They also found that they were going to have to pay twice what was budgeted for the tile to get it when they needed it. However, by this time a decision was essential and the options of economy and choice had forced an imperative solution that was less suitable than desired. Several observations can be made from this case that demonstrate the need in the planning, design, and construction business for timely action.

1. Delays in selecting key materials located at critical points on a project will almost always force a less-than-desirable course of action to be followed.

2. During the up-front design period large amounts of money are committed but actual cash expenditures are usually small, normally less than 10% of the total project cost. Therefore simulation, experimentation, and testing should be done up-front before construction starts in the field.

3. Use partnering to identify weak spots in the job at all phases. The mission and problem statements will almost always pinpoint where lack of timely action is most likely to cause job trouble.

4. Don't delay making decisions just because you think you have a lot of time to make the decision. Remember, there are usually many

things that must be done that take considerable time to do after you have made up your mind on a course of action.

Problem 7: Planning and Scheduling (Appears 396 Times, 14% of Total)   Planning and scheduling of design and construction work has become very complex over the past 45 years. Many factors have contributed to the increased use of higher levels of planning and scheduling. One of these is the many advantages that a design and construction project gains from good planning. Competent planning and scheduling provide:

- An accurate simulation model of the project by which office and field decisions can be tested before a project moves into actual implementation.
- An early statement of intent by which the project team sets the conditions under which a project is being designed and built.
- Encouragement to maintain good communications on a project. As we saw above, communication with others was one of the most common problems encountered on design and construction projects, receiving 35% of the mentions of problem statements in 23 charter meetings.
- One of the major tools by which project chronological sequencing progress can be measured and used as a management-by-exception tool.
- A method by which the performance of the project team can be evaluated accurately.
- An accurate and invaluable project history if the project plan is used in conjunction with a valid project monitoring system.
- A reliable method by which others with a stake in the project can be apprised by the project staff of the actual status and resource use of the project.

It is little wonder that planning and scheduling plays such an important role in the minds of all those who are at risk on a project. The absence of a good plan or schedule invariably alerts those involved in a project that they must be careful about the job. It signals that they should be thinking about how they can individually show their intent and the conditions under which they are working. Further, a good design and construction professional knows that planning and scheduling are major responsibilities of the manager.

Case Study 14.7: Difficulties on the Clark Campus

Clark County has been the fortunate recipient of a private financial grant to remodel an existing building on the urban campus of Clark Technical College. CTC has about 15,000 students and is literally the educational hub of a farmbelt covering three contiguous states, one of which contains the college. The mechanical engineering curriculum at Clark is unusually strong and Clark graduates are in demand all over the world. Especially strong is their knowledge of farm machinery, crop sciences, and rural hydrology.

Clark has constructed seven buildings ranging in cost from $5 to $40 million, all built during the past 10 years. The new project is a testing facility and classroom costing $25 million. The grant for the project has been hard earned from a local foundation, the Harvester, established many years ago by a wealthy farmer, Wayne Larkin. Early in his life Mr. Larkin placed much of his money in the foundation. Harvester has been instrumental in funding four of the seven facilities and has always required that the projects they help fund stimulate good relations between the farm community and the other parts of the regional economic system. They also strongly encourage the use of positive personal interaction techniques developed by local business and trade associations to maintain a high level of performance on Clark College's campus projects.

The new Larkin Testing Laboratory is a complex building, designed to help teach the most up-to-date destructive and nondestructive testing methods known throughout the world. The design is intricate enough so that the owner, the Harvester Foundation, and the user, Clark College, have brought together a team of planners, architects, engineers, testing experts, and constructors to work as a team on the program and to design the building. The program manager for the total job is George Stanford, a bright young agricultural engineering graduate of Clark who is employed by the Harvester Foundation to oversee all capital assets in which Harvester has a financial or operating stake. Mr. Stanford worked for an engineering design firm for about five years after graduation from Clark. He spent the next five years with a general contractor with operations in the United States, England, and France. This contractor was a design-build firm specializing in farm-related manufacturing facilities.

After 10 years of design experience, travel, and construction site operations, George Stanford was ready for Harvester. He joined the foundation four years ago and quickly built a reputation for being innovative, competent, honest, and fair that has now brought him into the higher middle management of his employer's organization. One of the soft areas of design and construction that has impressed George over the past 10 years is the use of partnering in front-end program and design work, followed by its application to field construction. He has used partnering several times and believes that it could be invaluable on this job. However, he does not quite see how it can benefit the team effort until the active design process begins. Currently, the project team is still writing the program. So he has held off suggesting a partnering effort at this early stage.

In the early parts of their work, members of the project program and design team, a group of hand-picked key people from each organization, have been doing considerable individual research. This keeps their efforts highly compartmentalized without a great deal of interaction among the team members. However, there are now increasing needs and pressures for full-team conferences as the data collected are put into program form.

Some inevitable frictions have appeared among the data collectors, the designers, the testers, the users, and the builders. So far nothing specific has caused much trouble. However, it is obvious that the construction group has about all the information they need to start their early proforma estimates of equipment, materials, and labor required for the project. The testing people could go on collecting information forever. The designers are ready to translate the data everyone has into graphic models. The users are getting itchy about the need to get some hard understandable information about the new building.

George has been depending on Tom Trainor, the contractor representative on the team, for planning and scheduling the team's work to date. However, Tom has obviously reached his planning and scheduling outer limit and is having trouble drawing the loose ends together so that the team members know with certainty where they stand on the project. Also, Tom, although competent, willing, and personable, is the builder presence on the team and is viewed with mild suspicion by both the design group and the testers.

By now, many of the team members have heard about partnering from a variety of sources. Lyle Claiborne, the architectural designer, has been told by his professional liability insurance representative that he is missing out on considerably lowered premiums project by project by not specifying partnering. Tom Trainor's employer, Trent Construction, is using project partnering on three jobs currently. All are running well and Tom has become a partnering advocate.

The Clark College dean of engineering and his boss, the Clark president, are interested in anything that will keep the project moving. Particularly attractive is the fact that Clark College and the Harvester Foundation have both learned from Lyle Claiborne's insurance company that partnering could reduce the cost of any additional liability insurance if they use his firm to design the project. Clark's president has seen some of Clairborne's early work on the new building and liked what he saw. It has been excellent despite the frustrations and apparent delays in starting full design development. He wants to use the firm as the architect/engineer of record. Claiborne Associates cannot, however, afford to carry the amount of insurance deemed advisable by Clark College and Harvester. So the college or the foundation might have to pick up a substantial portion of the extra coverage.

All things being considered—minor frictions, a few scheduling dysfunctions, some impatience among the engineering faculty, and the inquisitive nature of the program and design team—George Stanford thinks that it is time to call for a partnering charter meeting. He wants to build a solid foundation for future team design and construction efforts on the project while there is partnering enthusiasm.

The partnering charter meeting brought together a few key members of the Clark engineering faculty, the original team members, and a few additional players from the design firm who would soon be starting active schematic and design development work. The two program writers from Claiborne Associates were also present at the charter meeting since their work overlaps and connects to nearly every step in the design process. During the partnering session one of the problems most discussed was that of planning and scheduling and how to do it. The major subject of the discussions was about how to plan the up-front program and design work and who was to do the planning and scheduling. Tom Trainor candidly admitted that he did not have the skills to plan and schedule such a complex program; his abilities, he said, were in estimating, constructibility, materials, and structures. He added he was a good construction planner and scheduler but that this early planning was out of his league. None of the others present volunteered their services since they were all technical and design specialists with very little management and programming planning backgrounds. So the charter objective dealing with planning and scheduling that was adopted read:

> *Charter objective 5.* We, the Clark programming and design stakeholders, commit ourselves to:
>
> a. Work cooperatively with a competent professional planning and scheduling consultant to be retained by the Harvester Foundation.
> b. Help prepare, publish, and regularly update a network model showing activities through the programming and design stages needed to produce the full set of construction documents for the project.
> c. Help prepare and publish accurate short-term work plans and schedules.
> d. Assist Harvester to monitor and report regularly on project progress.
> e. Help train and coach key stakeholders to prepare, use, and update effective design and construction planning and scheduling documents for the remainder of the project.
> f. Honor schedule commitments and deadlines and accept responsibility for meeting deadlines.

Twenty people attended the all-day charter-writing session and all stayed on to sign the charter. They each received a beautiful scale model of a farm tractor as their memento of the meeting and were suitably impressed by its quality and attention to detail. A social hour hosted by Clark's president capped off the hard-working day and did much to provide the celebration so essential to successful partnering milestones.

Soon after the charter meeting the partnering evaluation task force met and produced an excellent charter-monitoring system. Concurrently, George Perry, Lyle Claiborne, and Tom Trainor selected an expert network modeler, Jessica Conover, who had coincidentally worked for all of them on past projects and was very familiar with innovative methods of effectively planning for the full range of line-of-action tasks from project concept to facility turnover. Jessica began her work immediately, and by the time the partnering evaluation system was in place, she had produced a first draft of the network model up through the issue of the early construction document bidding packages.

The timing of the charter and project progress monitorings was such that the project could be reviewed thoroughly every three weeks from both the noncontract aspect of the charter provisions and with respect to the contract requirements for each stakeholders. It was not long before the program writers were working to specific, well-defined sequence and time objectives set by them with Jessica as she prepared the network model. The model also provided the architects and engineers with authentic dates when they could begin their schematic design studies. The program and design dates in the network model gave the construction consultant the dates he needed to begin progressive estimating. It also allowed him to start aligning the cost targets of the program with the actual construction costs anticipated by the schematics.

The stakeholders, project-wide, began to feel more comfortable with their work and the work of others through the improved communications stimulated by competent project planning, scheduling, and monitoring. This aspect of job management was also assured of use in subsequent job and partnering stages.

Problem 8: Organization, Authority, and Responsibility (Appears 371 Times, 13% of the Total)   One of the most frequent causes of low employee morale and high turnover has consistently been cited as being organizational confusion. Usually, this problem is coupled to an expressed lack of knowledge by employees of their duties, responsibilities, and authority. Poor definition of employees' positions can also lead to faulty decision making, creating gaps in the informational chain, and preventing prompt and helpful resolution of disputes and conflict.

In partnering charter meetings, lack of well-defined organizational structure, fuzzy lines of project-related authority and responsibility, and obscure communications paths are mentioned as causes for concern to almost every person, discipline, and organization. Most attempts to improve clarity and effectiveness of project management applications are concentrated on sharpening the focus at all levels of management on organizational relationships. The Metroplex Airport project is only one example of how relatively simple measures can often produce results of great benefit to the project participants.

Case Study 14.8: The Metroplex International Airport Complex

Metroplex is a bustling urban complex of more than 4 million people living in five moderately sized, politically autonomous communities. Metroplex is currently remodeling and renovating its entire airport facility. The budget for the project remodeling, extending over 10 years, is nearly $300 million. The Metroplex Airport Authority is the organizational structure created to plan, design, construct, and operate the airport. It is composed of a large and, for the most part, competent managerial staff concerned with all aspects of airport operations. Included in the expansion program are several key people who are just beginning to find their positions in the project chain of organization, command, and responsibility. The complexity of the relationships on the project is increasing by leaps and bounds as more and more people join the project team. At present some of the key managers involved include:

- Linton Fredericks, director of Metroplex Airport Authority
- Albert Yardley, director of airport operations
- Aileen Lindell, design and construction project manager for the authority
- Ron Carrolton, project manager for Tinzel Construction, the at-risk construction manager firm[7]
- Lars Kunzel, project manager for Lincoln and Associates, architects of record
- Tom Prince, director of airport security
- Lena Warren, airport properties and leasing manager

Many others are being involved as the project moves through various stages of planning, design, and construction. Project work is now in phase 1 of eight major design and construction phases. Phase 1 design has been completed, the construction manager, Tinzel Construction, has taken proposals for all phase 1 work, and 70% of the construction cost contracts have been awarded for phase 1 construction.

The management team has decided that it should attempt to implement a staged project partnering effort and has just held its first charter writing session. Charter objective 1 stipulated that the organizational structure and the chain of command should be clarified, identified, and published. The charter provision dealing with organization, authority, and responsibility reads as follows:

Charter objective 1. The stakeholders will maintain effective communications on the project by:
  a. Conducting project briefings for each phase with all parties affected.
  b. Preparing and publishing a project chart showing project communications flow, responsibilities, and authority.

---

[7] An at-risk construction manager is a consultant and contractor who is, or ultimately will be, liable for the cost of the work on which it is engaged.

    c. Preparing and publishing an organization chart showing the organizational relationships among the stakeholders.

During the charter meeting, when the subject of organization and authority came up, the airport director, Linton Fredericks, asked who among the stakeholders was best equipped to prepare a chart showing the information required. To his surprise the staff members attending recommended that Mr. Fredericks himself prepare the charts in conjunction with a task force of volunteer help. The stakeholders felt that Mr. Fredericks was the only person who really knew what everyone was supposed to be doing.

After the charter meeting Mr. Fredericks asked Al Yardley of airport operations, Aileen Lindell, Metroplex's project manager, Lars Kunzel, architectural project manager, and Ron Carrolton, project manager for Tinzel, the construction managers, to meet with him the next day to review the stakeholder's request. At this meeting the task force was assembled in the large airport conference room where Mr. Fredericks outlined a suggested agenda for the group. "The main item of our work" he said, "is to determine what it was the stakeholders wanted done." Then, he added, the task force could work most effectively at how to best do it.

After considerable discussion, Aileen said that the most important thing for her to understand was what she was expected to do by each of the people on the job. Of equal interest to her was how to communicate with the other stakeholders as to what she expected of them in their work with her. Aileen illustrated her ideas by a sketch on a flipchart of what interrelations might exist between her and Linton Fredericks. The matrix chart looked like this:

	Linton	*Aileen*
Linton	—	(a description of what Linton expects from Aileen in her job as it affects her)
Aileen	(a description of what Aileen expects from Linton in his job as it affects her)	—

Task force members liked the idea of the dual expectations and decided to put each of their relations with the others into matrix form, as shown below. Since they did not know yet what would go in the chart, they numbered the spaces in which the interrelational descriptions were to be put. The matrix chart included the numbers from 1 to 20 in parentheses, as follows:

	Linton	Al	Aileen	Ron	Lars
Linton	—	(05)	(09)	(13)	(17)
Al	(01)	—	(10)	(14)	(18)
Aileen	(02)	(06)	—	(15)	(19)
Ron	(03)	(07)	(11)	—	(20)
Lars	(04)	(08)	(12)	(16)	—

The task force next began defining the text that fit each numbered space in the matrix, remembering that the vertical columns represent what is expected of the person named at the head of that column by the people named on each horizontal line. To begin the process, Linton Fredericks asked all of the task force members what they expected of him in doing their jobs on the airport improvement program. The individual responses were placed in spaces (01), (02), (03), and (04). Responses were quick to come and are listed below.

Linton Fredericks is expected:

- *Space (01):* By Al Yardley "to provide management and resource support for mutually agreed on program work."
- *Space (02):* By Aileen Lindell "to funnel his management requirements to her through Al Yardley and to be available when needed."
- *Space (03):* By Ron Carrolton "to support agreed-upon recommendations at all higher management levels."
- *Space (04):* By Lars Kunzel "to inform the design team of his design requirements through Al Yardley and to support approved design concepts in his work with the authority."

This went so well that Mr. Fredericks requested they continue the process and let him tell them what he expected of them. The stakeholders agreed. These responses were what would appear in spaces (05), (09), (13), and (17).

- *Space (05):* Of Al Yardley "to manage the total airport expansion to completion within the design program and mission to which the authority and its management have agreed."
- *Space (09):* Of Aileen Lindell "to manage the expansion at the project level so that the program and mission are achieved and to keep him informed through Al Yardley."
- *Space (13):* Of Ron Carrolton "to manage construction operations so that the program and mission are achieved and to keep him and the authority informed through Aileen Lindell."

- *Space (17):* Of Lars Kunzel "to design the new and remodeled airport facilities so that the program and mission are achieved and to keep him and the authority informed through Aileen Lindell."

The experimental process of determining what the five members of the task force expected of each other brought out many important points in the management of the job at various levels. One of the more significant was that in the organizational structure, Mr. Fredericks would be available to stakeholders as they needed him. However, he wants to manage this program primarily through the efforts of Mr. Yardley and Aileen Lindell. This management desire conveyed a clear message to Ron Carrolton and Lars Kunzel—"work through Aileen and Albert—don't come to me with any micro problems until you've talked through them."

The task force meeting went so well that Mr. Fredericks asked the group to implement the program immediately with the full complement of thirty-five stakeholders who attended the charter meeting and signed the charter. The organizational matrix resulting provided a springboard for several other organizational innovations on the project. It also gave the stakeholders a good working insight into the total management structure of the airport authority and helped them manage on a need-to-know basis. Charter objective 1 consistently received high marks in the monthly partnering evaluations and was considered by most stakeholders to be a major factor in the successful completion of phase 1.

15

TWENTY-SIX RECOMMENDATIONS FOR IMPROVEMENT

Partnering provides a unique opportunity to reach to the core of the planning, design, and construction process and to see how the building business is affected by destructive and unresolved conflict. In Chapter 14 and in Appendix D are tabulated the results of dozens of workshops in many partnering efforts (see Appendix D for a detailed explanation of problem types). The nature of our problems is surprisingly consistent with what most practicing construction professionals already know—that many of our wounds and hurts are self-inflicted.

> We have met the enemy and he is us.
>
> —Walt Kelly and Pogo

The 45 classifications of 2855 problems discussed in Chapter 14 and Appendix D exhibit a broad range of concerns about frequently mentioned problems that cause difficulties in the planning, design, and construction business. For those who truly want to improve their professional and business abilities, the problem statements and the recommendations contained in the charters offer a sound beginning. To summarize briefly, the eight most frequently mentioned construction problems areas encountered are:

- Job management
- Communicating with others
- Staff morale and attitudes
- Personnel quality and problems
- Being a good on-site neighbor

- Timely action
- Planning and scheduling
- Organization, authority, and responsibility

These elements have consistently proved to be among the most important factors in the success or failure of a project. Even without a statistical verification of their critical nature, most experienced design and construction managers recognize the need to give these items a high priority in their project management work. That given, some summary recommendations based on the detailed problem statements from which the list above was abstracted might prove of benefit to both new and experienced project managers. The recommendations that follow do not necessarily fit into a specific problem category. They are derived from a variety of problem statements that have been merged with the observations of practitioners who have years of experience. The surprising feature of the recommendations is that they are not necessarily high-cost items. Most of them merely require that we do our jobs better.

Recommendations for Planners, Designers, and Constructors to Help Make Them Better Technical and Business Professionals

Recommendation 1: Manage the project as if all team members are working toward the same project end objectives. Stakeholders may have different methods of working, different trades to install, different styles of managing, but they all must be playing the same music and must all end up together, and in tune. The examples that management sets on a design and construction project are often reflected in the actions of those being managed. If the owner fails to attend job meetings regularly, the architect and engineer might rightly or wrongly feel that they are entitled to follow the same management mode— indifference. If a prime general contractor sends third- and fourth-tier managers to the construction meetings, he can expect his subcontractors to do the same.

Recommendation 2: Set a good example for other managers on your project. Understand which duties and tasks require your active participation to move the project in the direction that you and the others have set as essential and desired. Then give the full measure of your skill and abilities to their accomplishment. Failing to take responsibility for actions has become a common management deficiency, seen most frequently in projects that drift aimlessly, without apparent leadership. Reluctance to make critical decisions or to be frightened of risks that must be taken or rejected are managerial missteps that create severe hardships for others. They often subject subordinates or other low-tier managers to unfair situations that the lead manager should have resolved. This destroys confidence and weakens leadership.

Recommendation 3: If it is your job to do it, do it! If you must take an action without having the authority and it is crucial to the project, take the action and stand the consequences. Needed actions taken by discerning managers, exercising sound judgment, will usually have positive consequential outcomes. Consistent exercise of good management policies and principles will allow those being managed to know where they stand most times and under most conditions. Confusion of communication, lack of adequate leadership, and vacillating direction will destroy the ability and the will of those being managed to follow and to work effectively.

A foolish consistency is the hobgoblin of little minds. . . .

Ralph Waldo Emerson

Recommendation 4: Exercise intelligent, consistent decision making, tempered with good judgment, and empathy for others; and then ensure that your decisions are communicated clearly to those people who must do the work, and that they do the work. The excellent manager must learn how to make good decisions. There are a variety of methods available to help make decisions. Once these are learned, decision making becomes an art in which the data, the risks, the pros, the cons, and the action alternatives all become a colorful and integral part of the total picture.

Recommendation 5: Position your home base mental working platform within the project macro boundary, making certain that it provides you with a view of the entire program and project. Then learn mentally to move comfortably away from the platform to the outer project boundaries for macro viewing, and to zoom back from the macro position for a micro view. From these macro and micro journeys you must be able to return quickly and easily to your home base mental working platform. Many problem comments made in partnering meetings concern the scale of management efforts rather than a lack of quality. The good manager must be able to expand his or her view beyond the immediate situational boundaries, yet return to the normal mental working location with ease. Moving easily from the macro to the micro and being able to stop anywhere in between helps ensure that the manager viewing the scene gets a full look at what's going on in *and* around the situation. Such an ability, intelligently applied, will usually keep a good leader from becoming either a generalist or a myopic micro manager.

Recommendation 6: Plan well, know what your plan says, know what the product of every action in your plan is supposed to produce, and then decide how to lead others through your plan while allowing them to accomplish what they want and need to be successful. One of the most important functions of design and construction managers is to see that the project is planned and scheduled properly. In the case of lead managers responsible for the work of their own staff and of others who are working under them, provision of effective planning and scheduling becomes a necessity. Time after time in partnering workshops stakeholders say that lead contractors do not maintain the schedules; or that the design team does not adhere to its approval commitments in accordance with agreed-to submittal dates shown in the contractor's schedules; or that the subcontractors fail in their adherence to agreed-to delivery dates shown in the schedule. Planning is not incidental to good management—it is an essential ingredient of good management.

Recommendation 7: Listen well and be certain to understand what other people are saying, thinking, and doing that affects your work and your management of that work. Often, we are so enthusiastic in what we are saying and so confident of its accuracy that we fail to understand the view that others might have of the management bases from which we direct our work. The perceptions of others, particularly those experienced in generic construction and knowledgeable about its management and execution, may be able to bring insight and resolution if we will only listen to them patiently and with genuine interest. Listen to those who advise—listen with ears that hear, hearts that have understanding, and minds that know how to be discerning—but listen!

> The clearsighted do not rule the world, but they sustain and console it.
> —Agnes Repplier

Recommendation 8: Avoid using emotional words when writing or talking about management situations in which different people with different viewpoints must interact. Emotional words are often interpreted differently by different people. If misunderstood, they can cause trouble. When emotional words such as *hate, love, prejudice, stubborn, uncaring, stupid,* or similar nouns, verbs, and adjectives are used incorrectly, intentionally or accidentally, and are misunderstood, the reaction can seriously injure normally clear lines of communication. Like action effectiveness, language effectiveness depends to a great extent on the intent of the sender and the perception of the receiver being in tune with each other. Emotional words are used to convey subjective feelings of a speaker or writer and are of a nature that if stated in an objective and unbiased manner would usually convey little or none of the speaker's or writer's emotional entanglement with the words. Choose your words carefully, and if you have doubts as to whether you should use an emotional word or phrase, don't!

Recommendation 9: Remember the worlds we live in—we live in a world of words when we speak, write, or draw; and we live in a world of nonwords when we act. When your nonword world matches your world of words, you can be said to have planned and simulated well. Many of the construction professional's working hours are spent in the world of words. If we fail to convey the meaning of words, and phrases produced in this world of words, the potential value of the simulations and modeling done through them disappears.

Recommendation 10: If there exists a possibility that abatement of any type for any materials whatsoever may be required, be certain to define clearly what work is liable to be affected by taking or not taking abatement action, what is to be done, where is the work to take place, when it is to be done, how it is to be accomplished, who is to do it, and who is to certify that the work is complete. Job problems caused by the need for abatement are so numerous and some are so disastrous that singling out any for praise or criticism is impossible. I have found that the most frequent cause of regulatory confusion on projects is failure to define clearly the conditions surrounding the situation. When you answer the six classic questions—what, where, when, how, why, and who?—you are well on your way to resolving an abatement problem.

Recommendation 11: Submit properly prepared pay requests at the time required, in the format specified, and to the proper parties. Prompt payment in the design and construction business is a lifeline of strength that can preserve the vigor and integrity of a project, or if ignored can cast the project adrift. The payment process is a two-way street—to be paid properly you must request payment properly. For instance, in a recent charter meeting a subcontractor was heard to complain to the general contractor project manager about slow approval of his latest pay request. The project manager responded saying "that pay request of yours was just submitted a few days ago, and it is not in the format required by the owner. We can't process it."

Recommendation 12: Learn to appreciate the importance and complexity of closing out your job quickly and cleanly. Then go to work and learn how. Closing out a job is very difficult. It is like trying to get rid of a piece of sticky flypaper or a scrap piece of mending tape. The importance of closing out properly is indicated by the number of mentions in charter preparation— 112 times in 2855 problem statements. The owner, architect, engineer, and contractor should all prepare their own check list of close-out items to do. It should cover such subjects as certificates of completion and occupancy, punch-out procedures, operating and maintenance manuals, training, construction record documents, and dozens more. Time spent early in the job on planning how to close it out will pay large dividends.

Recommendation 13: List the submittals you know that you and others who affect your work will have to prepare on the job. Then develop a system of insuring that these submittals are prepared, tracked, and acted on promptly and fairly. Submittals can include shop drawings, manufacturer's cuts, samples, mockups, test reports, and other documents containing critical information about materials, equipment, and systems that are important to maintaining job progress. Providing such submittals is only the first step, but it is a critical one. If the architect does not have the hollow metal shop drawings, he or she cannot approve them. Like most other problem actions, "it takes two to tango." Stay current on your submittals.

Recommendation 14: Know what submittals you are going to get, know what you have to do with them to keep the project running well, then do it! This is the corollary to recommendation 13. The progress of most projects depends on getting shop drawings submitted and returned promptly. Delay in approval of properly prepared and accurate submittals is inexcusable. It can easily cause cost overruns, schedule disruptions, loss of confidence by the submitting party, and severe disagreements that could lead to destructive conflict. Work hard at processing submittals. Many organizations have made their reputations by being good at this.

Recommendation 15: Consider the utility companies, the building departments, the zoning boards, the health departments, the fire marshals, and all the other public and quasi-public regulatory organizations as important participants in your project. They can help you get the job done. Occasionally, project participants consider others on the job as either friends or enemies. In conflicts or arguments it is often "we" and "they" who are involved. The appearance of this sharp division almost always signals management deficiencies that if not corrected will lead to a decline in project staff effectiveness. Developing a team spirit to reach plan objectives depends on getting rid of we–they attitudes. An example of we–they thinking can often be seen when design and construction practitioners fight with public agencies charged with protecting public health, welfare, and safety. A resulting request heard when the occasional public or utility official is invited to a partnering charter meeting is that stakeholders treat the official as a member of the project team. This request has considerable merit since the regulations will usually be enforced whether or not you agree with them. If the rules are reasonable and you are not, you will probably lose whatever argument you might have. Try to make the public officials who regulate your job friends, not enemies.

Recommendation 16: Educate and train your staff in the principles and basic ingredients of partnering. Partnering is more a way of thinking than it is a technique or methodology. The commitment of stakeholders as they write a partnering charter is to use their already acquired skills in a better manner than is usual. This is possible in a partnering arrangement only because the other stakeholders have agreed to try to do the same thing. They all risk exposing areas of uncertainty and vulnerability. The mutual agreement to improve protects them. Educating and training your staff members in partnering is primarily getting them to understand why the technique is important to them and the organization. When that is understood you can teach them the methods of preparing a charter, of resolving issues, and of evaluating performance.

Recommendation 17: Take the charter mission and objectives seriously and work hard to achieve both. The charter is the fundamental document that signifies a moral agreement exists between and among the stakeholders. It is a document that requires hours of discussion, argument, evaluation, and decision making, and culminates in a serious commitment that is not to be taken lightly. Once a stakeholder demonstrates that he or she is not going to abide by the charter, the spark of spirit that is so vital to a partnering effort is extinguished. If commitment is high among the other stakeholders, those who give it lip service only are liable to find themselves working in a vacuum.

Success depends on two things—finding the right people and being the right person.

Recommendation 18: The partnering system gives you a wonderful opportunity to set a good example to industry newcomers by displaying the talents you have. Show your best abilities with the intent that what the freshmen learn from you will benefit the design and construction profession. Most experienced professionals in the building business are always looking for ways by which they can make a positive contribution to their industry. We see this in the amount of time the active participants spend in working in their professional and trade associations. Partnering has provided another window of opportunity to coach and mentor young and upcoming project managers, tradespeople, and superintendents in how teamwork pays off on the job. As it always has, teaching others gives the instructor as much or more benefit as is gained by the student. The partnering effort is a training and educating opportunity that benefits all the stakeholders and their organizations. Use it to help others.

Knowledge exists to be imparted.

—Ralph Waldo Emerson

Recommendation 19: Use the "1 to 10" method to measure problems, situations, opportunities, performance, change, and the myriad other things that surround us and make us successful or doom us to failure. Get in the habit of considering 10 as the best and 1 as the worse. Then make your evaluations and base your actions on improving in increments, not all at once. You'll have more fun becoming successful and the people you work with will appreciate your increased tolerance for their efforts to succeed. So often in the world of sum-zero games such as baseball, hockey, politics, and hard-money bidding, we find that you either win or you lose. Wise people know that there are always tempering shades of gray in between the losing and the winning. Partnering helps us find these middle grounds, the 2's through the 9's, that mark most of our actions. The system then works to bring the extremes up or down to a compromise that benefits all participants—that prevents the overreaction of one or two views from destroying hope of resolution. Use the "1 to 10" method, not the "1 or 10" attitude.

Recommendation 20: Be available to those who depend on you for your advice, help, decisions, and leadership. Being available is one of the most critical attributes of good leaders and excellent managers. After all, the reason a leader or manager exists is so that those being led or managed can benefit from the leader's visibility and availability. Proper delegation, attention to a suitable span of control, and the scale of the situation (remember the "1 to 10" method) enters into a decision about just how available you must be. However, most good professionals just seem to be where they are needed when they are needed. This is more than an accident: It is people skills at work. Be available.

Recommendation 21: Start out a project by believing that others on the job want to do well. Next, consider the fact that if you don't care who gets the credit, you can accomplish anything. Finally, determine that you are going to help people do well and accomplish all that you and they wish to accomplish. If you have done these three things, you are one-third of the way to project success. People are full of surprises. Recently, a project with a tough time schedule seemed hopelessly behind. There were only three months to go and the general's superintendent, Frank, did not really believe he was going to complete on time. Only three months earlier the job had been a model job. Frank had said at the project kick-off partnering meeting that he was looking to partnering as providing a positive tool to help improve job performance. What had happened? His boss, a wise man, suggested that he get the project stakeholders together and see what suggestions they had to bring the project back in line. Frank did.

The stakeholders were gratified to be consulted—they wanted the job to be successful. They also made it clear that the real problem was poor scheduling, inadequately monitored. Frank got the message, assembled a blue ribbon planning group from the stakeholders, and they quickly revised the schedule of remaining operations to suit what had to be done. Now Frank's boss stepped back in—"Frank, he said, "remember that you and your people are also committed. You also have to abide by this schedule if it is to be used successfully." It worked—the job was completed on time at an acceptable cost.

Three things help you to succeed: people who want to do well, unclaimed credit, and helping others succeed.

Recommendation 22: Keep the job clean and the site well organized. Working on the site day after day and being constantly barraged with problems that threaten progress and costs, many field managers give up trying to keep their jobs clean and well planned. Many years of field monitoring experience have proven to me that in most cases, good housekeeping almost invariably signals good field management. Cleanup becomes a problem only when a working resident of the site does not choose to do it. Failure to clean up then becomes a matter of poor management and deficient leadership. Both of these qualities are usually found to cause other disruptive job problems, such as poor scheduling, cost overruns, careless paperwork, and a multitude of other problems. These other problems are not so easy to see as a well-kept job site, but rest assured that they are probably there if the site is a mess. A clean job is a valuable asset.

> Penny-wise. Pound-foolish.
>
> —Ben Franklin

> A penny will hide the biggest star in the universe if you hold it close enough to your eye.
>
> —Samuel Grafton

Recommendation 23: Try to keep good trades managers and journeymen on the job until their work is finished. Do this by making them want to stay on your project. All too often a contractor's home office will pull one of its key managers or journeymen off a project as it begins to wind down. This practice is disruptive, even though the reasons for it are valid. Good people, it appears, are more valuable to start up a new job than to finish an old one. Conventional wisdom may be faulty when the new management and trades crews have to do a quick-learn on the job. This often is done after the original group has left and frequently occurs during the closing out of the project. That means that the finishing touches are often make-dos, not a crowning achievement.

Recently, a little squib in a trade paper made a statement about paying attention to closing out. It said that close out must begin at the 85% point in the job because the last 5% of project work takes 15% of the total job effort to complete successfully. The actual numbers may be difficult to determine accurately, but the difficulty of closing out properly is well known. Make your people want to stay on your job by good partnering practices and developing a feeling for what people need to succeed in their work. Your turnover will diminish and you'll find that people will fight to stay on your project to the successful end.

Recommendation 24: Be certain to determine all the actions needed to start work on any given action during the job. Delays to, or untimely starts of, any project action, particularly very early or very late in the job, can be disastrous. Frequently, in job meetings and partnering workshops the remark is made that timely action is critical. Typical fragmentary comments are: "shortage of installation time due to previous work delays," "not having the work ready to punch out," "lack of well-defined up-front schedules," and the list goes on and on. Design and construction consists of two major kinds of actions: supportive and ex'e'cutive (see Appendix A for full definitions of *supportive* and *ex'e'cutive*). Support actions supply the job with resources, including information and approvals. Executing actions use the resources to put the job together in the field. In most of today's project delivery systems the project manager is responsible for support actions, the field superintendent is responsible for ex'e'cutive actions. Whichever you are, be certain that your work meshes with the work of the other. Intelligent action taken in a timely manner is the hallmark of a good manager.

Recommendation 25: Be honest and open with other stakeholders about your plans and schedules. It is a constant source of surprise how some project mangers and supervisors in planning, design, construction, and owner offices use their plans and schedules as weapons of force rather than as tools of persuasion. Just like statistics, a work schedule can be molded to convey any impression the manager wishes. An example: "We prepare two schedules: one to give the subcontractors, the other to run the job with." This dual use of planning and scheduling certainly contradicts good partnering practice and can easily backfire. For instance, a mechanical contractor might add unneeded time into the delivery schedule of an air-handling unit and receive it on time by his schedule but a month earlier than his published schedule date. The general contractor meanwhile could easily have constructed the mechanical floor slab and equipment bases to match the actual delivery, but did not since he was following the inflated time schedule. This foolish gamble incurred additional risk and charges for ground storage and protection of the unit, and unnecessary costs for double handling the early-arrival unit. Don't play games with schedules and don't use them as weapons.

> Construction Russian roulette is a high-risk game, usually not worth winning or losing.

Recommendation 26: Determine early in the project the profit motive held most important by each of the parties and try to work to help them achieve that specific profit through your efforts. The seven types of profit are financial, social, self-actualization, value system, technical, enjoyment, and educational (see Appendix A for definitions of profit types). Each is dear to the hearts of those who are trying to achieve them. If you are able to achieve your own profit while helping others on the project team to be profitable, all of you will soon be working together in an effort to accomplish the other noncontract objectives defined in the charter. Those who have worked with charters on successful jobs say that the cohesiveness of purpose that is produced among people on the project is one of the major benefits of the partnering concept. Partnering works.

APPENDIX A

MASTER GLOSSARY OF TERMS

Abatement   The process of correcting a perceived and/or hazardous condition at a geographic location: for instance, the removal of a hazardous spill of toxic chemicals.

Acceleration   Contract work performed in a time period shorter than that originally contemplated by the contract; or contract work performed on time when the contractor is entitled to an extension of time for his or her performance.

Administration   Those activities considered to be supportive of the ex'e'cutive operations in an organization. Administrative costs may be considered the cost of management.

Administrative operations   Actions performed by those persons who provide the support services that make possible the production of products or performance of services by the line operations staff of an organization or business.

Administrative settlement   A resolution of a dispute through discussion between the disputing parties and agreement upon a mutually satisfactory settlement.

Adversarial   Taking the position of an opponent or enemy. Opposing another's interests or desires.

Advisory arbitration   An abbreviated hearing before a neutral expert or a group of neutral experts acting as arbitrators. The neutral arbitrator or arbitrators issues an advisory award and renders a prediction of the ultimate outcome if the matter is adjudicated.

Advisory opinion   An opinion usually issued by an unbiased neutral expert on what the outcome of a dispute issue might be if subjected to various forms of settlement, such as binding arbitration, litigation, or other methods of dispute resolution.

Advisory relations The interaction of parties related to each other by an obligation, either contractual or informal, where the service performed is of an advisory nature only.

Agency authority A relation in which one person or organization acts on behalf of another with the other person's or organization's formal authority.

Agent A person or firm whose acts are asserted by the third party to bind the principal.

Agreement, partially qualified An agreement made based on a moderately broad range of measuring values used somewhat consistently by the principal.

Agreement, totally negotiated An agreement made based on a very broad range of measuring values applied as desired by the principal. The negotiated selection of an agent or contractor is usually made with very little visible competition.

Agreement, totally qualified An agreement made based on a very narrow range of measuring values (e.g., price), but used consistently by the principal. The totally qualified selection of an agent or contractor is normally made with fully visible competition.

Alternative dispute resolution In its generic form, a method of resolving disputed construction claims outside the courtroom. Alternative dispute resolution may make use of nontraditional combinations of conventional dispute methods.

Apparent authority A situation in which one person or organization acts on behalf of another person or organization without the other person's or organization's formal authority.

Approval An official or formal consent, confirmation, or sanction.

Arbitration The process by which the parties to a dispute submit their differences to the judgment of an impartial person or group, usually called arbitrators or neutrals and who are appointed by mutual consent or statutory provision.

Architect, engineer ruling The ruling of the architect or engineer in an issue or dispute on a construction project on which he or she is the design professional of record. Where specified, the ruling may be binding if accepted as stated in the contract.

Articulate To express oneself easily in clear and effective language.

Assigned contractual relations The interconnection of those parties bound by subsequent assignment of a contract to other than the initial parties.

At-risk A position or action that puts an individual or organization in the position of possibly suffering harm, loss, or danger. Often, the hazard poses an uncertain but potential danger.

At-risk construction manager   A manager of a construction program who takes the responsibility for paying for the construction of the project and then collecting the costs from his client under a contract with the client.

Authority   The prerogatives, either vested or acquired over a long period of time, that allow a person to carry out his or her responsibilities and duties. This includes the right to determine, adjudicate, or otherwise settle issues or disputes; the right to control, command, or determine.

Bench trial   A trial before a judge without the benefit of a jury.

Binding arbitration   A process in which opposing parties submit disputes to binding determinations by a neutral third person or panel.

Binding resolution   A third-party-imposed solution to a contested claim in which the conditions are legally binding on the parties.

Breach of contract   Failure to perform all or part of a contract where there is no legal excuse for such failure.

Building components   The basic units into which the most building construction projects can be divided. Usually, the components represent distinct construction and construction-related actions that have common characteristics.

- **Design work (des)**   Project-related work that concerns production and issuing of contract documents.
- **Exterior skin (esk)**   All elements required to close the building to weather.
- **Front-end work (few)**   All nonconstruction project–related work concerning real estate, financing, and preconstruction leasing.
- **Interior finish work (ifw)**   All interior building components that must be protected totally or in part from weather.
- **Interior rough work (irw)**   All interior building components that can be exposed totally or in part to weather.
- **Off-site work (ofs)**   All work outside the property or construction site boundary line that is included in the project contract scope of work.
- **On-site work (osi)**   All project work outside the building line and inside the property or contract boundary line.
- **Procurement (pro)**   Work related to solicitation of proposals, award of subcontracts, preparation of submittals, approval of submittals, and fabrication and delivery of materials and equipment to the job site.
- **Substructure work (sbw)**   All foundation work upon which the superstructure bears directly or indirectly. Includes site preparation for start of field work on the building area.
- **Superstructure work (ssw)**   All major structural load-carrying components that bear on the substructure directly or indirectly.
- **Unit systems work (usy)**   All work that can be installed as a unit and is somewhat isolated during construction from other components of the building.

Bulletin An official notice that a change is being considered and that it is desired that those affected parties to the contract provide an estimate of the cost of the proposed change.

Cardinal change A change that is outside the scope of the contract.

Change Any revisions to the contract documents that alter the scope of work agreed to.

Change order An official notice that the changes specified in the change order are to be done. A properly executed change order is a revision to the scope of work and the contract documents.

Charter A document prepared and agreed to by the project partnering stakeholders and containing a set of informal guidelines to successful performance in the execution of noncontract project matters. The charter is normally signed by the stakeholders and is used in conjunction with a mission statement from which the guidelines are derived, a partnering evaluation system by which noncontract practices are periodically evaluated, and an issue resolution system containing guidelines to the settlement of contested disputes about project matters.

Charter monitoring The process of evaluating achievement of an established set of performance standards as measured against a partnering charter mission and objectives.

Claim A demand for something as due; an assertion of a right or an alleged right. In construction, generally a demand for something as due or in which the demand is disputed.

Claim potential The measure of potential that any project has to encounter disputes during its implementation.

Claim-prone job A design and construction project that has a relatively high potential for the generation of contested claims by or against any of the at-risk parties to the project.

Close out The process of completing a construction project. Usually extends from the start of preparing the contractor's punch list through receipt of final payment to the designers and constructors. May occasionally extend through the warranty period.

Closed system A system in which there is no import or export of information or physical materials and in which, therefore, there is no change of components.

Committed costs Committed costs are promised funds for purposes that if aborted require that a penalty must be paid. Penalties may include such items as:

- Option costs
- Right-of-first-refusal costs
- Legal fees
- Early engineering fees

- Early planning fees
- Displeasure of political entities
- Staff time expenditure lost
- Loss of credibility
- Loss of opportunity

Communicate   To convey information about, to make known, or to impart knowledge, ideas, or thoughts.

Conceive and communicate   To mentally form and develop an idea for construction of a facility, to initiate the effort to provide resources needed for design and construction of the facility, and to translate the concept of the facility into a common language from which the project can proceed through to completion and occupancy.

Conceiver   Those who conceive the idea and provide the wherewithal to bring the environmental program to a successful conclusion. The conceiver may be the owner but might also be a governmental agency, a financial source, an architect, an engineer, a contractor, a vendor, or a potential tenant looking for space. We identify the conceiver since he usually is the key person driving the project to completion.

Conflict   A state of disagreement and disharmony.

Construct   To convert a concept and its related plans and specifications into an actual physical environment.

Construction management   A system of attempting to better manage the construction process by providing expert construction knowledge and resources throughout all phases of the project. The goal of the process is to make information available to all participants that is best provided by an expert skilled in construction practices.

Construction services contract   A legally enforceable oral or written agreement between two or more parties specifying construction-related services to be provided by one or more of the parties to other contract parties. The services generally relate to services that directly concern the relation, nature, cost, performance, or installation of specified work into specific facilities construction.

Constructive change   An owner's action or inaction that has the same effect as a written directive.

Constructor   Those who interpret the construction language and convert it to an actual physical environment. Occupying this role are general contractors, specialty contractors, vendors, suppliers, manufacturers, artists, and others who put the materials into place in the field.

Consulting services contract   A legally enforceable oral or written agreement between two or more parties specifying design- and construction-related services to be provided by one or more of the parties to other contract parties.

Contested claim   A demand or claim in which the demand is disputed.

Contingency   A program of action set out against the possibility that an unlikely or unintended event may occur.

Contract   A legally enforceable oral or written agreement between two or more parties specifying goods or services to be provided by one or more of the parties to others of the contract parties.

Contract document matrix   A two-dimensional grid in which the rows contain action items for the various project components and the columns usually designate the geographic location of the item. At the intersection of a row and a column is inserted the designation of the contract document package in which the information is contained.

Contract documents   Usually considered to be the documents that provide the full definition of the scope of work for which the parties are legally responsible. Could include the agreement, the drawings, the specifications, instructions to bidders, addenda, and any other material included by mutual agreement and clearly identified as part of the contract.

Contract provisions   Those elements of the legal contract between the parties that contain the mandatory directions and rules of behavior for parties to the contract.

Contractor   The party, where there is a principal and a contractor, who agrees to the doing or not doing of some definite thing for a stipulated sum.

Control   Maintaining firm, competent managerial direction of any given situation. Control leads to achievement. It is usually accomplished by the invisible use of leverage.

Coordinate   To harmonize in a common action or effort. Many design and construction consultants recommend that the word not be used in contracts since it has indistinct meanings as related to management in design and construction.

Cost/benefit   A comparative measure of benefits to be gained at a cost. A cost/benefit analysis usually establishes standards by which the benefits are given a value, and standards by which value added is measured against what is desired and what can be afforded. This allows the highest benefit/cost ratios to be identified within the standards adopted.

Cost growth   An increase in project costs from the expected costs, occurring during the planning, design, construction, and occupancy phases of the line of action.

Credentials   A formal certification for a qualified person to do something for which special talents, training, and education are required.

Critical path method   A mathematical modeling technique that allows the user to establish ranges within which resources can or must be used.

Critical transition points   Points in a project line of action where the nature of the job undergoes a significant change in responsibility, authority, staffing, construction sequencing, or other activity performed along the road to project completion. A critical transition point often indicates a time where project management and policies may need revisiting and revision.

Culture, business   A way of doing business that has been generated by a group of human beings and is passed along from one business generation to another, generally by unstructured communication.

Cuts   Excerpts from catalogs, drawings, or flyers that depict a configuration to be used in the construction process.

Daily reports   Daily technical reports about the project, containing data on personnel, weather, major activities, equipment on job, and other job-related statistical information. Usually, the daily report form is preprinted and in loose-leaf format.

Decision-to-action time span   The amount of time required from the point at which a decision is made to the point where the decision is implemented. In a management structure it is important to ensure that the full span of time from decision to action at all management levels is covered, from shortest to longest.

Defective or deficient contract documents   Contract documents that do not adequately portray the true scope of work to be done under the contract.

Delay   A problem or situation beyond the control of the contractor, and not resulting from the fault or negligence of the contractor, which prevents him or her from proceeding with part or all of the work.

Deposition   A written record of sworn testimony, made before a public officer for purposes of a court action. Usually, the deposition is in the form of answers to questions posed by a lawyer. Depositions are used for the discovery of information or as evidence at a trial.

Design   Generically, to conceive in the mind, to form a plan for, and to create in an artistic and highly skilled manner.

Design–build   A method of providing total design and construction services under one cost and liability umbrella. Usually a design–build contract is based on a scope of work performance specification prepared by the owner or user. The ultimate aim of the design–build system is to provide single-source management and liability for the total facility program.

Destructive conflict   Animosity or disagreement which results in lowering the potential for a person or organization to succeed.

Development   A business operation in which the primary goal is to locate and produce profitable and marketable real-estate assets.

Diary   Similar to a log but dealing more with personal observations of the person writing it relative to his or her feelings about the job and the people.

Differing site conditions   Where actual site conditions differ materially from those indicated in the contract documents; or where unknown physical conditions at the site differ materially from those ordinarily expected to be encountered in work of the nature contemplated by the contract.

Direct negotiations Conflict in which the matter in dispute is taken immediately to those that have the authority to make a final binding decision in any project-related matter. These are called the ultimate decision makers.

Direct question A question asked in a group and directed to one or more specific people in the group.

Directed change A written or verbal change that falls within the scope of the contract. The owner has the responsibility of paying for the change.

Discovery The act of disclosing or being compelled to disclose data or documents that a party to a legal action is obligated to disclose to another party, often an unfriendly one, either prior to or during a legal proceeding.

Disincentive A penalty imposed on a contract party for less-than-satisfactory performance on a project. The disincentive is usually coupled to a bonus or incentive.

Dispute To engage in argument or discussion. To quarrel or fight about.

Dispute resolution board (DRB) A method of dispute resolution where project participants establish procedures, by formal contract or informal agreement, to settle disputes proactively as they arise during the course of the project.

Documentation An organized collection of historical records that describe the events comprising a project or program. Also, the act of preparing or supplying documents or supporting references in a project or program for future reference.

Dysfunction, organizational An organizational problem that hinders or prevents achieving objectives. May be temporary or permanent.

Early finish (EF) The earliest possible date by which a task can finish in a network model.

Early start (ES) The earliest possible date at which a task can begin in a network model if all tasks immediately preceding it have been completed by their early finish dates.

Education The teaching and learning process by which the principles of doing things are conveyed to the learner.

Effective Of a nature that achieves identifiable goals and objectives in accordance with an action plan and achieves worthwhile peripheral goals through intermediate accomplishments.

Efficient Exhibiting a high ratio of output to input.

Empathy Identification with and understanding of another's situation, feelings, and motives.

Engineer or architect of record The legally licensed architect or engineer who oversees the production of drawings and specifications from which something is to be built. The architect or engineer of record is usually required to sign and seal the documents and is liable for their correctness.

Enrichment Adding to the scope of work originally contracted for with the intent to avoid being charged or paying for the extra work. Often seen in as-noted remarks on submittals.

Ethical   In accordance with the accepted principles of right and wrong that govern the conduct of individuals in a profession and in their relationships with others.

Everyone-must-know communications   An organizational communications system based on the managerial belief that if everyone in the organization knows what all or most other people in the organization are doing and working on, the organization's overall output quality will be superior.

Ex'e'cutive   A spelling of the word *executive* used to stress the executing nature of this kind of management action. This provides a stylized but accurate picture of the word.

Executive   Of, relating to, capable of, or suited for carrying out or executing. The executing arm of the organization is that closest to the flow of expense and income experienced in achieving the organization's prime objectives. Closely related to line operations.

Facilitator   A chairman or chairwoman who leads and guides a group through production of a mental or physical object of value.

Facility   Something created to serve a particular function.

Fact-finding   A process in which one or more third parties investigate and make factual determinations regarding issues in dispute.

Feedback loop   The loop of communication around and through a project in which information is conveyed to the various components of the project.

Field order   An official notice that the actions or changes described in the field order are to be done. The field order is usually issued only in emergency situations where the time between decision and action does not permit issuance of a bulletin followed by a change order. A method of payment is usually specified in the field order.

Financing   Providing the funding either or both interim and permanent for planning, designing, and constructing a facility.

Force Majeure   An unexpected or uncontrollable event.

Front-end work   Nonconstruction project–related work usually concerning real estate, financing, and preconstruction leasing. May in some cases include design work.

Functional, as related to continuous management   A business operation designed or adapted to perform a specialized activity or duty usually exerting a direct influence on the continuous operations of the company. Examples are departments of estimating, accounting, legal, office administration, and similar ongoing functions.

Gantt chart   A visual management control tool with time shown on the horizontal axis and the tasks to be done arrayed in a vertical listing, usually showing the tasks in order according to when they start. The earlier the task starts, the closer it is to the top of the task listing. A bar showing the length of the activity in a suitable time scale is placed on the chart in a position with its start at the planned start date and its finish at the planned end of the activity.

General conditions The portion of the contract agreement that contains legal requirements for the work.

General requirements The portion of the contract agreement that contains overall technical support specifications governing work on the job.

Generic construction The field of business practice that encompasses all phases of the construction industry, including programming, planning, designing, building, operating, and maintaining facilities.

Goals The unquantified desires of an organization or individual expressed without time or other resources assigned.

Grapevine The communication line for informal transmission of information, gossip, or rumor from person to person. The grapevine is often more accurate and rapid than formal transmission lines.

Guaranteed maximum price (GMP) The price for a specified scope of work to be provided by a contractor that contractually binds his or her performance to a specified guaranteed maximum price. Often, the guaranteed maximum price is tied to a time-and-material performance with the price not to exceed the agreed-upon maximum.

Hard-money A construction contract type in which a specified amount of work is to be done for a specified cost.

Histogram A graph showing a quantity on the vertical axis measured against equal intervals of time shown on the horizontal axis. In construction, often a depiction of the resources required per day over a period of time.

Incentive A bonus paid to a contract party for performing its work in a manner superior to that specified. The incentive is usually coupled to a penalty or disincentive.

Incentive–disincentive system A payment system used in construction to pay a bonus or incentive to a contract party for performing their work in a superior manner to that specified. The bonus may relate to cost, time, quality, safety, or other such measurable component of the total job performance. If the standards set are not reached by a measurable point on the project, a disincentive is triggered where the contract party is penalized for inferior performance on the project.

Independent advisory opinion An opinion rendered by a qualified neutral of what outcomes can be expected if certain courses of action are followed.

Industrial Revolution A complex of socioeconomic changes, such as the ones that took place in the United States in the nineteenth century and which were brought about by extensive mechanization of production systems and the use of large-scale factory production.

In-house work Relating to activities that are managed and directed by a permanent staff of an organization.

Interfaces Points at which different but related activities exert direct influences upon each other. Interfaces are often the points where direct objective activities contact dependent objective activities. Poor management of interface situations usually causes problems and dysfunctions.

Issue A point or matter of discussion, debate, or dispute.

Issue resolution A method of reaching agreement and closing out disputes and problems among the disputants at the lowest possible management level, in the shortest possible time, and with the lowest potential for residual damage.

Judicial system Of, relating to, or proper to courts of law or to the administration of justice. Decreed by or proceeding from a court of justice that is vested with the authority for such action by a set of legally dictated processes established by laws enacted by a legislature.

Late finish (LF) The latest allowable date by which a task can be completed in a network model without forcing those tasks that follow past their latest-allowable start dates.

Late start (LS) The latest allowable date by which a task can be started in a network model without forcing those tasks that follow past their latest-allowable starting dates.

Laundry list A list of items, usually at random, that are to be classified, rearranged, and used to build specifically sequenced tabulations, network models, narrative schedules, or other systems of which the items in the laundry list are a component.

Law The actions or processes by which the rules of a society are enforced and through which redress for grievances is obtained.

LCD projection panel A liquid crystal display panel that sets on the light bed of an overhead transparency projector. The panel is wired to a computer and shows the computer screen image on the LCD display. This, in turn, is projected to a large screen by the overhead projector for viewing and discussion.

Leadership The process of persuasion or example by which a person induces a group to pursue objectives held by the leader or shared by the leader and his or her followers.—John W. Gardner   The art of getting someone else to do something you want done because he wants to do it.—Dwight D. Eisenhower

Leverage The effective use of vested and earned authority to solve problems and achieve goals and objectives.

Liable Legally obligated or responsible.

Life cycle cost The total cost of a system over its entire defined life.

Limited agent The individual or organization acting as an agent and authorized to do only what is specified or what it is reasonable to believe the principal wants done. A contract can be used to define the amount of authority to be granted an agent.

Line of action   A sequential statement of activities necessary to conceive, design, build, and operate an environment. Related to the generic construction process.

Line operations   Actions performed by those persons who actually make the products or perform services which are the prime end offering of an organization or business.

Liquidated damages   The amount established by the parties to a contract which must be paid, by one or either of the parties, in the event of a default or a breach. Related to the damages suffered by late performance.

Litigate   To subject to, or engage in, legal proceedings.

Log   A permanently bound, dated, handwritten record of job-related events that have occurred on a project.

Long list   The initial list of those participants offering professional planning, design, and construction services for a particular project. This list is usually prepared by the conceiver of a proposed project from those having qualifications to do the job. The long list is narrowed to a short list from which the final selection is made. *See also* Short list.

Macro matrix elements   The individual elements or components of a three-dimensional matrix that defines the actions needed, the skills that must be applied to do the action, and those who must take the action.

Maladministration   The interference of the owner with the right of the contractor to develop and enjoy the benefits of least-cost performance.

Manage   To define, assemble, and direct the application of resources.

Management by exception (MX)   A measuring and monitoring system that sounds an alarm to the manager when problems have appeared or are about to appear. The MX system remains silent when there are no problems. The system identifies the problem area, permitting the effective manager to manage the exception while leaving the smoothly running operations to continue.

Marketing   The process of conceiving, formulating, and implementing a system by which the ultimate service or product of an organization can be successfully sold.

Matrix   A multiple-dimensioned display of related data.

Matrix management   A management technique that employs a multiple command system. Usually, it results in one employee having two or more bosses on a time-to-time basis.

Mediation   A private, informal process in which the parties are assisted by one or more neutral third parties in their efforts toward settlement. Mediators often do not judge or arbitrate the dispute. Rather, they advise and consult impartially with the parties to bring about a mutually agreeable resolution of disputes. However, where authorized, a mediator can render a nonbinding decision.

Minitrial   A process in which counsel for the opposing parties present their cases in condensed form in the presence of representatives for each side who have authority to settle the dispute. Usually, a neutral third-party advisor is also present. The third-party advisor may offer certain nonbinding conclusions regarding the probable adjudicated outcome of the case and may assist in negotiations.

Mission   A statement of the most important result to be achieved by the project being completed successfully.

Mockup   A full-sized scale model of a structure, used for viewing, demonstration, study, or testing. Generally used in construction to obtain approval of a system, materials, or a product.

Money flow   The flow of income and expense measured against time.

Monitoring   Measurement of current project conditions and position against the standards of performance set for the job.

Motivation   The elements of a given situation that encourage and make effective, successful, and meaningful the activities of those engaged in the situation.

Multiplier   A number usually applied to a direct cost by someone providing a service. The product of the multiplier and the direct cost determines the actual charge to be billed for the service. The multiplier adds the overhead and profit to the direct cost.

Must list   Those items that must be included in the scope of work to make the project a go. If any of the items in the must list cannot be included, the project is a no-go.

Need-to-know communications   An organizational communications system based on the managerial belief that information should only be offered and provided to those who truly need it and can use it to add value to the product they are responsible for producing.

Negotiated contract   A contract obtained through offering multivalue benefits in addition to cost benefits to the prospective client. Usually, conditions of the final contract are negotiated after an offer has been conditionally accepted.

Network   A system of interconnected, interacting components.

Network plan   A graphic statement of the action standard of performance to be used in achieving project objectives.

Neutral   An unbiased outside expert capable of objectively listening, analyzing, and evaluating construction-related demands or claims that are in dispute and rendering an opinion or decision as to their disposition.

Nonbinding arbitration   A hearing before one or more third parties who draw conclusions regarding issues in dispute. The third party renders a decision, but the decision is not binding on the parties. The intent is to predict the probable adjudicated outcome of the case as a stimulus to a settlement.

Nonbinding resolution A suggested solution to a contested claim or problem in which the conditions are not legally binding on the parties but are an expert's recommendations for resolution.

Objectives Quantified targets derived from established goals. The most commonly used resources in converting goals to objectives are money, time, human abilities, human actions, equipment, and space. *See also* goals.

Objectives, dependent Objectives to be achieved that are affected by major influences beyond the manager's direct control. The dependent goal may be predictable or unpredictable. Dependent goals, although usually beyond the manager's control, may well be within the organization's ability to reach. Lack of correlation between organizational and individual efforts to achieve a manager's goals that are affected by others may cause severe dysfunctions.

Objectives, direct Objectives that can be achieved by managing conditions within the manager's direct influence.

Objectives, end Objectives realized from and upon total completion of the defined project work.

Objectives, intermediate Objectives achieved at specific and identifiable stages of the project (e.g., partial occupancy of a building, turnover of a mechanical system for temporary heat, or completion and issuance of foundation plans for early start of construction).

Objectives, peripehral Objectives realized on an ongoing basis through the life of the project and achieved as an indirect result of project activities. Peripheral objectives may be personal, professional, technical, financial, or social. Peripheral objectives might include staff promotion, profitable subcontractor operations, specialized experience, or achievement of design excellence in a special field.

Off site Located outside the contract site boundaries.

On site Located within the contract site boundaries.

Open system A system that exchanges energy, information, and physical components with its environments.

Operator Those who operate and maintain the completed physical environment on a continuing basis. Usually, the party responsible for this function is an owner or tenant working through a plant or facilities manager.

Organization The management and operational structure through which individuals and groups work systematically to conduct their business.

Over-the-wall management A management style that subscribes to the actions of participants completing their work responsibilities and duties, and then passing the work product along to others (or throwing it over the wall) without adequate briefing for the successors to do their work effectively. Often identified by statements such as "We did our job and now they can do theirs" or "That's not my job."

Par   An amount or a level considered to be average; a standard.

Par performance   A rating, usually numerical, that expresses the level of performance that will be accepted as the normal degree of competence expected of an individual or organization in the performance of an action.

Partnering, a base statement   A method of conducting business in the planning, design, and construction profession without the need for unnecessary, excessive, and/or debilitating external party involvement.

Partnering, Associated General Contractors   A way of achieving an optimum relationship between a customer and a supplier. A method of doing business in which a person's word is his or her bond, and where people accept responsibility for their actions. Partnering is not a business contract but a recognition that every business contract includes an implied covenant of good faith.

Partnering, Construction Industry Institute   A long-term commitment between two or more organizations for the purpose of achieving specific business objectives by maximizing the effectiveness of each participant's resources.

This requires changing traditional relationships to a shared culture without regard to organizational boundaries. The relationship is based upon trust, dedication to common goals, and an understanding of each other's individual expectations and values. Expected benefits include improved efficiency and cost-effectiveness, increased opportunity for innovation, and the continuous improvement of quality products and services.

Partnering, project or tactical   A method of applying project-specific management in the planning, design, and construction profession without the need for unnecessary, excessive, and/or debilitating external party involvement.

Partnering, strategic   A formal partnering relationship that is designed to enhance the success of multiproject experiences on a long-term basis. As each individual project must be maintained, a strategic partnership must also be maintained by periodic review of all projects currently being performed.

Partnering charter   The basic manual for operating a partnering system. Contains at a minimum, the mission of the project team, and their objectives for the project. Usually signed by those writing the document.

Perception   The process of becoming aware of something through any of the senses. To become aware of in one's mind; to achieve an understanding of.

Planning, in the management sense   Establishing and arranging necessary and desired actions leading to end, intermediate, and peripheral objectives.

Positive conflict   Hostility that is managed so that its resolution raises the potential for well-intentioned individuals or organizations to succeed at being excellent.

Prepare and publish   A phrase often inserted into the partnering charter to direct the stakeholders to write, publish, and implement a policy, procedure, or guideline for accomplishing a performance that may be required by contract but whose detailed nature is not specified. An example of such a charter provision might be: "Prepare and publish invoicing procedures for all levels of project operations."

Prescriptive   A document that provides detailed information as to the methods and means by which something is to be done or produced. The document explicitly identifies the material and equipment components of the finished product. Compare to performance-oriented documents, which describe the performance desired and the amount that is to be spent to achieve the performance in the finished product.

Preventive law   A technique for minimizing contract problems in the construction industry.

Prime contractor   A contractor whose business agreement is directly with the organization providing primary direction and financing for the project.

Principal   A person who authorizes another to act as his agent, or a person primarily liable for an obligation.

Problem   A deviation from an accepted and/or approved standard of performance.

Professional   Having great skill or experience in a special contributive field of work.

Profit, educational and training   Fulfillment of learning and teaching goals held by individuals and their companies.

Profit, financial   Fundamentally, the difference between organizational cash income and organizational cash expense. Further definitions of financial profit are complex and often unique to an organization or project.

Profit, self-actualization   Personal fulfillment realized after basic needs of shelter, safety, protection, love, and freedom from hunger are achieved.

Profit, socioeconomic   Organizational group, or individual achievement of social objectives within a financially profitable set of activities.

Profit, value system   Organizational and project fulfillment of personal, professional, technical, social, and financial values held important by individuals and groups related to the organization.

Pro forma   The term *pro forma* means according to form. It is often a financial model built early in a construction program to show, by projecting income and expenses, how the money flow will occur to and from the project. It is often used to establish the capital amount to be allocated to a project based on simulated operating conditions.

Program, as defining a step in the design process   A narrative-oriented statement of the needs and character of the proposed user operation, the requirements of the user and owner, the nature of the environment to be planned, designed, and built, and the corresponding characteristics of the space that will satisfy these needs and requirements. Sometimes called the brief.

Project, as a set of work actions A set of work actions having identifiable objectives and a beginning and an end.

Project delivery system A method of assembling, grouping, organizing, and managing project resources so as best to achieve project goals and objectives.

Project history A tabulation of the major events on the job, arranged chronologically for easy reference. The purpose of the project history is to give a quick, accurate look at past job events.

Project management The management process of establishing project objectives, planning how these objectives are to be reached, and then assembling and directing the application of resources needed to accomplish the objectives.

Project organization The arrangement and interrelations of people charged with actually achieving project objectives.

Punch list A list of contract construction items to be completed by the contractors and others, in order for the project work, as specified in the contract documents, to be certified as partially or totally closed out and complete.

Quality A characteristic of superior excellence.

Question, closed A question that can be answered with a yes or no or with a simple statement of fact.

Question, direct A question asked with a strong indication as to who or whom should answer.

Question, open Questions that cannot be answered with a yes or no or a simple statement of fact.

Question, overhead A question asked of a group without indication as to who or whom is to answer.

Question, relay A question passed along to someone else by the party originally asked.

Question, reverse A question returned to the questioner by rephrasing or rewording the original question.

Record Any retained information that can be used effectively in the future.

Regulators Those who fill a review and inspection position to help ensure protection of public health, safety, and welfare. This is usually done by enforcing regulations written and adopted by qualified public or private bodies. Examples of regulators include those who work for building departments, departments of natural resources, public health agencies, fire prevention organizations, technical societies, and other such groups.

Relations, formal functional Organizational connections that concern distribution and use of data, information, and decisions that flow along formally defined transmission lines. Formal functional communications usually are written and are normally both from and to individuals and groups. Formal relations are defined precisely and most day-to-day business is accomplished within the formal relation framework. The line expressing a formal functional relation usually has an arrowhead at each end to show a mutual exchange of responsibility and authority. If there is a higher authority to be implied, a single arrowhead can be used, pointing to the superior party.

Relations, informal The natural channels along which organizationally related material is most easily and comfortably transmitted. The informal relation exists by mutual consent of the parties to the relation and is stimulated to maximum effectiveness by mutual profit gained from the relation.

Relations, reporting The official channels through which each person conveys information; is given raises, appraisals, and evaluations; is fired, assigned, or is provided professional, vocational, and personal identity in an organization. The true organizational superior of an employee is usually that person with whom the employee maintains a reporting relation. The line expressing reporting relations has an arrowhead at one end pointing to the superior.

Relations, staff The business patterns through which a person or group provides consulting services necessary to achieve goals and objectives. Staff personnel usually have little or no authority over those outside the staff group. The line expressing staff relations has an arrowhead at each end.

Relations, temporary Those relations created when extraordinary or unusual management demands must be met. The temporary relation is usually unstable and should be kept active for only short periods of time. The line expressing a temporary relation can have an arrowhead at one or both ends, depending on the nature of the relations. Extensive use of temporary relations creates business dysfunctions, breaks down morale, and causes internal tensions.

Resolution A course of action determined and acted upon that results in clearing conflict or dispute.

Resource allocation The assignment of project resources such as money, time, space, people, and equipment to activities that must be done to achieve project objectives.

Resources The tools of the supportive and ex'e'cutive manager. Resources include time, talent, tools, equipment, money, experience, space, and materials, as well as intangibles, such as enthusiasm, morale, and leverage.

Responsibility The assignment, spoken or understood, that a person in an organization has as his or her part in maintaining the organization's health and vitality.

Revisiting When applied to the partnering charter, revisiting means that the current project decision makers are assembled, and the present charter is reviewed, revised, and reissued as might be called for by changed project conditions.

Risk Any exposure to the possibility of harm, danger, loss, or damage to people, property, or other interest. To expose to a chance of loss or damage.

Risk management The management and conservation of a firm's assets and earning power against the occurrence of accidental loss.

Schedule A graphic or written tabulation of project activities showing where activities are to start and finish. The schedule is derived from the plan of action and the network model by locking the tasks and the resources they require into a specific time position.

Selling Establishing and implementing the strategy of achieving the objectives of the marketing plan. The physical process of closing the negotiation for services and products for a consideration.

Shared savings An arrangement by which a construction contractor and its client share in any savings realized by building a facility for less then the guaranteed maximum cost.

Shop drawing A submittal in the form of a drawing, usually made specially for the application shown. Shop drawings usually show details of fabrication and installation.

Short list The final selection list of those participants offering professional planning, design, and construction services, usually to the conceiver of a proposed project. The final selection is usually made from the short list. *See also* Long list.

Situational thinking The ability to evaluate a set of project influences accurately by mentally moving from a long overview (macro) of them to a detailed picture (micro) and back, being able to stop anywhere in between to consider other scale pictures of these influences and their relationships.

Span of control The number of organizationally related individuals a manager controls directly on a one-to-one basis.

Specialized construction The field of business practice that encompasses single phases of the construction profession. Examples of specialized construction organizations are architectural/engineering offices, mechanical contractors, plastering contractors, and planning consultants, among others. Includes nearly any single organizational unit active in design, planning, construction, or related fields.

Specification A narrative description of the various materials and systems to be incorporated in the work. The specification concentrates on identifying quality of materials, source of materials, allowable practices, and general requirements and conditions of the contract performance.

Sponsor, partnering In the partnering context, a person or organization that strongly supports or champions an activity and assumes responsibility for its implementation.

Staff A supportive unit of an organization whose basic function is to advise the organization's staff. Staff functions are occasionally defined as overhead or nonproduction. They are considered to be the organizational partner of line operations.

Stakeholders   The parties at risk financially and legally or in an extended sense, those affected and potentially put at risk during the execution of a planning, design, or construction contract. Stakeholders are also those who participate in writing a partnering charter and are a signatory to the charter.

Standard of performance   A well-defined, explicitly stated, approved, and accepted statement of the measurements to be used as a gage of performance and goal and objective achievement.

Standing neutral   A technically trained, educated, and credentialed professional who is active in the planning, design, and construction disciplines. The standing neutral must be capable of objectively listening, analyzing, and evaluating construction-related demands or claims that are in dispute.

Standing netural process   A method where neutral third parties are available to assist with resolution of disputes arising during the course of a contractual relationship. In this system one or more standing neutrals are on call to address disputes as they arise. It usually requires the neutral to render a nonbinding determination of the issues in dispute, although in some cases, and upon request, the neutral can act as a binding arbitrator.

Strategy   Applies to the management skills required to attain a macro result. Strategy is sometimes considered the action taken to plan, direct, and implement larger and longer-range programs, particularly in the military.

Subcontractor   A contractor whose business agreement is directly or indirectly with a prime contractor.

Submittal   Any document submitted by contracting parties to the owner's agents for review for accuracy, responsibility of design, general arrangement, and approval. Submittals are used by the fabricator and the installer to show adequate details so that the intent of the contract documents can be achieved. Generally, they are not considered contract documents but rather, aids to better fabrication and installation procedures.

Sum zero game   A contest in which there is a winner of all that somebody else loses.

Superior knowledge   The owner's withholding specific data on matters of substance not known to contracting parties during the precontract period.

Supportive   The administrative group of the project organization that is responsible for bringing resources to the point of use by the executive or operational project group.

Surety   One who has contracted to be responsible for another, especially one who assumes responsibilities or debts in the event of default.

Suspension   An owner's or owner's agent's action of stopping all or a part of the work.

Synergism   The action of two or more substances, organs, or organisms to achieve an effect of which each is individually incapable.

System   An assemblage or combination of things or parts forming a complex or unitary whole.

Tactics   The management skills required to attain a micro or current result. Tactics may be considered the actions taken to plan, direct, and implement the day-to-day action itself.

Task force   A temporary grouping of resources and people responsible for accomplishing a specific objective.

Technography   The action of preparing meeting notes and related material on electronic equipment as the notes and materials are generated. Often, the recorded material is projected on a screen for viewing by those in a meeting.

Tenant coordinator   The title usually given to a developer's owner representative. The tenant coordinator is responsible for integrating and directing the lease execution, construction process, tenant move-in, and operational startup of tenant spaces in the base building.

Tenant work   Work done by the landlord inside a tenant space, paid for by an allowance negotiated by the landlord with the tenant when preparing and executing the lease for the space.

Termination   The dismissal of a contractor from a project for convenience, resulting from factors beyond the contractor's control, or for default when the contractor's performance is not acceptable.

Time-and-material contract   An agreement in which payment for services and material is made only for those services and materials actually furnished. A not-to-exceed amount on the total cost may or may not be imposed.

Total float (TF)   The amount of discretionary time available to a task. The total float is the difference between the early and late starts or finishes. Formally, it is defined as the duration of the task, subtracted from the difference between the late finish (LF) and the early start (ES) [i.e., $(LF - ES) - \text{duration} = TF$].

Total quality management (TQM)   The managing process which helps ensure that the quality of all components and of the final product in the planning, design, and construction of any facility are maintained at a level that meets the client's program performance requirements.

Traditional   Pertaining to those qualities of an organization, civilization, or other culture that are handed down from generation to generation. Usually, the transfer is by word of mouth or by practice.

Training   The teaching and learning process by which specific, explicit methods and systems of doing something, usually by rote, are conveyed to the learner.

Translation   Recasting standard of performance information into graphic, narrative, mental, oral, or other forms, to ensure optimum use of the information by those involved.

Translator   Those who translate the environmental program into construction language. Designers, subcontractors, suppliers, vendors, manufacturers, contractors, and the conceiver may all play a role in translating.

Trust   Reliance on an organization or individual to apply integrity, justice, fairness, good judgment, and other relational qualities for the benefit of those affected by the actions of the organization or person.

Turnkey   A project delivery system in which a single contractor is given the total responsibility to plan, design, construct, and turn the key over to the owner upon its completion. Often, a turnkey contractor will provide land and financing and in some cases, operate the facility for a specified time after construction.

Turnaround time   The amount of time required to process submittals.

Ultimate decision maker (UDM)   The person or group at the lowest management level that has the authority to make a final binding decision in any job-related matter.

Updating   The process of revising and reissuing a project network model to bring it into conformance with a current desired and necessary plan of action. Often, but not always, updating results from monitoring and evaluating the project. Usually, updating is done when it is found that the current plan of work does not adequately depict the actual conditions under which the project is being executed.

Upset price   A guaranteed maximum price agreed to in a time-and-material contract.

Users   Those who occupy and use a completed facility to conduct their work, their recreation, their domestic living, or other activities for which the facility was specifically designed and built.

Value   The increase in worth of an open system to which an item of value has been added. Often multiplied by the weight of a factor to give the weight and value rating of a factor to help determine a choice of alternatives.

Value added   The improvement in the worth of anything that results from the efforts, contribution, and involvement of people, processes, materials, and ideas.

Value engineering   An engineering and architectural cost analysis process designed to achieve minimum total cost while maintaining maximum product quality within the price constraints.

Vested authority   The endowing of privileges, strength, and leverage from a superior usually to a subordinate. Generally gained quickly rather than being earned by long and proven service in a related field within the organization.

Want list   Those items that are wanted and can be included in the scope of work, over and above the must list items, and which provide a definable and acceptable rate of return on their cost.

Warranty   A legally enforceable assurance of the duration of satisfactory performance or the quality of a product, a piece of equipment, or of work performed. Often, the warranty period begins when the installation is turned over to the owner.

Win–win   A situation in which there are no losers. Usually, some parties win more than other parties win.

Wish list   Those items that the owner and the user wish they could include but might not be able to, for budgetary or other reasons. Wish list items are best added, not deleted, as the project moves into construction.

Working drawings   The set of contract drawings that graphically show the intended appearance of a job when complete.

World of nonwords   The world in which we live by our physical actions.

World of words   The world in which we live by simulating, through words and other symbols, what might happen in the world of nonwords.

APPENDIX B

BIBLIOGRAPHY

Allen, Richard K. *Dispute Avoidance and Resolution for Consulting Engineers.* New York: ASCE Press, 1993.

American Consulting Engineers Council and The American Institute of Architects. *A Project Partnering Guide for Design Professionals.* Washington, D.C.: American Institute of Architects, 1993.

Associated General Contractors of America. *Partnering: A Concpet for Success.* Washington, D.C., AGCA, 1991.

Bachner, John P., ed. *ADR: Alternative Dispute Resolution for the Construction Industry.* Silver Spring, Md.: Association of Engineering Firms Practicing in the Geosciences, 1988.

Beck, Emily Morison, ed. *Bartlett's Familiar Quotations.* Boston: Little, Brown, 1980.

Bittel, Lester R. *The Nine Master Keys of Management.* New York: McGraw-Hill, 1972.

Blake, Robert R., and Jane Srygley Mouton. *The Managerial Grid.* Houston, Texas: Gulf Publishing, 1964.

Boeckh, E. H. *Boeckh's Manual of Appraisals,* 5th ed. Washington, D.C.: E. H. Boeckh and Associates, 1956.

Brooks, Hugh. *Illustrated Encyclopedic Dictionary of Building and Construction Terms.* Englewood Cliffs, N.J.: Prentice Hall, 1976.

Business Roundtable. *Summary Report of the Construction Industry Cost Effectiveness Project,* Vols. A1-7, B1-3, C1-7, D1-5, and E1. New York: Business Roundtable, 1983.

Carnes, William T. *Effective Meetings for Busy People.* New York: McGraw-Hill, 1980.

Carr, A. H. Z. *How to Attract Good Luck.* New York: Cornerstone Library, 1965.

Carzo, Rocco, Jr., and John N. Yanouzas. *Formal Organization: A Systems Approach.* Homewood, Ill.: Dorsey Press, 1967.

Center for Public Resources, Inc. *Preventing and Resolving Construction Disputes.* New York: CPR, 1991.

Chandler, Howard M, ed. *Means Repair and Remodeling Cost Data,* 15th ed. Kingston, Mass.: R. S. Means, 1993.

City and County of Denver. *Practical, Profitable Partnering: Denver's Team Approach to Urban Reconstruction—A Guidebook.* Denver, Colo.: Department of Public Works, 1993.

Construction Industry Institute. *In Search of Partnering Excellence.* Austin, Texas: University of Texas at Austin, 1991.

————. *Partnering: Meeting the Challenges of the Future.* Austin, Texas: University of Texas at Austin, 1989.

Coonradt, Charles A. *The Game of Work,* 2nd ed. Salt Lake City, Utah: Shadow Mountain, 1985.

Dale, Ernest. *Management Theory and Practice.* New York: McGraw-Hill, 1965.

Dorman, Peter J. *Dictionary of Law.* Philadelphia: Running Press, 1976.

Douglas, Clarence J., and Elmer L. Munger. *Construction Management.* Englewood Cliffs, N.J.: Prentice Hall, 1969.

Drucker, Peter F. *The Effective Executive.* New York: Harper & Row, 1966.

————. *Management: Tasks, Responsibilities, Practices.* New York: Harper & Row, 1973.

————. *Managing for Results.* New York: Harper & Row, 1964.

Feinberg, Mortimer R. *Effective Psychology for Managers.* Englewood Cliffs, N.J.: Prentice Hall, 1965.

Fitch, James Marston. *American Building: The Historical Forces That Shaped It,* 2nd ed. Boston: Houghton Mifflin, 1966.

Frisby, Thomas N. *Survival in the Construction Business: Checklists for Success.* Kingston, Mass: R. S. Means, 1990.

Gardner, John W. *On Leadership.* New York: Free Press, 1990.

Geneen, Harold. *Managing.* Garden City, N.Y.: Doubleday, 1984.

Gordon, William J. J. *Synectics: The Development of Creativity Capacity.* New York: Harper & Row, 1961.

Greater Phoenix, Arizona, Chapter 98 of the National Association of Women in Construction. *25th Anniversary Construction Dictionary,* 8th ed. Phoenix, Ariz.: NAWC, 1991.

Gutman, Robert. *Architectural Practice: A Critical View.* New York: Princeton Architectural Press, 1988.

Haviland, David. *Managing Architectural Projects: Case Studies,* Vols. 1 to 3. New York: American Institute of Architects, 1981.

————. *Managing Architectural Projects: The Effective Project Manager.* New York: American Institute of Architects, 1981.

————. *Managing Architectural Projects: The Process.* New York: American Institute of Architects, 1981.

Heyel, Carl, ed. *The Encyclopedia of Management,* 2nd ed. New York: Van Nostrand Reinhold, 1973.

Karrass, Chester L. *Give and Take: The Complete Guide to Negotiating Strategies and Tactics.* New York: Thomas Y. Crowell, 1974.

Leeds, Dorothy. *Smart Questions: A New Strategy for Successful Managers.* New York: McGraw-Hill, 1987.

Mackenzie, R. Alec. *The Time Trap.* New York: AMACOM, 1972.

National Economic Development Office. *Before You Build,* 2nd ed. London: HMSO, 1979.

Papageorge, Thomas, E. *Risk Management for Building Professionals.* Kingston, Mass.: R. S. Means, 1988.

Pilcher, Roy. *Principles of Construction Management.* London: McGraw-Hill, 1966.

Pollock, Ted. *Managing Creatively,* 2nd ed. Boston: CBI Publishing, 1982.

———. *Managing Yourself Creatively,* New York: Hawthorn Books, 1971.

Radcliffe, Byron M., Donald E. Kawal, and Ralph J. Stephenson. *Critical Path Method.* Chicago: Cahners, 1967.

Smith, Terry C. *How to Write Better and Faster.* New York: Thomas Y. Crowell, 1965.

Steiner, George A. *Top Management Planning.* New York: Macmillan, 1969.

Sweet, Justin. *Legal Aspects of Architecture, Engineering, and the Construction Process,* 3rd ed. St. Paul, Minn., West Publishing, 1985.

Uris, Auren. *The Executive Deskbook.* New York: Van Nostrand Reinhold, 1970.

U.S. Army Corps of Engineers. *Construction Partnering: The Joint Pursuit of Common Goals to Enhance Engineering Quality.* Omaha, Nebr.: The Corps, 1991.

Walker, Mabel. *Business Enterprise in the City.* Princeton, N.J.: Tax Institute, 1957.

Walker, Nathan, and Theodore K. Rohdenburg. *Legal Pitfalls in Architecture, Engineering, and Building Construction.* New York: McGraw-Hill, 1968.

Wickenden, William E. *A Professional Guide for Young Engineers.* New York, New York: Engineers Council for Professional Development, 1949.

Whyte, William H., Jr. *The Organization Man.* New York: Simon & Schuster, 1956.

APPENDIX C

PARTNERING CHARTERS

- Veterans Administration Medical Center
- Michigan Millers Mutual Insurance
- Frankenmuth Mutual Insurance
- West Suburban Health Campus
- Webber Street Retention Basin Project
- Magnesium Products of America Plant
- Muskegon Community College Center for Higher Education
- Fitzhugh Retention Basin Project
- National Computer Systems
- Salt/Fraser Retention Basin
- MAC/GTC Staff
- MAC/GTC Construction
- Bavarian Inn Expansion
- Connecticut Consolidation Terminal, Yellow Freight System, Inc.
- Michigan Millers Mutual Insurance, Revisited
- Carson City Middle/High School
- L. L. Pelling Company
- New Federal Courthouse
- AAA Office Building

Appendix C consists of a collection of 19 charters from 20 charter meetings for buildings of various types, designs, costs, locations, and sizes.[1] Details of

---

[1] One of the charter meetings was to revisit the original charter. No changes were made in the original charter text.

the project delivery system used varied from project to project. About 543 people attended the 20 charter meetings. The charters are arranged in ascending order by the date on which they were written. This is merely because that arrangement serves as a convenient method by which the reader can see some of the evolutionary characteristics of the charter form and content.

Some charters have been slightly edited to improve their readability and eliminate major grammatical mistakes. However, for the most part, the text has been left as written. Charters were signed by all charter meeting participants except for two charters which lacked four signatures. Three participants in one of these meetings were prohibited by their employer, a major government agency, from committing the department they represented to any formal or informal agreement. The fourth participant, a contractor, had been told by his organization to attend the meeting but not to sign the charter.

Only those attending a major portion of the meetings were permitted to sign the charter, and all signatures except two were affixed by the genuine party. The two exceptions were allowed to have a designated attendee sign for them. In both cases the party had to leave the meeting for good reason and just before the signature copy was produced, after having been in the session for the full day. The objectives contained in the charters make good study material from which the conscientious planning, design, and construction practitioner can work to improve his or her professional skills.

Charter for Veterans Administration Medical Center, Detroit, Michigan

Date written: April 16, 1992
Project: Veterans Administration Medical Center
Location: Detroit, Michigan
Cost: Approximately $375,000,000
Owner: Veterans Administration
Contractor: Bateson/Dailey Joint Venture
Designer: Smith Hinchman and Grylls

Mission

We, the undersigned, recognize that we all have common objectives. We therefore agree to strive together to construct the Detroit VAMC safely, on time, and within budget to the highest-quality standards commensurate with its mission of serving veterans and the community. To achieve our mission we believe in the following principles:
- Commitment
- Mutual trust
- Integrity
- Personal pride

Objectives

1. Maintain open lines of communications.
   a. Recognize the need for quality information.
   b. Minimize submittal and response times in all matters.
2. Keep paper and administrative work to a minimum.
3. Develop and implement an alternative conflict resolution system.
   a. Promptly resolve conflicts at lowest possible level.
   b. Eliminate need for contracting officer decisions.
   c. Fairly interpret ambiguities.
   d. Be proactive (not reactive) in problem solving.
   e. Maintain objective attitude toward constructibility and practicality.
   f. Accept responsibility for our actions or inactions.
   g. Have empathy in all matters.
   h. Clearly describe changes to contract work.
4. Limit cost growth.
   a. Develop cost-effective measures.
5. Maintain clean, efficient, and secure work site.
   a. No lost time due to accidents.
   b. Properly staff project.
   c. Be a good neighbor.
6. Seek to maintain good job morale and attitudes.
   a. Promote partnering attitudes at all levels of contract administration.
   b. Have fun.
   c. Have pride in our product.
7. Commit to quality control in all project-related matters.
   a. Do it right the first time.
   b. Maintain proper work sequence.
   c. Meet design intent.
   d. Recognize owner's needs in occupation and operation of the facility.
8. Close out job in proper and timely manner.
9. Maintain and implement a partnering evaluation system.

Charter for Michigan Millers Mutual Insurance, Lansing, Michigan

Date written: August 19, 1992

Project: Michigan Millers Mutual Insurance office addition and renovation
Location: Lansing, Michigan
Size: Addition, 45,000 square feet; renovation, 65,000 square feet
Owner: Michigan Millers Mutual Insurance
Contractor: Christman Construction Services
Designer: MBDS, Inc.

Mission

We, the project team, commit to construct a quality facility on time and within budget, maximizing safety, communication, and cooperation so that all participants can be proud and profitable in their accomplishments.

Objectives

To accomplish our mission, we recognize a need to work to the following goals and objectives:

1. Submittals.
   a. Clarify objectives and expectations of the submittal process.
   b. Minimize submittal and approval times.
   c. Provide accurate, prompt, clear, and concise approvals.
2. Payments.
   a. Make payments in accordance with the published flowchart process.
3. Information processing and paperwork.
   a. Expedite all information and indicate desired response times.
   b. Maintain open lines of communication among project team members.
   c. Be available.
   d. Attempt to offer possible solutions to questions within a proper scope.
   e. Provide clear responses to requests for information.
4. Legal matters.
   a. No litigation.
   b. Settle disputes at originating level.
5. Abatement.
   a. Establish, approve, and publish a plan of abatement.
   b. Abate promptly.
6. Planning and scheduling.
   a. Provide, obtain, and use accurate activity information.
   b. Clearly monitor the project against the plan and schedule.
   c. Commit to and fulfill man hour projections.
7. Decision making.
   a. Architect/engineer team to regularly inspect work and advise compliance.
   b. Define and clearly communicate quality expectations.
   c. Properly empower those at all decision-making levels.
8. Policies and procedures.
   a. Prepare, review, approve, and publish policies and procedures that will serve as guidelines to manage the project.
9. Site layout and management.
   a. Formulate and publish a trash removal and parking plan.
   b. Properly establish and maintain benchmarks and control lines.
10. Processing revisions.
    a. Provide written authorization prior to work proceeding.
    b. Respond to requests for information, bulletins, and change orders promptly.
    c. Prepare, approve, and publish a flowchart for processing revisions.
11. Be a good partnering neighbor.
    a. Commit to protecting your work and the work of others.
    b. Show all participants due respect and acknowledgment.
    c. Maintain proper work sequences.
12. Total quality management.
    a. Prepare, approve, publish, and commit to a TQM program.

Charter for Frankenmuth Mutual Insurance, Frankenmuth, Michigan

Date written: January 28, 1993

Project:	Frankenmuth Mutual Insurance Company headquarters office building addition and renovation
Location:	Frankenmuth, Michigan
Size:	115,000 square feet in additions
Owner:	Frankenmuth Mutual Insurance Company
Contractor:	R. C. Hendrick & Son, Inc.
Designer:	Wigen, Ticknell, Meyer & Associates

Mission

Our mission is to work together in a trustworthy and professional manner to produce a quality project completed within budget, safely, and on time.

Objectives

To accomplish our mission, we recognize a need to work to the following goals and objectives:

1. Maintain lines of effective communications.
   a. Hold regular team progress meetings and prepare and publish minutes.
   b. Prepare and publish organizational chain of command (with phone and fax numbers).
   c. Continually communicate a spirit of cooperation through actions.
   d. Prepare and implement a partnering evaluation system.
   e. Prepare and publish progress schedule and update regularly.
2. Paper and administrative work.
   a. Prepare and submit complete and accurate submittals and shop drawings in a timely manner.
   b. Prepare and publish standard procedures for payment, changes, questions, and other documentation.
   c. Prepare and publish close-out procedures for all trades.
   d. Prepare and publish submittal processing procedures.
3. Prepare and implement an effective alternative dispute resolution system— general contractor to appoint resolution task force.
4. Cost management.
   a. Encourage value engineering.
   b. Identify and resolve cost-growth problems early.
   c. Hold changes to a minimum.
5. Good work site.
   a. Plan, organize, and publish site layout and organization.
   b. Keep disruptions to owner's operations at a minimum.
   c. Maintain a clean, safe, secure site and surrounding area.
   d. Hold regular safety meetings to be attended by all workers.
6. Job morale and attitude.
   a. Stress and encourage pride and workmanship.
   b. Respect other trades.
   c. Address the problem, not the person.
   d. All construction employees maintain professional relationship with Frankenmuth Mutual employees and the public.

7. Quality control.
   a. Prepare and publish program to regularly monitor and report on job quality.
   b. Use qualified personnel.
   c. Treat this project as if you were the owner.
8. Payment.
   a. Pay promptly.
   b. Prepare and publish accurate schedule of the value of subcontracts.
   c. Make timely release of retainage.
9. Legal matters.
   a. Avoid litigation.

Charter for West Suburban Health Campus, Plymouth, Minnesota

	Date written: March 5, 1993
Project:	West Suburban Health Campus, Outpatient Neighborhood Facility
Location:	Plymouth, Minnesota
Developer:	Tobin Real Estate Company
Developer consultant:	Kirk Program Management
Designers:	BWBR, architect of record
	HKS, design architect
Construction advisor:	Kraus-Anderson, preconstruction services contractor

Mission

Design an effective and flexible community-based outpatient-centered facility that provides for present and future quality health care services.

Objectives

1. Maintain control of design costs and construction budgets.
   a. Prepare and publish design development-based total target cost.
   b. Prepare and publish must, want, and wish list.
   c. Prepare and publish FFE budget.
   d. Prepare and publish life-cycle costing guidelines.
   e. Prepare and publish preconstruction costing guidelines.
2. Properly document project activities.
   a. Prepare and publish guidelines for single-source documentation.
   b. Make decisions promptly.
   c. Prepare and distribute glossary.
   d. Prepare and publish payment policies.
   e. Prepare, publish, and periodically update schedule for entire project.
   f. Prepare and publish submitting, reviewing, and approving process guidelines.
3. Maintain an effective mode of communication on project
   a. With medical and nonmedical staff.
   b. With surrounding community.
   c. With regulatory agencies.
4. Provide approvals promptly from proper management level.

5. Define standards of performance expected to achieve program conformance.
   a. Provide forum for periodic total project review by entire preconstruction team.
   b. Do it right the first time.
   c. Define community image to be projected by project team and the facility.
6. Establish issue resolution process.
   a. Prepare and publish conflict resolution guidelines.
   b. Resolve issues promptly and at originating level.
   c. No litigation.
7. Generate and maintain high levels of project team morale.
   a. Exhibit and expect others to exhibit good partnering practices.

Charter for Webber Street Retention Basin Project, Saginaw, Michigan

Date written: April 22, 1993

Project:	Webber Street Combined Sewer Overflow, retention and treatment facility
Location:	Saginaw, Michigan
Size:	3.9 million gallons
Owner:	City of Saginaw
Contractor:	Spence Brothers
Designer:	Spicer Engineering

Mission

Beyond the contract requirements the project partners will achieve a quality project, mutual success, and avoid litigation by a commitment to:
- A safe workplace
- Effective communications
- Trust
- Timely action

Objectives

To accomplish our mission we recognize a need to work to the following goals and objectives:
1. Maintain partnering effectiveness.
   a. Prepare and publish a partnering effectiveness measurement system.
   b. Meet regularly and evaluate partnering effectiveness.
   c. Take prompt steps to correct any deterioration of partnering effectiveness.
2. Maintain effective project communication.
   a. Be available.
   b. Minimize response times.
   c. Maintain an appropriate level of documentation.
3. Submittals.
   a. Prepare and publish submittal processing guidelines.
   b. Process submittals in a timely manner.
   c. Insure proper distribution of submittals.

4. Planning and scheduling.
   a. Prepare, issue, and maintain current project schedules.
      (1) Long term.
      (2) Short term.
5. Maintain a clean and well-managed workplace.
   a. Minimize time lost due to accidents.
   b. Be a good neighbor to adjoining area residents.
   c. Use good construction site housekeeping practices.
6. Close out project in a proper and timely fashion.
   a. Prepare and publish acceptable close out guidelines.
   b. Establish clearly defined punch-out procedures and standards early in the project.
7. Maintain good job morale and attitudes.
   a. Promote partnering attitudes at all levels of contract administration.
   b. Have pride in your work.
8. Resolve problems effectively.
   a. Prepare and publish a responsive conflict resolution system.
   b. Promptly resolve conflicts at lowest possible levels.
   c. Attempt to anticipate and prevent damaging problems.

Charter for Magnesium Products of America Plant, Eaton Rapids, Michigan

Date written: June 25, 1993

Project: Magnesium Products of America magnesium die-casting plant and office
Location: Eaton Rapids, Michigan
Size: 98,000 square feet
Owner: Magnesium Products of America
Contractor: Christman Construction Service
Designer: Hobbs and Black

Mission

As a team, design and safely construct a world-class magnesium die-casting facility on time, within budget, while developing positive and profitable relationships among all team members.

Objectives

1. Alternative dispute resolution.
   a. Resolve problems in an effective, timely, and fair manner at the lowest level possible.
   b. Address all issues in a professional, nonpersonal, timely manner.
   c. Maintain an active, open-issue log.
2. Close-out and final payment.
   a. Implement rolling punch list technique.
   b. Prepare and publish acceptable close-out guidelines.

3. Communicate effectively at all levels.
   a. Attend meetings and be accessible.
   b. Maintain flow and tracking of documents and information.
   c. Maintain effective written communication and documentation.
4. Control cost growth.
   a. Document approved changes and forecast potential changes.
   b. Define and resolve outstanding design issues quickly.
   c. Recognize owner's desire to control operations and maintenance costs.
5. Expedite project payments.
   a. Maintain invoicing procedures so as to facilitate prompt payment.
6. Quality control.
   a. Define quality objectives.
   b. Conduct preinstallation meetings.
   c. Keep all field testing current.
   d. Perform quality review throughout design and construction.
7. Job morale, attitudes, good neighbor, good work site.
   a. Maintain a clean and safe work environment.
   b. Maintain a teamwork attitude with openness and respect for other contractors' work.
8. Leadership, responsibility, and authority definition.
   a. Identify and communicate organizational responsibilities to all team members.
   b. Issue timely goals/objectives for weekly activities progress meetings, field activity.
   c. Educate team to specific needs/design of all process equipment.
9. Maintain partnering effectiveness.
   a. Prepare and publish partnering effectiveness procedures.
10. Planning and scheduling.
    a. Publish an exception report highlighting critical path, behind-schedule activities, and changes.
    b. Monitor, update, and issue project schedules.
    c. Communicate schedule changes and requirements.
    d. Issue timely goals and objectives for weekly activities, progress meetings, and field activity.
11. Revisions and submittals.
    a. Interface shop drawings among trades.
    b. Define submittal priorities, turnaround commitments, and lead times.

Charter for Muskegon Community College Center for Higher Education, Muskegon, Michigan

Date written: July 23, 1993

Project: Muskegon Community College Center for Higher Education
Location: Muskegon, Michigan
Owner: Muskegon Community College and State of Michigan
Contractor: Muskegon Construction Company
Designer: WBDC Group

Mission

We will strive to construct the Muskegon Center for Higher Education in a quality, safe, professional, and profitable manner within the limits of the project budget and the schedule while satisfying all participants.

Objectives

1. All partners (stakeholders) agree to submit, review, and process in a timely manner all shop drawings, samples, requests for payment, revisions, and other important documentation.
   a. Set submittal priorities and establish time frames for processing.
   b. Submit bulletin pricing as soon as possible.
   c. Take time to review shop drawings and answer questions.
   d. Be available.
   e. Let those involved know about problems immediately.
   f. Make timely notification to subs of expectations (schedules, submittals, manpower).
2. Prepare a proper and reasonable construction schedule with involvement and commitment from all participants and use it for proper management of project.
   a. Identify shutdowns in schedules.
   b. Limit use of construction site until project is substantially complete.
   c. Consider preparation of a critical path plan and schedule for the project.
   d. Properly assign priorities to operations in all work areas.
   e. Provide regularly updated issues of the project schedule.
3. Establish and identify effective lines of communication.
   a. Clarify chain of command and identify single-source responsibility.
   b. Follow through on decision making.
   c. Maintain effective communications within the project team.
   d. Be willing to cooperate with other trades.
   e. Keep paperwork to a minimum.
   f. Respect other trades' needs.
4. Resolve problems effectively.
   a. Be reasonable in resolving problems.
   b. Provide prompt answers to field problems by partners.
   c. Let partners know about problems immediately.
5. Limit cost growth
   a. Limit unnecessary change requests.
   b. Expedite contracts, bulletins, and change orders.
6. Establish and implement a proper project close-out procedure as specified.
   a. Limit use of new facility until project is substantially complete.
   b. Provide appropriate facility training to the right people.
   c. Establish and follow clearly defined close-out procedures.
7. Maintain a clean, safe, and orderly work site.
   a. Plan for proper delivery and storage of materials and equipment.
   b. Be a good partnering neighbor.
   c. Be a good neighbor to the adjoining community.
   d. Provide skilled and trained personnel.
   e. Provide adequate and safe equipment and tools.

    f. Continue to find methods of improving site storage and parking space availability.
8. Establish and maintain a partnering evaluation system.
    a. Prepare and publish a partnering effectiveness measurement system.
    b. Meet on a regular scheduled basis and formally evaluate partnering effectiveness.
    c. Take prompt steps to correct any deterioration of partnering effectiveness on the project.

Charter for Fitzhugh Retention Basin Project, Saginaw, Michigan

Date written: August 16, 1993

Project: Fitzhugh Retention Basin
Location: Saginaw, Michigan
Owner: City of Saginaw
Contractor: Gerace Construction
Designer: Spicer Engineering

Mission

We commit to profitably construct this project within contract bid amounts in a safe and timely manner, providing for owner requirements and quality workmanship as defined by contract, and through:
- Cooperation
- Communication
- Trust
- Respect

Objectives

1. Maintain effective, timely, and directed communications throughout the project.
    a. Each organization to prepare and submit a project directory and an organizational chart for this project to the general contractor.
    b. Periodically review communications systems and update as required.
2. Maintain effective, timely, and accurate documentation of project activities and issues.
3. Prepare and publish accurate long- and short-term work plans and schedules.
    a. Maintain plans and schedules so as to properly achieve mission.
4. Make timely project payments within the requirements of the contract provisions.
5. Maintain owner's operations as required by contract provisions.
    a. Owner will communicate operational needs promptly and accurately.
6. Maintain a clean, safe project site.
    a. Be a good neighbor to adjacent businesses.
7. Close out project cleanly, quickly, and effectively.
    a. Owner, engineer of record, and general contractor prepare and publish close-out guidelines.

8. Prepare, submit, and process submittals accurately, fairly, and promptly.
    a. Provide schedule of required submittals.
9. Be a good partner.
    a. Follow through on commitments.
    b. Be available.
    c. Treat each other fairly.
10. Establish and implement an issue resolution system.
    a. Promptly resolve problems at the originating level of management.
    b. Involve affected parties in problem discussions.
    c. Avoid litigation.
11. Employ intelligent and timely use of cost/benefit concepts on the project.

Charter for National Computer Systems, Iowa City, Iowa

	Date written: September 9, 1993
Project:	National Computer Systems Facility
Location:	Iowa City, Iowa
Owner:	National Computer Systems
Owner Consultant:	The Wilkinson Project Group, Ltd.
Designer:	RSP, Minneapolis, Minnesota
Contractor:	McGough Construction Company

Mission

Through mutual trust and cooperation, we will strive to recognize and satisfy the owners need's and provide a quality project on schedule, within budget, safely, profitably, and to the satisfaction of all concerned.

Objectives

1. Establish and maintain an effective and timely decision-making process.
    a. Set, define, and follow appropriate communication paths and methods.
    b. Provide timely and meaningful information to make proper decisions.
        (1) Program statements.
        (2) Design documents.
        (3) Schedules.
        (4) Budgets.
    c. Prepare and process submittals promptly and fairly.
2. Develop and implement an alternative dispute resolution system.
    a. No litigation.
    b. Promptly resolve conflicts at lowest possible levels.
    c. Be proactive in problem solving.
3. Payments.
    a. Pay properly submitted invoices promptly.
4. Maintain good job morale and attitudes.
    a. Have fun.
    b. Encourage partnering attitudes.
    c. Be proud of your contribution to the project.

5. Close out project promptly and properly.
   a. Prepare and publish close out items of work and guidelines to accomplish this work.
6. Properly manage cost and schedule.
   a. Employ intelligent and timely use of cost/benefit concepts on the project.
7. Remain open and receptive to the ideas and needs of other project partners.
   a. Be sensitive to the special space and functional needs of the owner.
   b. Seek, respect, and consider input from other team members.
   c. Strive to educate and communicate to employees, all project team members, and the public regarding safety and access during construction.
8. Define and maintain quality standards expected by the owner within budget constraints.

Charter for Salt/Fraser Retention Basin, Saginaw, Michigan

Date written: October 13, 1993

Project:	Salt/Fraser Combined Sewer Overflow–Retention and Treatment Facility
Location:	Saginaw, Michigan
Size:	2.8 million gallons
Owner:	City of Saginaw
Contractor:	Spence Brothers
Designer:	Hubbell Roth & Clark

Mission

Complete this project to meet the expectations of the owner in a safe and timely manner, within budget, so as to be profitable for all parties, and without third-party intervention.

Objectives

1. Job morale and attitude.
   a. Behave toward others on the project as you would have them behave toward you.
   b. Maintain an open mind and healthy attitude toward resolving problems.
2. Communications.
   a. Communicate effectively with others.
   b. Prepare and publish a project organization chart.
   c. Provide prompt notification of expected delays to the project.
3. Provide adequate manpower to meet job commitments.
4. Use alternative dispute resolution (ADR).
   a. Prepare and implement an issue resolution system for the project.
   b. Resolve disputes quickly and at the lowest level possible.
5. Submittals.
   a. Process submittals expeditiously.
   b. Assign processing priorities to all submittals.
   c. Where deemed reasonable, minimize resubmittal requirements.

6. Planning and scheduling.
   a. Provide and regularly update project schedules.
   b. Provide required information to update project schedules.
7. Keep paperwork to a minimum and process all paperwork promptly.
   a. Recognize the importance of project documentation.
   b. Execute contracts, change orders, and field orders for project work promptly.
   c. Promptly process all requests for quotes and requests for information.
8. Be a good neighbor.
   a. Commit to and implement effective partnering practices on project.
   b. Be a good community neighbor.
   c. Be a good project neighbor.
9. Policies and procedures.
   a. Establish project guidelines for policies and procedures.
      (1) Back charges.
      (2) Cleanup.
      (3) Chain of command protocol.
      (4) Close out.
10. Payment.
    a. Establish and implement prompt payment to and from all project participants.

Charter for MAC/GTC Staff, Minneapolis, Minnesota

	Date written: November 22, 1993
Project:	Minneapolis–St. Paul International Airport, GTC middle/ west/valet parking and roadway expansion, staff operations
Location:	Minneapolis, Minnesota
Cost:	Approximately $22,000,000
Owner:	Minneapolis Airport Development
Construction Manager:	Kraus-Anderson Construction Company
Designers:	The Alliance

Mission

Emphasizing teamwork, we will construct this project on time, within budget, and with the least disruption to all customers.

Objectives

1. Maintain effective communications on the project.
   a. Conduct project briefing on each phase with impacted parties.
   b. Prepare and publish a project chain of command showing communications flow, responsibilities, and authority.
   c. Solicit and receive MAC staff input prior to making major project decisions.
   d. Establish and maintain an internal and external public relations program that anticipates issues that affect achieving the project mission.

2. Be a good partner.
    a. Build a shared vision with others on the project.
    b. Be available and have staff available.
    c. Understand others' problems and concerns.
    d. Be open to ideas for improvement.
    e. Maintain a clean, safe work site.
3. Make effective decisions in a timely manner.
    a. Provide decision alternatives from which to choose.
4. Process submittals fairly and promptly.
    a. Prepare and publish submittal processing guidelines.
5. Prepare and publish conflict issue resolution guidelines.
    a. For internal MAC issues.
    b. For contract parties.
    c. For the customers of MAC.
    d. In resolving conflict, address the problem, not the people.
    e. Resolve problems promptly and at the originating level where possible.
6. Prepare and publish partnering evaluation guidelines.
7. Maintain operations at minimal customer inconvenience during construction.
8. Planning and scheduling.
    a. Clearly define deadlines and responsibilities for meeting deadlines.
9. Keep project meetings timely, well structured, and directed toward a defined agenda.
10. Control cost growth by effectively applied budgetary controls.
11. Close out project effectively, on time, and completely.
12. Seek to maintain good job morale and attitudes.
    a. Encourage pride in workmanship.
13. No litigation.
14. Have fun.

Charter for MAC/GTC Construction, Minneapolis, Minnesota

	Date written: November 23, 1993
Project:	Minneapolis–St. Paul International Airport, GTC middle/ west/valet parking and roadway expansion, staff operations
Location:	Minneapolis, Minnesota
Cost:	Approximately $22,000,000
Owner:	Minneapolis Airport Development
Construction Manager:	Kraus-Anderson Construction Company
Designers:	The Alliance

Mission

To deliver to the owner's satisfaction, a safe, profitable project, while maintaining professional quality, in which the community can take pride and satisfaction

Objectives

1. Be a good project partner.
   a. Maintain a mutual respect for all individuals involved.
   b. Be cooperative and open with your partners.
   c. Listen well and strive to understand others' concerns.
   d. Address the issue, not the person.
   e. Be an active participant in project-related matters.
2. Communicate effectively.
   a. Ask how your intentions and your actions might affect others—don't assume.
   b. Communicate in a timely, forthright, and thorough manner.
   c. Maintain an appropriate level of documentation.
3. Manage your work well.
   a. Be available.
   b. Ensure that capable staff is available.
4. Prepare and maintain an accurate and timely planning and scheduling system.
   a. Honor schedule commitments.
   b. Prepare and process submittals in a timely and fair manner.
5. Maintain ongoing airport operations.
   a. Be aware of the public's needs.
   b. Make the public aware of our needs.
   c. Remember—MAC customers may be you, your family, or even your future customers.
6. Maintain a good work site.
   a. Effectively allocate and manage available storage areas.
   b. Keep site clean and well organized.
7. Strive to avoid litigation.
8. Process project revisions promptly.
   a. Conduct management walk-throughs as needed to identify potential project revisions.
   b. Promptly and thoroughly document and disseminate changes through agreed-upon channels.
9. Prepare and publish a project chain of command showing communications flow, responsibilities, and authority.
10. Maintain partnering effectiveness.
    a. Prepare and implement a partnering evaluation system.
    b. Prepare and implement an issue resolution system.
11. Control cost growth through controlled revisions.
12. Close out the project in a proper and timely manner.
13. Maintain good project attitude and morale.
    a. Encourage pride in workmanship.
    b. Have fun.

Charter for Bavarian Inn Expansion, Frankenmuth, Michigan

Date written: May 3, 1994

Project: Bavarian Inn Motor Lodge addition
Location: Frankenmuth, Michigan
Size: 102,000 square feet, five stories, 156 rooms
Owner: Bavarian Inn
Contractor: R.C. Hendrick and Sons
Designer: Manyam & Associates

Mission

To deliver our project safely, on time, within budget, while maintaining quality, a spirit of cooperation, enjoyment, profitability, and a mutual respect for others' interests.

Objectives

1. Establish and maintain an effective approval process.
   a. Prepare and publish submittal guidelines.
   b. Keep resubmittals to a minimum.
2. Be a good on-site and off-site neighbor.
   a. Be a good neighbor to each other, to guests and staff, and to the community.
   b. Respect the work of other trades.
   c. Provide proper resources to maintain work progress.
3. Closing out.
   a. Project task force prepare and publish a close-out guideline.
   b. Complete your work promptly and well.
   c. Document and communicate changed conditions.
4. Communications.
   a. Prepare and publish project communications guidelines.
   b. Clearly communicate expectations.
   c. Respect and be receptive to contractors' suggestions and input.
5. Control cost growth.
   a. Avoid or minimize back-charges.
6. Maintain a clean and effective work site.
   a. Prepare and publish construction staging and storage guidelines.
   b. Mutually establish guidelines for a clean and effective work site.
7. Job staff morale and attitude.
   a. Prepare and publish partnering guidelines for project staff (orientation).
   b. Conduct orientation and training sessions on partnering and the partnering charter for on-site and off-site employees.
8. Avoid legal entanglements and resolve issues effectively.
   a. Make a good-faith effort to resolve all project disputes quickly, on site, and at the originating level.
   b. Establish and maintain an issue resolution policy.
9. Mutually establish clear and acceptable organizational authority and responsibility guidelines.
10. Keep project-related paperwork accurate and timely.
11. Project task force establish and stakeholders maintain a monthly partnering evaluation system.

12. Accurately prepare, submit, and process payment requests at all project levels.
13. Maintain proper plans and schedules.
    a. Prepare and periodically update and publish a master project plan and schedule.
    b. All parties will work together to establish schedules early and often.

Charter for Connecticut Consolidation Terminal, Yellow Freight System, Inc., Middletown, Connecticut

---

	Date written: June 22, 1994
Project:	Connecticut Consolidation Truck Terminal
Location:	Middletown, Connecticut
Owner:	Yellow Freight System, Inc.
Contractor:	Newfield Construction
Designer:	Warner Nease Bost Architects

Mission

Through teamwork, cooperation, commitment, and communication, we will build a truck terminal with quality workmanship, on schedule, safely, within budget, and with profitability for all partners.

Objectives

1. Do it right the first time!
2. All parties recognize and understand that the success of this project construction rests to a great extent on proper handling of site conditions.
3. Prepare and publish organization, authority, and responsibility guidelines for management of the project.
4. All parties will strive to reduce excessive and unnecessary resubmittals of shop drawings and other submittals.
5. Implement specific approval guidelines for all submittals.
6. Develop, approve, and implement a responsive conflict resolution system.
    a. Identify and resolve disputes and conflicts at the originating level and as quickly as possible.
7. Implement an acceptable revision process.
8. Experience no lost time from accidents.
9. Be a good on-site and off-site neighbor.
10. Use good construction housekeeping practices.
11. Review, agree upon, and implement an acceptable payment process at all contractor levels, including change orders and close-out processes.
12. All parties will strive to close out the project in a timely and efficient manner.
13. Define and maintain an acceptable and effective protocol for project communications.
14. Promote partnering attitudes and awareness at all levels.
15. All partners will remain receptive to new ideas and fresh approaches.
16. Prepare, issue, execute, and update project schedules in a timely, accurate manner, with full project partner participation.

17. Strive to keep project management staff changes to a minimum.
18. All project participants will consider the regulatory bodies and public utilities as team members.
19. All project team members will accept responsibility for their actions.

Charter for Michigan Millers Mutual Insurance, Revisited, Lansing, Michigan

Date written: September 13, 1994

Project: Michigan Millers Mutual Insurance office addition and renovation
Location: Lansing, Michigan
Size: Addition, 45,000 square feet; renovation, 65,000 square feet
Owner: Michigan Millers Mutual Insurance
Contractor: Christman Construction Services
Designer: MBDS, Inc.

Mission

We, the project team, commit to construct a quality facility on time and within budget, maximizing safety, communication, and cooperation so that all participants can be proud and profitable in their accomplishments.

Objectives

To accomplish our mission, we recognize a need to work to the following goals and objectives:

1. Submittals.
   a. Clarify objectives and expectations of the submittal process.
   b. Minimize submittal and approval times.
   c. Provide accurate, prompt, clear, concise approvals.
2. Payments.
   a. Make payments in accordance with the published flowchart process.
3. Information processing and paperwork.
   a. Expedite all information and indicate desired response times.
   b. Maintain open lines of communication among project team members.
   c. Be available.
   d. Attempt to offer possible solutions to questions within a proper scope.
   e. Provide clear responses to requests for information.
4. Legal matters.
   a. No litigation.
   b. Settle disputes at originating level.
5. Abatement.
   a. Establish, approve, and publish a plan of abatement.
   b. Abate promptly.
6. Planning and scheduling.
   a. Provide, obtain, and use accurate activity information.
   b. Clearly monitor the project against the plan and schedule.
   c. Commit to and fulfill man-hour projections.

7. Decision making.
   a. Architect/engineer team to regularly inspect work and advise compliance.
   b. Define and clearly communicate quality expectations.
   c. Properly empower those at all decision-making levels.
   d. Avoid using the partnering charter as leverage in unfair negotiations.
8. Policies and procedures.
   a. Prepare, review, approve, and publish policies and procedures that will serve as guidelines to manage the project.
9. Site layout and management.
   a. Formulate and publish a trash removal and parking plan.
   b. Properly establish and maintain benchmarks and control lines.
10. Processing revisions.
    a. Provide authorization, preferably written, prior to work proceeding.
    b. Respond to requests for information, bulletins, and change orders promptly.
    c. Prepare, approve, and publish a flowchart for processing revisions.
11. Be a good partnering neighbor.
    a. Commit to protecting your work and the work of others.
    b. Show all participants due respect and acknowledgment.
    c. Maintain proper work sequences.
12. Total quality management.
    a. Prepare, approve, publish, and commit to a TQM program.
13. Long-term partnering results.
    a. In doing this, we improve our and others' professional and technical abilities and help insure an ongoing relationship with other partners.

Charter for Carson City Middle/High School, Carson City, Michigan

Date written: October 19, 1994

Project: Carson City/Crystal Area Middle/High School
Location: Carson City, Michigan
Owner: Carson City/Crystal Area School District
Contractor: Granger Construction Company
Designer: Kingscott Associates

Mission

This partnering team commits to safely deliver a quality project that is on time, within budget, and profitable for all through continual communication and cooperation.

Objectives

The stakeholders signing this charter will strive to adhere to the following objectives:

1. To do it right the first time.
2. To print and distribute a payment release requirement process.
3. To attend all progress meetings held while my work is proceeding on site or as required by the construction manager.

4. To assign a resolution date and attempt to meet the date for all decisions required.
5. To process all submittals and deliveries to maintain the current project schedule.
6. To prepare, publish, and maintain a current project construction plan and schedule.
7. To provide and maintain material, equipment, and skilled worker levels adequate to meet the current schedule requirements.
8. Each contractor to perform their own punch list prior to the owner management group (owner, architect/engineer, construction manager) writing its punch list.
9. To strive for zero punch lists at completion.
10. To promptly submit operating and maintenance (O&M) manuals and to follow close-out procedures as required by the contract and in a timely manner.
11. To be proactive toward job site storage areas, cleanliness, security, layout, and sequencing of stored materials.
12. All stakeholders will inform and enforce the conditions of the partnering charter on their employees and their subcontractors.
13. To respect the rights and property of all stakeholders (follow the Golden Rule).
14. To establish and maintain a regular partnering evaluation system.
15. To commit to open dialogue.
16. To identify ultimate decision makers (those at the lowest management levels who can make final binding decisions in any project-related matter).
17. To distribute meeting minutes to all ultimate decision makers.
18. As stakeholders if we don't know for sure, we will find out—we will not assume.
19. To resolve potential interferences with other trades early enough to avoid installation conflicts or delays.
20. To process contract revisions promptly.
21. To respect the no-tobacco provisions per school policy.
22. To be available.
23. To attempt to avoid issue settlement through litigation.
24. To establish and maintain an issue resolution policy that encourages dispute resolution promptly and at the originating level wherever possible.
25. To maintain good job morale and attitudes.

Charter for L.L. Pelling Company, Cedar Rapids, Iowa

Date written: October 21, 1994

Project: L.L. Pelling Company office and maintenance building
Location: Cedar Rapids, Iowa
Owner: L.L. Pelling Company, asphalt paving contractor
Contractor: Merit Construction Company
Designer: Shive-Hattery Engineers

Mission

To construct a functional, state-of-the-art L.L. Pelling company facility, combining total quality, craftsmanship, and safety in a timely manner, within budget, to the satisfaction of all partners.

Objectives

All partners on this L.L. Pelling job agree to:
1. Maintain a safe, clean, and well-organized work site.
2. Do it right and completely the first time.
3. Treat each other fairly and honestly.
4. Maintain mutual respect for other partners and their work.
5. Be proactive in seeking solutions to problems.
6. Train all personnel to use and adhere to the partnering charter.
7. Prepare and publish submittal processing guidelines.
8. Prepare and publish a billing and payment process.
9. Review and resolve actual and potential discrepancies among trades.
10. Prepare and publish weather-related contingency plans.
11. Prepare and publish formal lines of communication and responsibility.
12. Prepare and publish all change approval processes with price limits.
13. Promptly notify other partners about potential off-site material fabrication problems.
14. Provide and maintain a current schedule of all project work, including rough-in and finishes.
15. Maintain project material, labor, and equipment quality at the levels specified in the contract documents.
16. Prepare, publish, and commit to a set of punch-list and close-out guidelines and conduct user orientation as part of the close-out process.
17. Recycle waste material whenever possible.
18. Maintain accurate construction record documents on site.
19. Make timely decisions on finish material and colors.
20. Be available!
21. Resolve problems effectively.
    a. Develop, approve, and implement a responsive conflict resolution system.
    b. Resolve disputes and conflicts at the originating level if at all possible.
    c. Resolve disputes and conflicts as quickly as possible.
    d. Eliminate the need for third-party legal involvement.
22. Prepare and publish a partnering evaluation process.
23. Have fun and take pride in your work.

Charter for New Federal Courthouse, Fan Pier, Boston, Massachusetts

	Date written: October 26, 1994
Project:	New Federal Courthouse, Fan Pier
Location:	Boston, Massachusetts
Cost:	Approximately $165,000,000
Owner:	General Services Administration

Construction services: Parsons Brinckerhoff Construction Services, Inc.
Designers: Pei Cobb Freed & Partners and Jung/Brannen &
                          Associates
Contractors: George Hyman Construction Company

Mission

Recognizing the enduring contribution the Boston federal courthouse will make to the system of justice and to the community, we will produce this building in accordance with the established design intent, to the highest standards of construction, within the established parameters of contract, schedule, and budget. We will work together in a manner that will allow each member of the project team to take pride and feel personal satisfaction in his or her contribution to the successful completion of the building.

Objectives

In furtherance of this mission, the stakeholders on the Boston federal courthouse project agree to:

1. Respect design and construction excellence as a fundamental goal to be achieved.
2. Be available.
3. Make timely decisions in all project-related matters.
4. Prepare, publish, and implement a project procedures manual that provides all stakeholders with guidelines for:
   a. Submittal processing.
   b. Payment processing, including retention.
   c. Revision and change order processing.
   d. Time commitments for procedures.
   e. Giving priorities to assignments.
5. Establish and implement procedures acceptable to all stakeholders, to expeditiously process requests for information (RFIs).
6. Mutually prepare, publish, implement, and keep current a project schedule that is useful to all stakeholders.
7. Prepare and publish a chart of channels for communication, responsibility, and authority.
8. To establish and implement close-out guidelines that provide direction for:
   a. Punching out the job.
   b. Prompt issuance of the certificates of substantial completion.
   c. Setting intermediate occupancy dates.
   d. Maintenance and transmission of contract record documents.
9. Maintain a safe, orderly, well-organized work site.
10. Do it right the first time.
11. Identify and remedy incorrect performance in a timely manner.
12. Plan for and meet the human resource needs of the project.
13. Use human and technological resources to their maximum effectiveness.
14. Keep paperwork to a minimum.
15. Meet individual and organizational obligations.
16. Respect financial profit as an incentive for private-sector stakeholders.
17. Establish a trustful work environment with other stakeholders.

18. Establish and maintain good informal working relations on the job.
19. Minimize disputes and resolve conflicts quickly and at the lowest possible management level.
20. Prepare, issue, and implement a dispute resolution system.
21. Prepare, issue, and implement a partnering evaluation system.
22. Maintain key personnel continuity on the project.
23. Respect others and their work.
24. Extend the spirit of partnering to all project participants.
25. Strive to maintain high job morale and cooperative attitudes among all project participants.

Charter for AAA, Dearborn, Michigan

	Date written: December 14, 1994
Project:	AAA Insurance Association office renovation, addition, and expansion
Location:	Dearborn, Michigan
Size:	Existing building, about 375,000 square feet; addition, about 240,000 square feet
Contractor	A.J. Etkin Company
Designers:	Giffels Associates

Mission

To plan, design, and construct a facility that meets or exceeds the project team's goals to:
  • Satisfy AAA Michigan's needs and expectations
  • Minimize operational disruptions
  • Achieve on-time, in-budget performance
by optimally using team members' skills through a process that will be recognized as an industry benchmark.

Objectives

In recognition of the importance of achieving their mission, all stakeholders agree to:
1. Prepare and publish an owner-approved project budget in a timely manner.
2. Make timely decisions and abide by them.
3. Prepare, publish, and monitor a comprehensive design, construction, and move-in plan and schedule. This plan and schedule is to be reviewed and updated as required by project conditions.
4. Prepare and publish invoicing procedures for all levels of project operations.
5. Make prompt, full payment of all properly submitted pay requests.
6. Respect the ideas, needs, expertise, and work of others on the project.
7. Prepare and publish project commissioning and project close-out guidelines and procedures.
8. Provide adequate resources to properly maintain and complete the project work.

9. Approve and publish final project program documents in a timely manner.
10. Prepare and publish procedures for bidder selection and award of all construction subcontracts.
11. Strive to reduce the time required and improve the results achieved by project meetings. This to be encouraged through agenda planning and prompt issue of meeting minutes.
12. Be fair and reasonable on pricing adds and deducts for proposed changes.
13. Strive to minimize cost creep, time creep, and disruptive revisions to the project work.
14. Prepare and publish an approved quality control system for preparation of contract documents.
15. Maintain and publish current project cost estimate and cost accounting.
16. Establish effective project management and technical communications channels with all other stakeholders.
17. Prepare and publish procedures for tracking and resolving open issues.
18. Prepare and publish procedures for approval and release of design systems, packages, and features.
19. Prepare and publish procedures for processing project changes.
20. Prepare and publish procedures for processing change orders.
21. Establish value engineering/enhancement procedures.
22. Prepare and publish bid packaging procedures for all trades and disciplines.
23. Prepare, approve, and publish a mobilization and construction phasing plan.
24. Prepare and publish submittal processing procedures.
25. Prepare and publish substitution procedures.
26. Conduct preconstruction meetings with subcontractors.
27. Avoid litigation.
28. Prepare and publish procedures for issue resolution.
29. Prepare, publish, and implement a partnering evaluation procedure.
30. Strongly consider the construction, review, and approval of prototypes for selected systems.
31. Be available.
32. Do it right the first time.

APPENDIX D

PARTNERING PROBLEM TYPE LISTINGS

SECTION A: INTRODUCTION

Appendix D is a tabulation of 45 major classes of design and construction problems identified by stakeholders who participated in 23 partnering charter meetings over a three-year period.[1] Project construction costs of the partnered facilities designed, built, and occupied by these men and women range from $4 million to nearly $400 million. The projects are located in 14 cities and in six states. They included a broad range of building types, from a magnesium extrusion plant to a new federal courthouse. Data are reported from detailed meeting notes taken during the sessions.

Approximately 685 men and women attended the 23 meetings and participated actively as partnering stakeholders. Those that took part represent nearly every management level, technical discipline, and age group active in the industry today. Stakeholders ranged from key trades workers to presidents of construction companies, from design firm principals to elected public officials. Their work in the partnering meetings provides a comprehensive view of what design and construction professionals believe are the problems our industry faces in the near and distant future.

Study and analysis of these conflict and dispute-potential areas can help point the way to:

[1] Three full-scale simulated partnering seminars conducted by me at the University of Wisconsin are included in the sampling. The problem statements resulting from these seminars were written by professional practitioners who are active in the design and construction business.

411

1. Reducing design and building costs
2. Improving organizational, business, and personal relations in the offices and at the job sites
3. Encouraging project participants to apply resources to the actual design and building process more productively than in the past

The problem statements resulting from the 23 sessions were actually encountered in writing the partnering charters. They were taken directly from the results of workshops 1 and 2 held in each of the meetings. Workshop 1 was devoted to answering the question, "What actions do others take that cause problems for us?" Workshop 2 was concentrated on answering the question, "What actions do we take that cause problems for others?" It is interesting to note that workshop 2 usually takes about one-half the time it requires to complete workshop 1. However, the problem statements from the two sessions are similar, and both are included in the tabulations.

The information analysis was done as fairly and objectively as possible by following a specific procedure:

. All 2855 problems mentioned by stakeholders attending the charter meetings were coded in accordance with a special classification system. Problem statements resulting from the charter meeting fell broadly into 45 major classes. These classifications are listed alphabetically in Table D.4 along with the three-letter identification code used to identify the problem in the database.

2. Problem statements and their accompanying characteristics were next entered in a database file containing the following information fields:
   • Problem statement
   • Problem codes applicable, not limited in number
   • Problem master code: the best and most descriptive code of those applicable
   • Project on which the potential problem was identified
   • Workshop in which the problem was identified

3. Problem statements were edited to clarify intent.

4. Problem statements were arranged both alphabetically and in descending order of total mentions.

5. Where problem duplicates were encountered the problem statement was kept in the file, but coded to indicate that there were similar problem statements in the file. One often-mentioned problem statement was kept in the active file. However, all mentions were counted in the tabulations.

TABLE D.1 Types and Number of Partnered Projects

New hospital, 1
Insurance office addition and expansion, 3
Neighborhood outpatient health clinic, 1
Stormwater retention basins, 3
National data processing center, 1
Magnesium extrusion plant, 1
Community college activity center, 1
International airport addition and expansion, 2 (internal staff and construction)
Resort hotel, 1
Major freight terminal, 1
Paving contractor maintenance and office facility, 1
Major urban federal courthouse, 1
Elementary and middle school addition and remodeling, 1
Revisited charters, 2 (hospital and one insurance office)
Seminar case studies on office buildings, 3

Where there was a doubt about the meaning or value of the apparent duplicate, it was kept in the records.

SECTION B: PROJECT TYPES AND LOCATIONS

Types and locations of the projects included are summarized in Tables D.1 to D.3.

SECTION C: MAJOR CLASSES OF PROBLEMS

The problems listed in Table D.4 have their genesis in practices and procedures within the construction process. Almost as soon as problems are raised in a partnering charter meeting, the distinction of whether they are "problems others cause us" or "problems we cause others" evaporates. A problem is a problem is a problem.

In the discussions below the abbreviation "tm" stands for total mentions . . . the total number of problems where this factor appears as an identifiable

TABLE D.2 State Locations of Partnered Projects

Connecticut
Iowa
Massachusetts
Michigan
Minnesota
Nebraska

TABLE D.3   City Locations of Partnered Projects

Detroit, Michigan
Lansing, Michigan
Plymouth, Minnesota
Saginaw, Michigan
Eaton Rapids, Michigan
Muskegon, Michigan
Omaha, Nebraska
Frankenmuth, Michigan
Minneapolis, Minnesota
Middletown, Connecticut
Cedar Rapids, Iowa
Boston, Massachusetts
Carson City, Michigan
Dearborn, Michigan

component in the 2855 specific problem statements. The "sm" abbreviation stands for single mentions. They are what I considered to be the most applicable descriptor of a problem statement. For instance, one participant mentioned as a problem that "owners side with engineers in almost any dispute." That problem included elements of cwo (communicating with others), pqp (personnel quality problems), jma (job management), and sma (staff morale and attitudes). Each of these problem types earned one of its total mentions in that example. As I studied this particular problem, the factor "staff morale and attitude (sma)" emerged as the overriding cause of the problem and was assigned as the single-mention (sm) master code. Each of the problem statements is made up of one or more of the 45 factors in the total mentions and has been further refined to a single-mention cause.

Table D.4 is a list of the 45 factors by which the design and construction problems referred to above were identified. Generally, a problem is associated with more than one factor which has caused the event to become a problem. Letters used in the abbreviation type follow no set formula. They are mnemonic device to help remember their meanings. Problem types are shown in alphabetical order in Table D.4. The problem type is listed first, followed by the problem type code and then by the number of times that code type was assigned by me to each of the problem statements made in the charter workshops. Since any given problem could receive more than one code classification, the totals for all statements add up to considerably more than 2855.

In Table D.5 the problem code material has been arranged in descending order of the number of times the problem was assigned that type. The rankings are shown in the left column, followed by the number of total problem code assignments, and then by the problem category.

In Table D.6 the problem code material has been arranged alphabetically by problem type. The total problem statement mentions are shown in the left numeric column, followed by the percent of these to the total of all statements,

TABLE D.4 Problem Types Listed Alphabetically

Approval processes (apv), 90
Back-charges (bch), 11
Being a good off-site neighbor (ofn), 88
Being a good on-site neighbor (onn), 475
Closing out the project (clo), 112
Communicating with others (cwo), 984
Constructibility (cbl), 20
Construction document quality (cdq), 267
Contract interpretation (coi), 97
Cost growth (cgr), 64
Decision making (dma), 133
Documents and documentation (doc), 141
Equipment and material problems (emp), 145
Financial matters (fin), 11
Inspecting and testing (ite), 69
Issue, conflict, and problem resolution (ire), 166
Job management (jma), 1146
Labor conditions (lab), 14
Legal matters (leg), 14
Maintaining regular project evaluations (mpe), 52
Organization, authority, and responsibility (oar), 371
Paperwork and administrative work (paw), 92
Payment processing (ppr), 95
Personnel quality and problems (pqp), 593
Planning and scheduling (pas), 396
Policies and procedures (pop), 70
Procurement of materials and equipment (prc), 125
Program conditions (prg), 233
Project cost structure (pco), 116
Quality management (qma), 97
Regulatory agency matters (reg), 49
Revision processing (rev), 268
Safety (saf), 52
Staff morale and attitudes (sma), 684
Staffing and personnel (stf), 69
Submittal processing (spr), 205
Substitutions and alternates (sal), 58
Time growth (tgr), 73
Timely action (tac), 467
Training (tng), 22
User-group interaction (ugi), 166
Value engineering (ven), 22
Warranty conditions (war), 5
Weather conditions (wea), 10
Work-site conditions (wsc), 288

**TABLE D.5 Total Assignments and Problem Types
Listed by Frequency of Appearance**

1.	1146	Job management
2.	0984	Communicating with others
3.	0684	Staff morale and attitudes
4.	0593	Personnel quality and problems
5.	0475	Being a good on-site neighbor
6.	0467	Timely action
7.	0396	Planning and scheduling
8.	0371	Organization, authority, and responsibility
9.	0288	Work-site conditions
10.	0268	Revision processing
11.	0267	Construction document quality
12.	0233	Program conditions
13.	0205	Submittal processing
14.	0166	Issue, conflict, and problem resolution
15.	0166	User-group interaction
16.	0145	Equipment and material problems
17.	0141	Documents and documentation
18.	0133	Decision making
19.	0125	Procurement of materials and equipment
20.	0116	Project cost structure
21.	0112	Closing out the project
22.	0097	Contract interpretation
23.	0097	Quality management
24.	0095	Payment processing
25.	0092	Paperwork and administrative work
26.	0090	Approval processes
27.	0088	Being a good off-site neighbor
28.	0073	Time growth
29.	0070	Policies and procedures
30.	0069	Inspecting and testing
31.	0069	Staffing and personnel
32.	0064	Cost growth
33.	0058	Substitutions and alternates
34.	0052	Maintaining regular project evaluations
35.	0052	Safety
36.	0049	Regulatory agency matters
37.	0020	Constructibility
38.	0022	Training
39.	0022	Value engineering
40.	0014	Labor conditions
41.	0014	Legal matters
42.	0011	Back-charges
43.	0011	Financial matters
44.	0010	Weather conditions
45.	0005	Warranty conditions

TABLE D.6   Total Assignments of Problem Types[a]

Approval processes (apv)	tm = 0090, % = 03, sm = 042
Back-charges (bch)	tm = 0011, % = 01, sm = 011
Being a good off-site neighbor (ofn)	tm = 0088, % = 03, sm = 007
Being a good on-site neighbor (onn)	tm = 0475, % = 17, sm = 087
Closing out the project (clo)	tm = 0112, % = 04, sm = 081
Communicating with others (cwo)	tm = 0984, % = 35, sm = 234
Constructibility (cbl)	tm = 0020, % = 01, sm = 013
Construction document quality (cdq)	tm = 0267, % = 09, sm = 196
Contract interpretation (coi)	tm = 0097, % = 04, sm = 024
Cost growth (cgr)	tm = 0064, % = 02, sm = 014
Decision making (dma)	tm = 0133, % = 05, sm = 073
Documents and documentation (doc)	tm = 0141, % = 05, sm = 028
Equipment and material problems (emp)	tm = 0145, % = 05, sm = 023
Financial matters (fin)	tm = 0011, % = 01, sm = 002
Inspecting and testing (ite)	tm = 0069, % = 02, sm = 041
Issue, conflict, and problem resolution (ire)	tm = 0166, % = 06, sm = 029
Job management (jma)	tm = 1146, % = 40, sm = 319
Labor conditions (lab)	tm = 0014, % = 01, sm = 003
Legal matters (leg)	tm = 0014, % = 01, sm = 004
Maintaining regular project evaluations (mpe)	tm = 0052, % = 02, sm = 011
Organization, authority, and responsibility (oar)	tm = 0371, % = 13, sm = 106
Paperwork and administrative work (paw)	tm = 0092, % = 03, sm = 026
Payment processing (ppr)	tm = 0095, % = 03, sm = 083
Personnel quality and problems (pqp)	tm = 0593, % = 21, sm = 034
Planning and scheduling (pas)	tm = 0396, % = 14, sm = 098
Policies and procedures (pop)	tm = 0070, % = 02, sm = 005
Procurement of materials and equipment (prc)	tm = 0125, % = 04, sm = 060
Program conditions (prg)	tm = 0233, % = 08, sm = 100
Project cost structure (pco)	tm = 0116, % = 04, sm = 033
Quality management (qma)	tm = 0097, % = 04, sm = 053
Regulatory agency matters (reg)	tm = 0049, % = 02, sm = 023
Revision processing (rev)	tm = 0268, % = 09, sm = 118
Safety (saf)	tm = 0052, % = 02, sm = 042
Staff morale and attitudes (sma)	tm = 0684, % = 24, sm = 299
Staffing and personnel (stf)	tm = 0069, % = 02, sm = 047
Submittal processing (spr)	tm = 0205, % = 07, sm = 123
Substitutions and alternates (sal)	tm = 0058, % = 02, sm = 040
Time growth (tgr)	tm = 0073, % = 03, sm = 008
Timely action (tac)	tm = 0467, % = 16, sm = 097
Training (tng)	tm = 0022, % = 01, sm = 012
User-group interaction (ugi)	tm = 0166, % = 06, sm = 026
Value engineering (ven)	tm = 0022, % = 01, sm = 019
Warranty conditions (war)	tm = 0005, % = 01, sm = 004
Weather conditions (wea)	tm = 0010, % = 01, sm = 008
Work-site conditions (wsc)	tm = 0288, % = 10, sm = 133

[a] Code definitions, number of total mentions (tm), % of total mentions to 2855 responses (%), and number of single mentions (sm). Percentages have been rounded up and down. Those below 01% are given as 01%.

and then by the number of master code assignments for that problem type (see Section A of Appendix D for explanation of the single-mention code).

SECTION D: PROBLEM CODE DEFINITIONS

The meaning of each problem category is described below. Problem categories all have a positive and a negative side. If certain elements of the problem type go well, it is possible that there is no problem. On the other hand, if the elements of a problem type go poorly, the situation caused by the elements going poorly will usually cause difficulties. Problem types are listed in alphabetical order. The problem type code follows the name. The total mention (tm) and single mention (sm) information is as described in Section C.

Approval process (apv)   tm = 90, sm = 42

The official acceptance of information or submittals needed on the project from regulatory agencies, governmental bodies, the user, the owner, the design team, or any of the members of the construction group is critical to job success. A delay in approval can seriously affect job planning, scheduling, and field progress.

Back-charges (bch)   tm = 11, sm = 11

Back-charges are charges for actions such as cleanup, hoisting, equipment use, damage to installed work, or similar items for which the party furnishing the item feels they are entitled to be paid. A back-charge is often deducted from a payment being made by the party providing the item to the party receiving the item. Problems arise when back-charges are deducted without prior negotiation or notification, especially when there appears to be insufficient cause for the charge.

Being a good off-site neighbor (ofn)   tm = 88, sm = 7

Being a good off-site neighbor is project participant behavior that relates to the people, organizations, or facilities outside construction site boundaries. When on-site actions cause off-site aggravation—noise or dust from a project; or when off-site actions interfere with off-site neighbors—dirt and other debris left on roadways—it is difficult to be an effective builder. Nearly everyone must get to the site by going through the neighborhood—be friendly to the people who live there.

Being a good on-site neighbor (onn)   tm = 475, sm = 87

The on-site behavior of project staff determines how well they are treated by other on-site people. Poor job behavior almost always damages the informal organizational and social relations so critical to healthy jobs. The best rule is still to treat others the way you want to be treated. It is the quickest way to learn the benefits of being a good on-site partner.

Closing out the project (clo)   tm = 112, sm = 87

Closing out means properly finishing the project totally or in part. Factors related to close-out problems affect owners through delayed occupancy, and contractors and subcontractors by delays in completing their work. Improper close-out also adversely affects payment of retainage and often increases costs difficult to associate with any specific party to the job.

Communicating with others (cwo)   tm = 984, sm = 234

Information exchanges between or among individuals, groups, or organizations can be oral or visual and may express a new thought or a commonly understood policy. Problems caused by the inadequate exchange of thoughts, messages, or information in construction makes communication with others an important factor in design and construction.

Constructibility (cbl)   tm = 20, sm = 13.

Constructibility is the degree to which the design of a facility is found to be buildable. Often when there is a constructibility problem, the project or a component of the project cannot be built as called for by the contract documents. This may lead to serious delays, extra costs, redesign, and hard feelings on the job.

Construction document quality (cdq)   tm = 267, sm = 196

Problems are caused by poor quality control in the preparation of working drawings and specifications. Difficulties are usually caused by unclear or contradictory notes, drafting errors, poor workmanship, incomplete information, dimensional errors, or similar detractions.

Contract interpretation (coi)   tm = 97, sm = 24

Any contract is open to interpretation. Serious problems may arise from substantial differences in those interpretations, especially in the understanding of various parties as to what their work scope is and what they are entitled to claim when they are hurt by a unilateral contract interpretation. Contracts being legally binding, this factor can quickly escalate from a simple problem into a disaster if not resolved promptly.

Cost growth (cgr)   tm = 64, sm = 14

Changes in project cost, either greater or less than expected, often affect the program or project. Growth may be positive for some participants and negative for others. Problems considered here often produce damaging impacts through unfair risk assignment.

Decision making (dma)   tm = 133, sm = 73

Wise decisions at the proper time are much sought after. Inadequate, improper, or untimely decision making on project-related matters by those not competent or authorized is frequently a cause of trouble.

Documents and documentation (doc)   tm = 141, sm = 28

Every construction job requires documentation from conception to occupation. Improper, inadequate, unneeded, or excessive paperwork that blocks effective management and implementation is likely to result in long-standing and difficult problems.

Equipment and material problems (emp)   tm = 145, sm = 23

You cannot build a job without equipment and materials. Problems with procurement, storage, installation, or functioning of equipment and materials used on the project can create a nightmare.

Financial matters (fin)   tm = 11, sm = 2

Financing is at the heart of any building project. Problems related to the methods, amount, availability, or reliability of project funding are difficult to discern early and are even more difficult to resolve before they do their damage.

Inspecting and testing (ite)   tm = 69, sm = 41

Safety and quality are the hallmarks of good construction. Inspection and testing are designed to guarantee safety and quality. That means that someone qualified must inspect and test. Problems generated by poor or untimely inspections and poor testing methods, personnel, management, or interpretation can have a serious impact on the project.

Issue, conflict, and problem resolution (ire)   tm = 166, sm = 29

Problems are meant to be solved. The best course of action is to agree in advance how the parties will resolve emerging issues fairly and speedily. Prompt settlement of conflicts, contested claims, and other disruptive or destructive action between or among the project participants is essential to conserving profit. Unresolved issues cost dearly and create hard feelings.

Job management (jma)   tm = 1146, sm = 319

Good leadership and knowledge in depth of the total project or of its components constitute 80% of job management. The proper use of skills in planning and scheduling, assigning resources, and assembling and effectively utilizing resources enhance the prospect of job success. Conversely, bad management can doom a design and construction project before it begins.

Labor conditions (lab)   tm = 14, sm = 3

Conditions, rules, laws, and obligations exist under which project participants work on any project. The term *labor* usually refers to tradespeople of all skills located at the job site. Problems arise when there are poorly managed union–nonunion disputes, ineffectual communications between management and tradespeople, financing problems, or any of the multitude of conditions that adversely affect the lifeline of the project – financial health for all.

Legal matters (leg)   tm = 14, sm = 4

The construction practitioner operates under the rule of law but cannot afford to become preoccupied by it. Adverse legal actions expected or taken on a project can reduce or destroy potential for good project performance.

Maintaining regular project evaluations (mep)   tm = 52, sm = 11

Competent monitoring, analyzing, and acting on information derived from a plan of work is an integral part of managing. In partnering, evaluation is often implemented by regularly measuring actual partnering performance against standards set by the stakeholders in the charter. Problems arise when the process is ignored by the stakeholders or when subjective evaluations replace objective measurements.

Organization, authority, and responsibility (oar)   tm = 371, sm = 106

Organization, authority, and responsibility patterns spring from a functional need for competence and responsible leadership. The pattern may be assigned or assumed and will generally govern project and program actions on the job. Problems follow when the organization, authority, and responsibility needs are disregarded or unfilled. The results will often be a disrupted project, uninformed participants, and frayed tempers.

Paperwork and administrative work (paw)   tm = 92, sm = 26

Documents, letters, and other communications, whatever the media, must flow quickly and accurately among, between, from, and to project participants. Paperwork frequently creates a love–hate relationship. Imposing too much communication without a corresponding value added is a distraction and annoyance. Too little communication may produce a value-subtracted situation by encouraging management by blindfold, where stakeholders run their work by guessing and assuming. There is a proper amount of paperwork for each job.

Payment processing (ppr)   tm = 95, sm = 83

The methods, practices, and timing of payments due to or from project team members are usually spelled out in contract documents. Problems arise when one party disregards that agreement or when practices in billing and paying become sloppy. Prompt payment is a great stimulator of good work.

Personnel quality and problems (pqp)   tm = 593, sm = 34

The labor pool, wages, and the press of business will determine who works on what job. Variations in personnel abilities, qualifications, desires, skills, attitudes, and honesty of the project staff working in the interests of the project can give rise to any number of conflicts and problems.

Planning and scheduling (pas)   tm = 396, sm = 98

Competent design and construction sequencing, resource assignment, scheduling, and procurement planning for project actions are some of the easiest roads to a successful job. Failure to plan and schedule will lead to failing to do the job well. The job of the manager is to plan the work, and then work the plan.

Policies and procedures (pop)   tm = 70, sm = 5

Policies and procedures are detailed statements of expected behavior, sequences, courses of action, and principles that help determine decisions, actions, and other matters for the participants on a planning, design, and construction program. Usually, policies and procedures are set both for the firms involved in doing the project work and for the project. Problems arise when those policies and procedures are unrealistic or when involved firms cannot or will not conform to agreed-on policies and procedures.

Procurement of materials and equipment (prc)   tm = 125, sm = 60

Procurement is the process of detailing, approving, fabricating, and delivering materials, equipment, and other physical elements to be installed in the facility. Intelligent, experienced management and a strong interest in excellent performance is at the core of successful procurement. Procurement problems cause frustration and delays.

Program conditions (prg)   tm = 233, sm = 100

The quality of the project program has a sizable effect on the design, construction, turnover, and use of the facility. A good program helps design and build a good facility. Poor programs hinder work and often lead to damaging project surprises.

Project cost structure (pco)   tm = 116, sm = 33

The characteristics of the project relative to how funding is determined, allocated, and disbursed to the project participants determine the project cost structure. Cost structure is usually established during early programming of a project. It can begin there as a problem or it may rear its ugly head later if there is an unwelcome change.

Quality management (qma)   tm = 97, sm = 53

Quality management concerns factors in project success or failure that are related to the quality of people, workmanship, materials, equipment, or organizations being used on the project. Quality, as used here, means of a nature that meets contract requirements and produces results that satisfy or exceed expectations. Anything less may be a problem.

Regulatory agency matters (reg)   tm = 49, sm = 23

Rules and guidelines are often set by regulatory agencies in the public or private sectors. Regulations can be maintained by voluntary compliance or by compliance dictated by law. Intelligent compliance with legitimate, well-interpreted regulations helps a job. If the rules and guidelines are misused or poorly interpreted, problems will surface. Regulatory difficulties often occur because regulators are sometimes not considered as a participant in the project. The result of this is an us–them mentality that produces potentially damaging conflicts between the regulators and the stakeholders.

Revision processing (rev)   tm = 268, sm = 118

Revision processing includes steps taken to produce proper and effective project revisions from formulation to implementation of the change. As a supportive action, good revision processing is almost invisible. Continued poor performance in this critical part of a design and construction project leads to progressive deterioration of nearly all job management functions.

Safety (saf)   tm = 52, sm = 42

Provision and maintenance of safe working conditions on the job site are crucial to job success. Safety problems usually result in damage or injury. Both harm job quality and progress.

Staff morale and attitudes (sma)   tm = 684, sm = 299

Individual and collective morale and attitudes of people can heavily influence and shape working conditions and outcomes on a project. Often, morale and attitude problems are matters of perception that may or may not correspond with reality. Good morale and constructive enthusiasm on design and construction projects are always welcome contributions to project health.

Staffing and personnel (stf)   tm = 69, sm = 47

Staffing and personnel items relate to the number of staff resources on the project and their quality, competence, and abilities. When resources are available, the job moves well; when resources are lacking, frustration and confusion result.

Submittal processing (spr)   tm = 205, sm = 123

Submittal processing concerns preparing, delivering, reviewing, approving, and returning shop drawings, specifications, designs, samples, cuts, and other documents or objects that must be approved as required by the contract provisions. Done well, submittal processing makes a job support system function well. Lack of competent attention to the procedure causes problems and delays.

Substitutions and alternates (sal)   tm = 58, sm = 40

Often, suggested or actual substitutions or alternative materials, equipment, methods, or systems are considered for use in place of those specified or shown on the contract documents. Problems arise when substitutions and alternatives degrade quality, present a false cost saving, or unfairly shift profit or loss among project participants.

Time growth (tgr)   tm = 73, sm = 8

Time growth refers to a change in time either greater or less than expected that produces an impact upon the project or program. This impact, particularly when time is extended, almost always indicates a problem will or has appeared.

Timely action (tac)   tm = 467, sm = 97

Timely action can mean action taken at the correct or effective time, or action taken for a correct or effective duration. Problems can be related to taking, or failing to take, timely action on any project or program-related matter.

Training (tng)   tm = 22, sm = 12

Adequate training and education of the project team is a management necessity. Problems arise when training and education are inadequate.

User-group interaction (ugi)   tm = 166, sm = 26

To produce a successful project, project team members and stakeholders must maintain effective informational, technical, business, and professional relationships with the owner and the end user of the facility. When these relations are damaged or ignored, problems are almost certain to follow.

Value engineering (ven)   tm = 22, sm = 19

Cost and other cost-related benefits are often gained on a generic construction project by improving the means, methods, materials, and sequences of architectural and engineering systems used. Without striving to improve value within the target cost restraints, a job remains a static system. Value engineering is best applied before construction contracts are awarded.

Warranty conditions (war)   tm = 5, sm = 4

Warranty conditions are those construction guarantees placed in effect subsequent to completion of the work and usually upon acceptance by the owner. Warranty problems arise when their starting or expiration dates are unfairly assigned or unilaterally imposed for the benefit of one party and to the detriment of the other.

Weather conditions (wea)   tm = 10, sm = 8

Weather and construction are either fighting or are friends, but weather will have its way. Bad weather at a poorly managed job can create insurmountable obstacles to good work. Weather is one of the best documented scientific occurrences that exist. The manager is not expected to change the weather. He or she is, however, expected to know the when, how, what, and where of weather in their locality so that the people on the job can maintain work continuity and profitability irrespective of poor weather conditions.

Work-site conditions (wsc)   tm = 288, sm = 133

Work-site conditions almost always affect a project. A poorly organized and badly maintained work site prevents people from doing their best work, even when they want to do well. A clean, safe, well-planned work site shows respect for those who earn their salaries by working there. It helps them do a good job. Poor site working conditions demotivate; good site working conditions motivate. One leads to trouble and danger; the other shows good faith and confidence.

INDEX